36th Ethylene Producers Conference 2024

Topical Conference at the 2024 AIChE Spring Meeting and 20th Global Congress on Process Safety

New Orleans, Louisiana, USA
24-28 March 2024

ISBN: 979-8-3313-1710-2

Printed from e-media with permission by:

Curran Associates, Inc.
57 Morehouse Lane
Red Hook, NY 12571

Some format issues inherent in the e-media version may also appear in this print version.

Printed with permission by Curran Associates, Inc. (2025)

For permission requests, please contact AIChE
at the address below.

AIChE
120 Wall Street, FL 23
New York, NY 10005-4020

Phone: (800) 242-4363
Fax: (203) 775-5177

www.aiche.org

Additional copies of this publication are available from:

Curran Associates, Inc.
57 Morehouse Lane
Red Hook, NY 12571 USA
Phone: 845-758-0400
Fax: 845-758-2633
Email: curran@proceedings.com
Web: www.proceedings.com

TABLE OF CONTENTS

CONTAMINANTS TUTORIAL

CIRCULAR ECONOMY/DECARBONIZATION

CONTAMINANTS AND IMPURITIES IN ETHYLENE PLANTS

EPC KEYNOTE:

ETHYLENE PLANT ENVIRONMENTAL AND SUSTAINABILITY

ETHYLENE PLANT MAINTENANCE AND RELIABILITY

ETHYLENE PLANT OPERATIONS

ETHYLENE PLANT ROTATING EQUIPMENT

ETHYLENE PLANT SAFETY

ETHYLENE PLANT TECHNOLOGY FUNDAMENTALS

INDUSTRY 4.0 IN ETHYLENE PRODUCTION

TUTORIAL: FLARE OPERATIONS IN A POST E-MACT WORLD

ETHYLENE PRODUCERS GENERAL TOPICS

Author Index

121a Applying Filtration and Coalescence Technologies for the Removal of Contaminants in the Ethylene Process - Contaminants and Impurities in Ethylene Plants Subcommittee Tutorial

Ali Arshad, Emmanuelle Biadi

The presence of contaminants plays a critical role in maintaining equipment reliability and process stability in the ethylene production. The critical areas where contaminant-free process fluid is essential include cracking furnace, low NOx burner, molecular sieve dryer, caustic scrubbing tower, process water steam stripper, and pygas hydrotreater. Contaminants can be detrimental to the process, resulting in such operational problems as accelerating coke formation, corrosion, catalyst poisoning, fouling of major equipment, over-consumption of utilities such as steam in the benzene stripper, among others.

Typical contaminants include various liquid impurities (of aqueous and organics), particulate solids (such as coke fines, salts, and corrosion products), deformable semi-solids (of tar-like high molecular weight compounds) and gases (including hydrogen sulfide, carbon dioxide, and air in the compressor lube oil).

An effective removal of contaminants at any point of entry into the process (such as feed stocks and burner fuels) as well as those generated within the process is of paramount importance to effectively protect the downstream equipment critical to the ethylene production and to prevent high OPEX, low productivity and energy inefficiency to ultimately meet the performance expectations and the production capacity.

This tutorial session reviews the criticality of fluid cleanliness in the ethylene process. It describes the various filtration and phase separation solutions that can be applied in various locations within the process to mitigate the operational problems associated with the presence of impurities by removing them from liquid and gaseous streams within the production process.

The most cost-effective options achieving a suitable removal of contaminants with the highest degree of separation and potential benefits will be highlighted. Furthermore, general guidelines will be discussed for effectively selecting and specifying an optimum filter for a particular application.

'Chemicals Used to Control Contaminants in the Ethylene Plant'*Contaminants and Impurities in Ethylene Plants Subcommittee Tutorial*

Debby Rossana, Sherri Elder

'Chemicals Used to Control Contaminants in the Ethylene Plant'

Contaminants and impurities in Ethylene Plants Subcommittee Tutorial

Debby Rossana, Ph.D., Retired, Nalco Water, An Ecolab Company

Sherri Elder, Ph.D., New Page Sustainable Chemistry

During the production of ethylene, there are a number of impurities the operator of the plant has to contend with. There are contaminations in the feedstocks and recycle streams as well as a multitude of ppm levels of impurities that are made during the cracking process. And if that is not enough, there are other contaminants that form during the separation/purification process. A large proportion of the latter become foulants. Many of these various contaminants and impurities can cause serious processing problems for the ethylene producer if they are not removed. These problems may be off spec products, catalyst poisons, corrosion, foaming, emulsions and fouling.

Because of issues these impurities bring to the production of ethylene, they need to be removed or controlled in some way. This paper looks at chemical means of dealing with some of the impurities in the system. There are situations where addition of a chemical can render the impurity harmless. These would be chemicals such as neutralizers to control acids (CO_2, H_2S and the organic acids - acetic primarily, but also propionic acid, butyric acid and higher analogs), scavengers to remove acetaldehyde or scavengers to remove CS_2. More often, however, chemicals are added to combat the problems associated with the impurity, rather than the impurity itself. These would include antifoams, filmers, emulsion breakers, dispersants and antifoulants.

Because these problems occur throughout the plant, in multiple pieces of equipment, an in-depth discussion of each area is impossible. However, a high-level look at the issues where chemical approaches can provide some benefit will be presented and this will benefit the engineer in solving processing problems.

121b Adsorbents for Contaminant Removal in the Production of Ethylene and Propylene

Jay Gorawara, David Radtke

Adsorbents, of various kinds, are used to help olefin producers manage their operations as contaminated feeds and increasingly sensitive polymerization catalysts continue to offer them challenges.

The most widely used adsorbents are the 3A based molecular sieves used to remove water from the cracked gas, cracked liquids, ethylene-ethane and hydrogen. Adsorbents are not only used for dehydration, but also to remove other contaminants such as, trace mercury, ammonia, arsine and phosphine from the cracked gas.

In the ethylene and propylene product streams, and as one moves downstream to the polymer plant, contaminants are a problem for sensitive polymerization catalysts and can impact polymer quality. Catalyst effectiveness can be greatly diminished by the presence of trace contaminants such as, Chlorides, Carbon monoxide, Oxygen, CO2, Sulfur and Nitrogen compounds. In this part of the plant, adsorbent beds of various combinations in a logical sequence, can be utilized to remove a variety of compounds. The key to trace contaminant removal from olefinic streams is the use of adsorbents with low reactivity towards the process stream. Although adsorbents are a cost effective option for olefin stream dehydration and purification, it is critical to select the right product.

Selection criteria for matching adsorbent products with specific streams and contaminants as well as tips for optimal operation will be covered.

AIChE Paper Number: 44a

Reformulated Feeds for Olefins Production

Kandasamy M. Sundaram

Ethylene Department
Lummus Technology

Ronald M. Venner
Chief Business Officer
Clean Fuels & Crude to Chemicals
Lummus Technology

and
Pedro M. Santos
Technology Director
Chevron Lummus Global
Lummus Technology

Prepared for Presentation at 2024 Spring National Meeting
Houston, Texas, March 24-28, 2024

Reformulated Feeds for Olefins Production

Kandasamy M. Sundaram, Ronald M. Venner and Pedro M. Santos

Lummus Technology, Bloomfield, NJ 07003

Abstract

The petrochemical industry is moving to increase sustainability and reduce greenhouse gas emissions. Instead of fossil feeds, the ethylene industry is looking for alternative feed sources. Bio feed in general and vegetable oil in particular is an excellent feed to produce olefins. In this paper, feed characterization and contaminants removal will be discussed using Chevron Lummus Global's (CLG) ISOTERRA process. Vegetable oil and other bio feeds can be converted to hydrogenated vegetable oil (HVO), a generic term used for bio feeds. Cracking HVO feed and its impact on existing ethylene plant performance will be discussed.

Introduction

Ethylene is one of the major petrochemicals produced in large quantities worldwide. It ranks number one in monomer production and is the basic building block of polyethylene. Recent annual ethylene production exceeds 220 million tons of ethylene [1]. The majority of ethylene is produced by thermally cracking petroleum-based hydrocarbon feeds such as ethane, LPG, naphtha and gasoil. A small quantity is produced by ethanol dehydration, MTO and other processes. There is an increased awareness to limit the fossil fuel consumption to reduce greenhouse gas emissions. Ethane is the simplest hydrocarbon used commercially to produce ethylene and it produces the highest amount of ethylene by thermal cracking in tubular reactors. Ultimate ethylene selectivity is more than 80 w% from ethane. All other methods such as dehydrogenation and oxydehydrogenation are proposed to reduce greenhouse gas emissions, but the ethylene selectivity is much less than that of thermal cracking. All other hydrocarbons produce significantly lower ethylene than ethane (20-40%). Significant energy is also spent in producing the feedstocks such as naphtha and gasoil from crudes. To minimize the energy and the greenhouse gas emissions, Lummus has introduced its Thermal Crude to Chemicals technology (Sundaram et.al, 2023; 2024). However, the feed is petroleum based and is not available everywhere. As oil supplies decline, alternate sources are required to secure the feeds. Nature has an abundant amount of biomaterials. By using these feeds, sustainability of feed sources can be maintained. Currently, biological sources are mainly used to produce food for human and animal consumption. In that process, quite a lot of biological materials are discarded as waste and often end up in landfills. Both non-foods and wasted bio feeds can be used to obtain feeds for ethylene production. Currently vegetable oils are produced and used for manufacturing of sustainable fuels. In this paper, how these oils can be used to produce olefins is discussed. Petrochemical plants produce a lot of consumer goods that after some time end up in landfills. So in addition to bio feeds, how waste materials can be used to generate feeds for ethylene plants will be discussed briefly. These new feed sources constitute reformulated feeds that are used to produce ethylene by conventional thermal cracking.

Recycling Chemicals to produce Olefins

Olefins such as ethylene and propylene are mainly used to produce plastics. The chemical industry is looking into recycling plastics to monomers. Though polyethylene and polypropylene can be directly cracked to produce olefins, the direct route produces low yield of olefins and high yield of other chemicals. Therefore, usually a two-step process is preferred. In the first step, the solid plastic is pyrolyzed to plastic oil (PyOil) and only a certain fraction of the plastic oil is cracked in tubular reactors to produce olefins. Unfortunately, this PyOil contains a significant amount of contaminants and sometimes clean-up of the feed is expensive. The general term plastics is used to denote PE, PP, PVC, polyesters and other polymers. In the production of polymers and consumer products, in addition to monomers other chemicals such as catalysts, inhibitors and additives are required. These added compounds and finished polymers may contain sulfur, nitrogen, chlorine, phosphorous and metals. They are potential contaminants in ethylene production. The ethylene plant produces 99.95% pure ethylene and hence any contaminant in feed requires attention. So, recycling all plastic oils is potentially not economical. Table 1 shows some of the properties obtained from different plastic materials and biomaterials. Tires are another commodity of interest. Unfortunately, tire derived oil (TDO) contains significant amount of contaminants and hence it is not worth to clean it up to produce feedstock for ethylene plant. Other uses such as fuel are currently considered the best solution. PE and PP based PyOil is currently cracked or co-cracked with other feeds. PyOil from other sources have low hydrogen content and are more expensive to upgrade to a suitable cracker feed.

Table 1: Characteristics of Different Pyrolysis Fuel Oils

	C2H4 Plant PFO	Plastic PyOil	Tire Pyrolysis Oil	Wood Pyrolysis Oil
Density, kg/L	1.086	0.809	0.92	1.15
IBP,C	175	103	69	100
10 vol%,C	200	160		120
50 vol%,C	407	249	243	300
EBP, C	735	640	514	480
% Recovery	>99			
BMCI	>120			
Sulfur, wt%	0.05	0.0032	0.7	<0.05
Nitrogen,wt%	0.0022	0.02	0.9	<0.4
Oxygen,wt%	0.12		8.6	
Chlorine, ppm	2	49		
Hydrogen, wt%	8		9.8	8
MCR,wt%	14			
Gross Heating Value, Kcal/Kg			10048	5671
Metals, ppm			20	
Asphaltene,wt%	>17	14.3		
Viscosity at 40 C,cst	200		10	15-35
% Aromatics	76		59.9	
PNA, wt%	42		27.2	

As seen from Table 1, all these feeds have significant amounts of contaminants and often the hydrogen content is very low. Typically, a feed that has less than 13 wt.% hydrogen will not produce a significant amount of ethylene. Only plastic PyOil has higher than 13% hydrogen, when the source plastic is carefully selected. With PE and PP only, the contaminant levels are also low. All these oils have to be upgraded to be suitable feeds for olefin plant. Some of the technologies such as plastic PyOil cracking has been discussed in webinars and in a crude to chemical paper [2,3]. Lummus and CLG over the last few years have invested a lot of research efforts in analyzing and characterizing the feeds. There are certain species when present in feed will have a direct impact on the process by either accelerating the coking in thermal cracking heaters and /or deactivating the catalyst in hydrocracking reactors. Some aspects were discussed in an earlier paper in relation to crude cracking [2]. This has been extended to other feeds shown in Table 1. In this paper we will mainly concentrate on hydrogenated vegetable oil as source for ethylene plant cracker feeds.

Hydrogenated Vegetable Oil (HVO) Feeds

Though vegetable oil can be obtained from many parts of plants, it primarily refers to oils produced from seeds. It is not the intention to replace food materials to produce chemicals. Used cooking oil (UCO) is often disposed as waste. How can we use those oils to produce olefins? In 2022, about 3.7 billion gallons of UCO was available globally and about 0.85 billion was available in the USA [4]. Majority of UCO is used to make fuels product only. However, for carrying out the experiments and understanding the chemistry, all types of fresh oils are used. Also, there are many wild plants (for example carnitas oil) that are not used as food materials. Oils derived from those plants can be used to produce HVO feeds. Cultivating those plants on a regulated basis will lead to a sustainable cracker feed source. In addition to vegetable oils, animal fats have similar characteristics. Therefore, in this paper, such fats are also included. Currently, these sources are used to produce jet fuel for commercial aviation and diesel marketed as bio diesel in the industry. Lummus uses these feeds and modified the process to produce feeds for olefin production instead of producing transportation fuels.

Lummus has two processes to make biojet/diesel from plant oil and animal fats: BIC ISOCONVERSION and ISOTERRA process. BIC ISOCONVERSION includes a process from ARA. ISOTERRA is a CLG process developed by Chevron. BIC ISOCONVERSION uses supercritical water as an agent to decompose oil and hence it is a thermal process. It is developed mainly to produce jet fuels and diesel from plant oils and animal fats. Being a thermal process, the product is rich in aromatics and hence the main products are not suitable for olefin production. However, naphtha produced in that process is suitable for olefin production, but the yield of naphtha is very small. Therefore, the BIC ISOCONVERSION process will not be considered further in this article. On the other hand, ISOTERRA is a catalytic process and produces mainly paraffins. As a part of a renewable fuel conversion project, Chevron's refinery at El Segundo in California started up successfully with the ISOTERRA process producing ASTM approved sustainable aviation fuel or renewable diesel in 2023. [5] This will be discussed at length in the following sections.

Vegetable oils are mainly derived from seeds. The oils or chemicals derived from barks, cellulose and from other plant parts contain significant number of heteroatoms and many functional groups. Vegetable oils are a class of bio compounds belonging to triglycerides. Theoretically, all these compounds contain hydrocarbons in the skeleton and hence some olefins can be produced. However, the cost of clean-up with some bio compounds is considerable. Olefin production requires high quality feeds. Otherwise, the polymer grade olefins production costs will be very high. Hence the products from barks and cellulose are attractive to produce biofuels and the presence of some heteroatoms is acceptable in some fuel grades but are not preferred for olefin production. Considering all these aspects for this article, we have limited the raw feeds discussion to vegetable oils only.

It is the aim of this article, to use these feeds in existing cracking heaters or new heaters which are suitable to crack current feeds such as ethane, LPG, naphtha and gasoils. The current olefin production is based on cracking coil technology where the feed is vaporized externally or internally in the convection section of the cracking heater, mixed with dilution steam, and then sent to radiant coils. The cracking reactions proceed via free radical mechanism in the coils placed in the radiant section in a short residence time (typically <0.5 sec). The products are quickly cooled in a transfer line exchanger and then sent to a recovery section for recovering the products. The reaction is endothermic and hence energy is supplied in the form of heat using fuel fired in the radiant box. The same reactor concept is applied to produce olefins from vegetable oils. The vegetable oil as such cannot be used in the current cracking heater since it contains some heteroatoms, which will contaminate the products. It needs to be conditioned before feeding it to a cracking heater. CLG's ISOTERRA process, developed for diesel production, essentially produces feed good for olefin production. However, ISOTERRA was developed to maximize bio diesel and hence as a fuel, the product has to meet fuel specifications. These are not required for olefin production. The ISOTERRA process can be modified to make it suitable for olefin production. To understand this, the chemistry is discussed first.

Vegetable oil contains mainly triglycerides and other biomasses such as cellulose, which contain 5 and 6 ring alcohol species. Figure 1 shows the structure of typical triglycerides present in vegetable oil and biomasses.

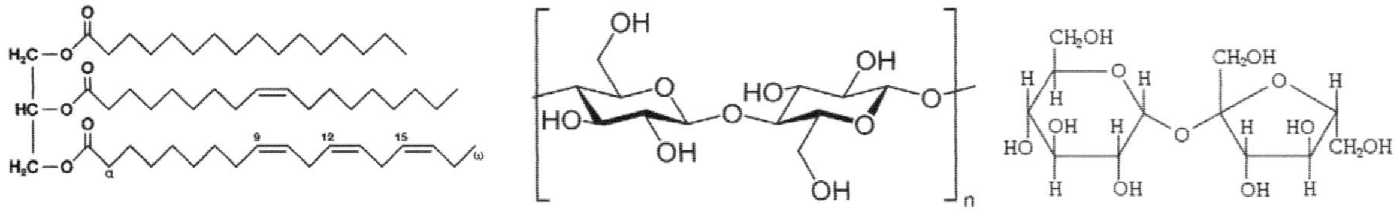

Figure 1: Typical Triglyceride, cellulose and sugar molecular structures

As seen from the chemical structures, all species contain oxygenates. However, triglyceride is a product of glycol and fatty acid reaction, known as esters. The structure of propylene glycol ($C_3H_8O_3$) and fatty acids contained in the above triglyceride are shown in Figure 2.

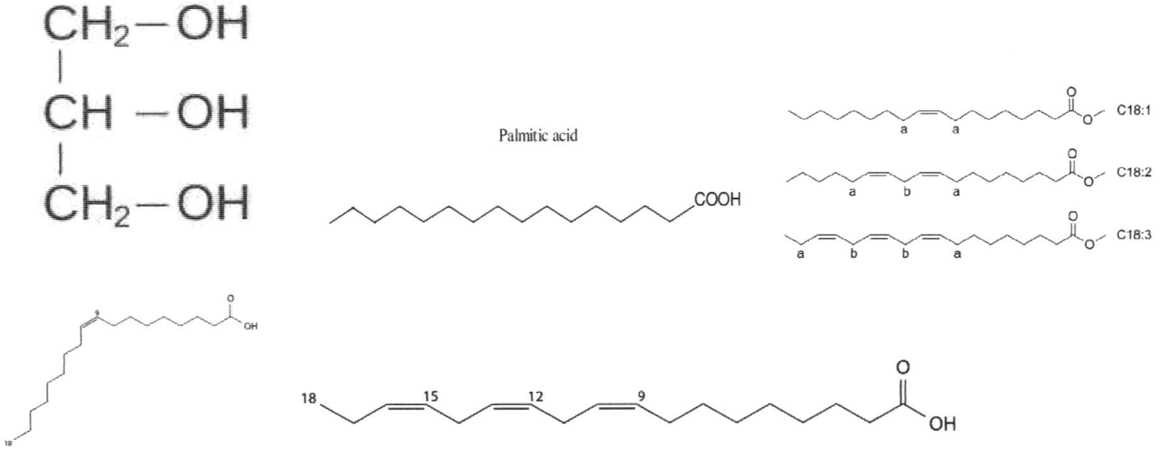

Figure 2: Structure of glycerol, palmitic acid (C16), oleic acid (C18:1) and linoleic acid(C18:3)

Note that triglycerides are esters of glycerol. So by hydrolysis, the triglycerides can be decomposed to glycerol (an alcohol) and fatty acids. These fatty acids contain mainly normal chain compounds with 0 to 3 double bonds. The carbon number of fatty acids varies from C16 to C24. Vegetable oil contains many such triglycerides. An example of camelina oil is shown in Table 2. This seed is mainly grown in Florida. The composition of many oils are available in the literature [6]. Some oils are shown in Table 3. As one can see, the molecular weight of triglyceride (>800) is much higher than that used in ethylene plant feeds. However, with hydrolysis, each triglyceride produces a fatty acid with a low molecular weight. Though vegetable oil cannot be cracked as such, triglycerides can be converted to fatty acids, which can be cracked.

Table 2: Camelina Oil Hydrolysis

Components	C	H	O	MW Triglyceride	Typical % in Oil
Glycerol segment	3	8	3		
Palmitic Acid (16:0)	16	32	2	806	6
Stearic Acid (18:0)	18	36	2	890	3
Oleic Acid (18:1)	18	34	2	884	17
Linoleic Acid (18:2)	18	32	2	878	17
Linolenic Acid (18:3)	18	30	2	872	35
Arachidic Acid (20:0)	20	40	2	974	1
Gondoic Acid (20:1)	20	38	2	968	15
Erucic Acid (22:1)	22	42	2	1052	2
(20:2)	20	36	2	962	2
(20:3)	20	34	2	956	2
Total					100

After hydrolysis, the total weight will increase to 106% due to consuming 6% water.

Table 3: Fatty acid distribution in some oils considered in ISOTERRA pilot plant studies

Fatty Acid Content by carbon Number and degree of Unsaturation													
	n.C12	**n.C14**	**n.C16**		**nC18**				**n.C20**		**n.C22**		**n.C24**
Boiling Point, F	421	489	459		603				649		695		736
Double Bonds	0	0	0	1	0	1	2	3	0	1	0	1	0
Palm Oil	<1	1-2	32-48	<2	4-6	37-53	6-13	<1	<1	<0.5	<0.2	-	-
Soybean	<1	<1	7-13	<1	1-6	12-27	52-63	3-12	0-4	<0.5	<1	<1	-
Tallow	0-2	2-8	23-28	0-5	14-29	35-50	1-5	<1	0.1	<0.1	-		-
Carinata	-	-	2-5	-	1-3	8-65	11-22	5-23	-	-	-		-

Whether the fatty acid is saturated or unsaturated, it contains an acid group. Therefore, straight forward cracking of hydrolysis products will produce additional acids, aldehydes, ketones and some oxygenated species. They can contaminate olefin plant products. Once the oxygen is removed then the acid produces only saturated normal paraffins or unsaturated normal olefins, diolefins or triolefins. Therefore, the process of removing the oxygen plays a vital role in the suitability as cracker feed. As mentioned earlier, the BIC ISOCONVERSION process is a thermal process using supercritical water after acid hydrolysis of the oil. Therefore, unreacted oxygenates and other oxygen compounds can be present like in any thermal process. Chevron Lummus Global developed a catalytic process to upgrade the oil suitable for bio diesel production. This is commercially known as the ISOTERRA process. The flow diagram is shown in Figure 3.

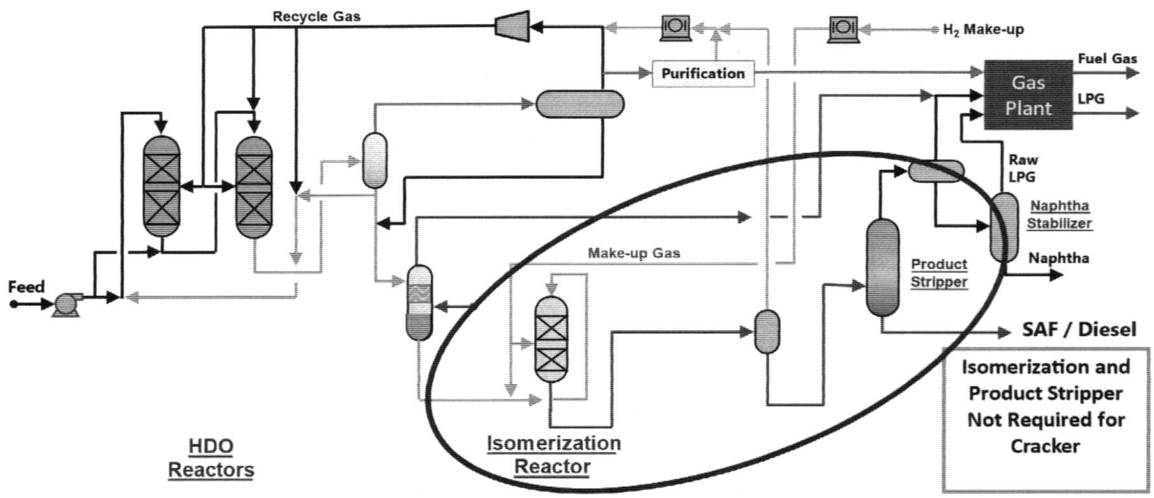

Figure 3: ISOTERRA Process for SAF/Diesel production

In this process, the vegetable oil is hydrotreated using a proprietary catalyst. The oil and the fatty acids formed are hydrotreated. Generally, glycerol is converted to propane and the fatty acids are converted to normal paraffins. In the presence of hydrogen, all of the double bonds will be saturated, and the liquid product will contain only normal paraffins (C15-C25). Oxygen in the fatty acids is converted to CO, CO_2 and H_2O and no additional aldehydes or oxygenates are formed. It is easy to separate CO/CO_2 and water from high molecular weight paraffins. The reactions taking place are summarized in Table 4.

Table 4: Reactions in ISOTERRA Process for Diesel

Hydrotreating reaction for triglycerides (typically vegetable oils)	
Hydrotreating reaction for fatty acids	
Isomerization	
Hydrogenation of olefins	$- 4 H_2$
Decarboxylation	$CO_2 - 3 H_2$
Decarbonylation	$CO, H_2O - 6 H_2$
Deoxygenation	$H_2O - 12 H_2$
Water gas shift	$CO_2 + H_2 \rightarrow H_2O + CO$
Methanation	$CO + 3H_2 \rightarrow CH_4 + H_2O$

In that table typical hydrogen consumption is also shown. The ISOTERRA reactor operates at moderate pressure and moderate temperature. The reactors are called hydrodeoxygenation reactors (HDO). The HDO reactors produce normal paraffins and hence the product will not meet diesel specifications. In the second reactor, an isomerization catalyst system is used to convert normal paraffins to isoparaffins. The ethylene plant prefers normal paraffins for ethylene production. Therefore, by removing the isomerization step, the complete products from the ISOTERRA process can be sent to an ethylene plant. A simplified block flow diagram for olefin production is shown in Figure 4.

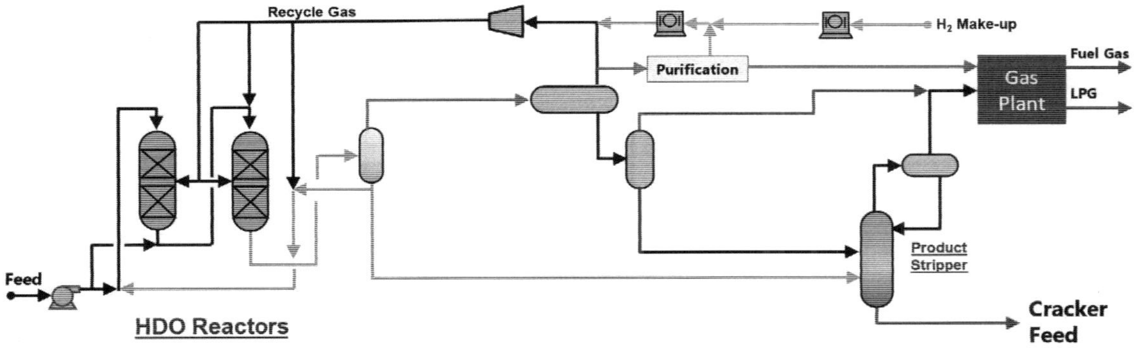

Figure 4: Modified ISOTERRA process for olefin Production

To meet fuel specifications, a complete fractionation column is required in ISOTERRA process. As will be seen, no hydrocarbons higher than C26 are present in many experiments and hence there is no need for a column and the PFD can be simplified significantly. The maximum carbon number depends upon the vegetable oil. Interestingly, animal fats share similar fatty acids range. The contaminants are different from petroleum feeds as shown in Table 5.

Table 5: Typical Contaminants Present in Bio Feeds

Feed Type	Used Cooking Oil	Palm Oil	Tallow (Animal fat)	Soybean Oil	Distillers Corn Oil (DCO)
Moisture,%	2.0	0.1	1.0	0.1	4.0
Insoluble,%	0.5		0.2		
Unsaponifiable,%	1.5	1.0	1.4	1.0	
Phosphorous, wppm	65.0	2.6	250.0	1.0	50.0
Total Metals (excl. P), wppm	170	5	300	3	
Total Nitrogen, wppm	100	5	500	1	95
Total Sulfur, wppm	50	5	80	1	100
Inorganic Chloride, wppm	110		32	1	10
Organic chloride, wppm	15		10	1	10
Total Chloride, wppm	125	10	43	2	20

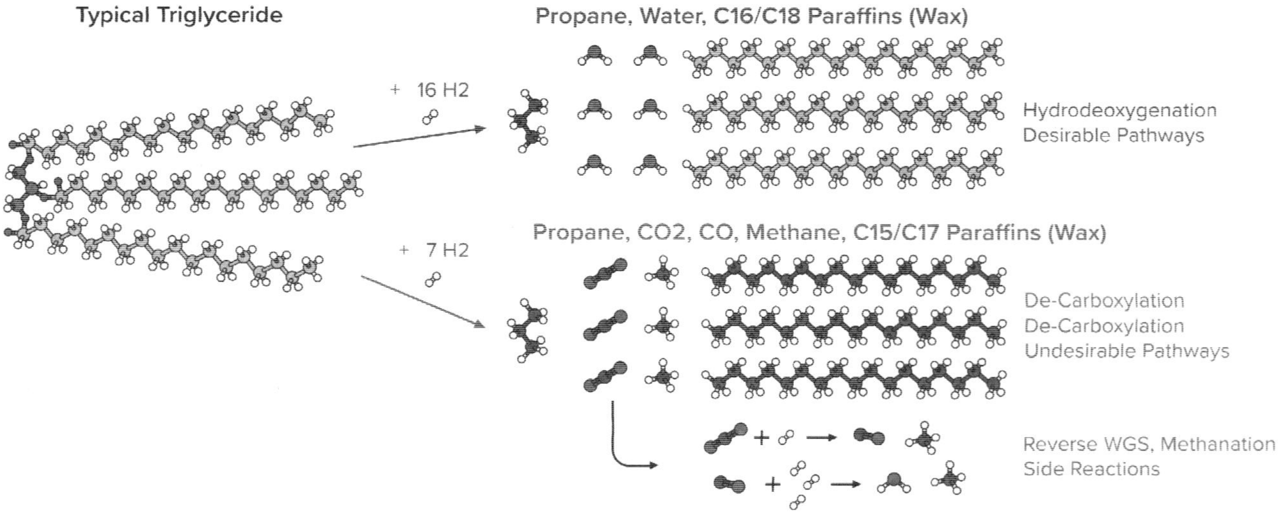

Figure 5: Different routes in catalytic process

As discussed above, in the first two reactors of the ISOTERRA process, hydrogen addition is carefully monitored so that it produces a maximum amount of paraffins, which will consume more hydrogen. As shown in Fig.5, hydrodeoxygenation consumes more hydrogen and produces water and paraffin as by product. This is the preferred pathway. With a lower hydrogen consumption, decarboxylation and decarbonylation are promoted and produce CO & CO_2. So operating parameters determine the selectivity to paraffin and CO_2 products. Reverse water gas reaction reacts with CO_2 and hydrogen to produce CO and methane. This route is not preferred since carbon is lost to methane or CO/CO_2. By cutting down the hydrogen flowrate, the process will consume less hydrogen, but it produces less valuable wax and more methane and CO/CO_2. With increased hydrogen consumption, oxygen is removed as water instead of as CO/CO_2. Through this route, the catalyst will also last longer. Certainly, for olefin production, high hydrogen consumption with wax (long chain normal paraffins) is the preferred route. Some of the feeds tested in the pilot unit are shown in Table 6. Some contaminants present in the feed and in the products are also shown in this table.

Table 6: Oils Tested and effect of hydrotreating

Oils Tested		
Crude Soybean Oil	Distiller's Corn Oil	Used Cooking Oil
Animal Fats	Yellow Grease	Brown Grease
Soapstock	Fatty acid distillate	Phosphate Gums
Algal Oils(70% phospholipids)		

	Feed			**Product**		
	P(ppm)	**Na(ppm)**	**K(ppm)**	**P(ppm)**	**Na(ppm)**	**K(ppm)**
Crude Soybean Oil	642	13.2	229	0.6	1.4	1.4
Packers Tallow	142	229	59.4	0.7	1.6	<0.2
Used Cooking Oil	41.2	14.2	33.9	1.4	0	0

Hydrotreating reduced phosphorus and all metals to extremely low levels in the product. The feed and products were rigorously analyzed for many contaminants. Note that the metals are organically bound and are not in elemental state. As mentioned earlier, oxygen is removed as H_2O, CO & CO_2. ISOTERRA produces much superior quality feed for an ethylene plant. It preserves the normal paraffins in the first stage.

Ethylene plant

Soybean was extensively tested in the pilot plant and hence its application as a cracker feed is discussed below as an example. However, all bio feeds mentioned above can be cracked in existing and/or new cracking heaters. For ethylene plant feed as mentioned previously there is no need for an isomerization section. Therefore, the liquid sample before isomerization was analyzed and the properties are shown in Table 7. The liquid product before isomerization is the optimal feed to be routed to cracker. Figure 6 shows the distillation curve of the feed to cracker and Figure 7 shows the carbon distribution.

Table 7: HVO feed characteristics obtained from Soy Oil

API	47.6
Viscosity@70C, cst	2.272
Br.#	1
Carbon, wt%	84.9
H, wt%	15.0
P-Total, vol%	97.2
N-Total	2.8
A-Total	0
Mono Naphthenes	2.8

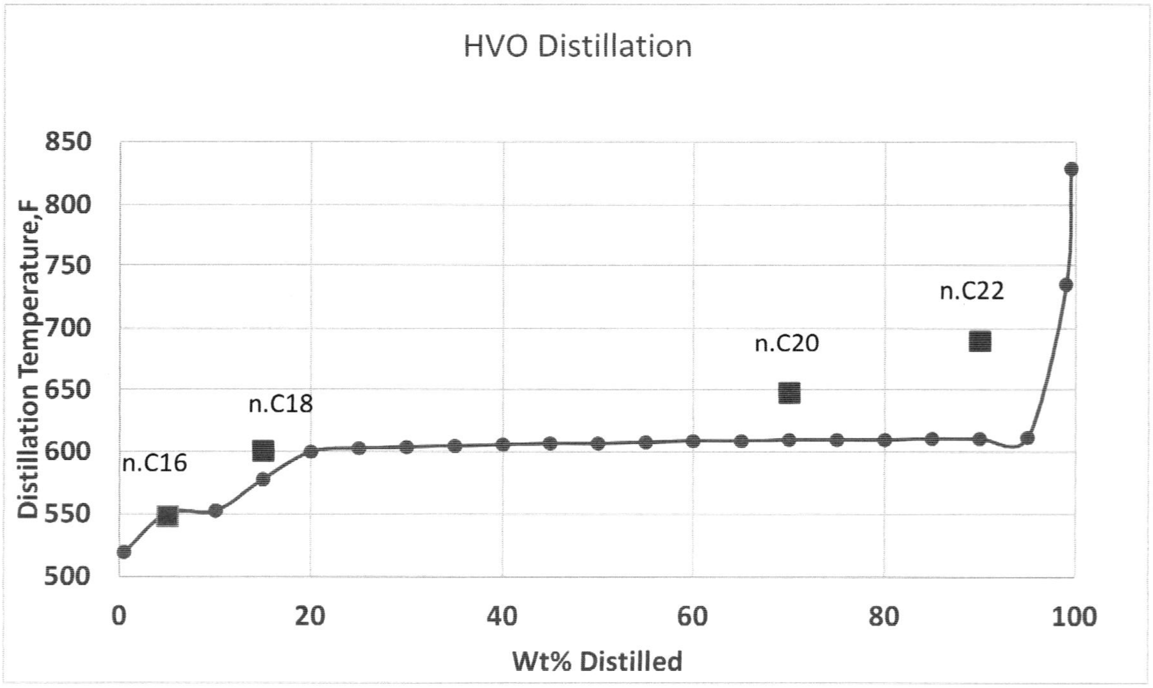

Figure 6: Distillation curve of HVO feed along n.C16-nC22 Paraffin Boiling Points

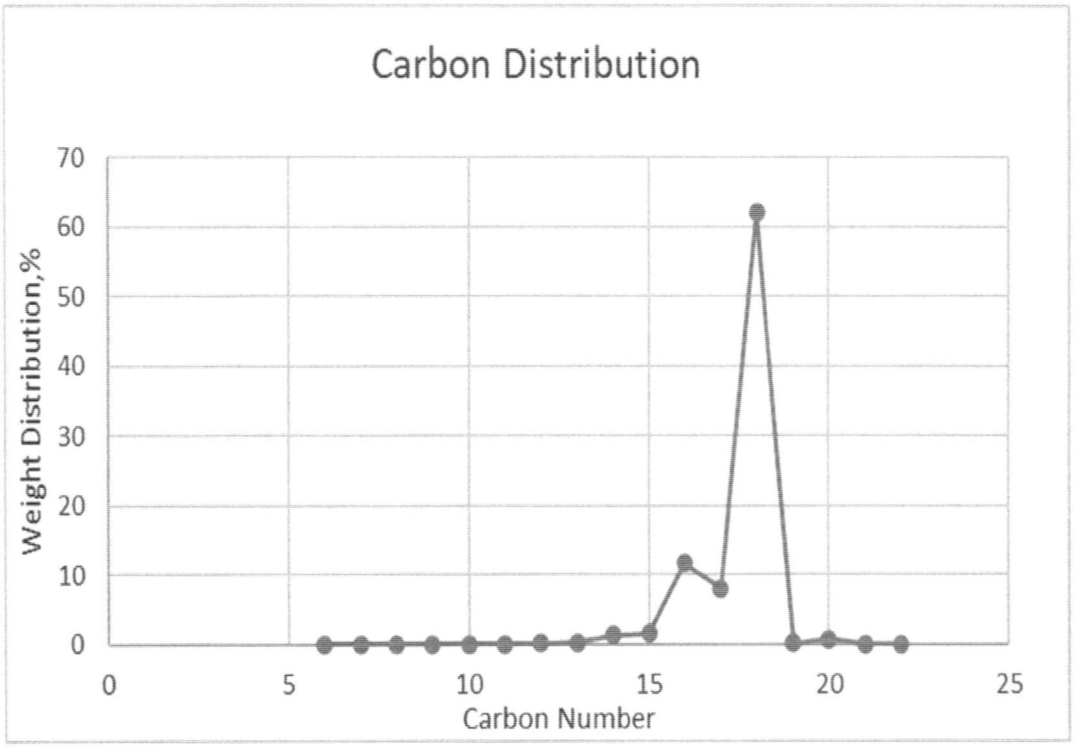

Figure 7: Carbon distribution of HVO feed before Isomerization

In Figure 6, boiling point of normal C16, C18, C20 and C22 paraffins are also shown. The carbon distribution is similar to other literature results [7]. The distillation curve is nearly flat for a major portion and boils around C18, normal paraffin, and that is also confirmed by the carbon distribution shown in Figure 7. This illustrates that the HDO reactor catalyst almost does not produce any carbon molecules lower than C14. Selectivity of the catalyst to preserve the original fatty acid is clearly seen in Figure 7. HDO reactor produces very low isomers and practically there are no aromatics and more than 97% is paraffins. The advantage of HDO reactor in the ISOTERRA process is very clear. It selectively removes the oxygen atoms and preserves normal paraffin chain, and no cyclization is performed. It produces an excellent feed to cracker. One familiar in the cracker industry knows that the normal paraffins are the best feeds to produce olefins and they produce the highest amount of ethylene for the given carbon number. The same HDO reactor removes all the contaminants as shown in Table 6. Therefore, there is no need to take any additional impurity clean-up in the cracker. Looking at the feed properties, the feed is mainly concentrated in C18 and only a small portion is above C20. This is the characteristics of many bio feeds (vegetable oils). It is very rare that naturally occurring fatty acid has higher (>5%) in the C22 range. Even n.$C_{22}H_{46}$ boils below 700°F (371°C). All of these molecules boil in the gasoil (or diesel) range. A typical gasoil from a petroleum source has less than 60% paraffins and the remaining are polynuclear naphthenes and aromatics. Therefore, when a gasoil is vaporized, condensed molecules have a high propensity to form coke in the convection section. It is imperative to vaporize the gasoil at lower temperatures since the coking rate increases exponentially with temperature. When the feed is completely normal paraffins, coking rate is also significantly reduced. Once vaporized, the performance in the radiant coil, whether it is a naphtha feed or in gasoil boiling range, is not significantly different. In fact, normal paraffins tend to coke less in the radiant coil also. In addition, fuel oil production is reduced significantly as shown later and hence TLE fouling rate is also reduced significantly though the HVO feed boils in the gasoil range. In summary, cracking HVO feed has less issues in the convection section, radiant section and in the TLE.

Table 8 shows once-through yields for a typical two pass coil for HVO cracking at high severity. The P/E ratio is 0.43 w/w. Since the hydrogen content of HVO is very high, especially compared with typical full range naphtha, it is not surprising the ethylene yield is also high. Even at such severity and with a high end point compared with a naphtha feed, PFO yield is low. So HVO cracking is better than naphtha cracking. Table 9 shows the overall material balance for a 100% full range naphtha feed (P= 69.2, N=21.2 and A=9.6 wt%) and 100% HVO feed for a typical 1,200 KTA plant at P/E=0.5.

Table 8: Typical HVO Cracking Yields

S.G	**0.788**
Yields, wt %	
Hydrogen	0.85
Methane	16.00
Acetylene	1.20
Ethylene	34.00
Ethane	2.80
MA+PD	1.20
Propylene	14.00
Propane	0.20
Butadiene	6.80
Butene	2.20
Butane	0.15
C5-400F	12.78
Fuel oil	7.82
Total	100.00
% Benzene in C5-400	44

Table 9: Material Balance For 1,200 KTA Ethylene production

Case	1	2
FEEDS, KTA		
Naphtha	3,449	0
HVO		3,283
Net steam reacted	4	3
TOTAL	3,453	3,286
PRODUCTS, KTA		
Hydrogen Product	22	20
Methane Rich Off-gas	560	528
PG Ethylene	1,200	1,200
PG Propylene from Cracker	593	593
BD Product	182	225
Butenes	147	126
Butanes	11	9
Benzene	190	149
C7-C8 Gasoline	298	72
C9-205 Gasoline	86	140
Pyrolysis Gas Oil	54	77
Pyrolysis Fuel Oil	105	146
Acid Gases	3	3
TOTAL	3,453	3,286

Case 1 is based on naphtha feed with ethane and propane recycle. The same coil and operating conditions are used for HVO feed, also with ethane and propane recycle. HDO reactor also produces some propane from glycerol oxygen removal and hence except for oxygen in the fatty acid, all carbon and hydrogen in the triglycerides are used in making olefins and only a small amount of carbon is rejected as CO/CO_2, usually much less than 0.6%. Note that to meet the same ultimate P/E of naphtha feed, HVO feed is operated at a much lower severity; and still the HVO feed requirement is about 5% less than naphtha feed. At high severity, the HVO requirement is still much lower. For this case, about 73,000 BPSD of raw vegetable oil is required, which is a large quantity. In 2022, the world production of soy oil was 59.3 million tons. Therefore, though HVO can give very good yields, the large quantity of oil will not be available as cracker feed. In addition, it is competing with biodiesel market, thus the supply of oil will be limited. Until the governments and industries work together, such large quantities will not be available in the near future. However, the source of the feed does not depend upon petroleum and hence with suitable program, the market is completely sustainable. All products (ethylene, propylene, butadiene, benzene etc.) obtained with HVO feeds are bio chemicals and qualify for government credit in many countries. Compared with ethanol to ethylene, which produces only bio ethylene, HVO directly produces bio olefins and other chemicals, and all are qualified for credits as bioproducts. Nature has given an excellent way to detect bio materials from fossil derived materials. In all natural (bio) oils, Carbon 14/Carbon 12 ratio determines amount of bio feeds. Fossil feeds do not contain Carbon 14 compounds and thus the chemical industry has used this marker to qualify for government credits. So even with slightly higher cost of HVO feed, it will be economically attractive. Moreover, this process uses UCO, which is currently disposed as waste. One report mentions that a total of less than 5 million metric tons of UCO were collected from all major industrialized world in 2020. They can be used to make olefins and then biopolymers.

As mentioned, such a large quantity of HVO feed is not easily available to supply all of the feed for an ethylene plant. However, based on recent enquiries from clients from all around the word for biodiesel production, 10,000 -15,000 BPSD is practical. So instead of one full plant, what is the performance if only one heater is operated with HVO feed? The heater performance is shown in Table 10.

Table 10: Heater Operating Conditions

Case	**1**	**2**	**3**	**4**
Feed	Naphtha	HVO	HVO	HVO
Flowrate, t/h	92.9	85.7	79.8	92.9
S/O, w/w	0.5	0.5	0.5	0.5
Severity	Ult P/E=0.49	Ult P/E=0.49	Max.C_2H_4	Max.C_2H_4
Cross Over Temp., C	619	593	596	610
Coil Outlet Temp., C	835	829	833	839
Fuel Fired, MMKCal/h	144.9	131.9	124.1	148.1
SHP Steam, T/h	101.5	91.9	86.3	104.6
Overall Th. Efficiency, %	94.7	94.3	94.3	94.1
C_2H_4 Production, KTA	217	217	217	253
C_2H_4 Production, % Naphtha case	100	100	100	116
Run-length, % Naphtha case	100	128	107	74
CO_2 emission, t/t-C_2H_4	1.16	1.06	1.00	1.02
CO_2 emission, % naphtha basis	100.0	91.0	85.7	102.3

In Table 10, Case is for an existing heater is designed to produce 217 KTA ethylene for naphtha operation. The same heater with the corresponding P/E ratio with HVO feed is shown as case 2. As discussed previously, since the HVO feed produces higher olefin yields than naphtha feed, the feed rate is reduced for the same ethylene production. Therefore, for the same ethylene production, fired duty is decreased and hence SHP steam production is decreased. The consequential benefit is reduced CO_2 emission and longer run heater length. There is no heater limitation at this condition. Generally, when the convection section is designed for a specific feed, the final vaporization location of the feed is critical. When the final vaporization occurs at high temperature (lower mixed preheat coil), there is a high probability of increased coke formation at that location. To avoid this, higher steam to oil ratio is used. This reduces the hydrocarbon partial pressure and shifts the vaporization point at low temperatures (upper mixed preheat coil). Fig.8 shows the dew point of naphtha-HVO blends for different weight ratios with and without steam addition at typical convection section conditions. Without steam addition 100% HVO boils well above 390°C, but with steam (S/O=0.5) the vaporization point is reduced by more than 100°C. At these conditions, HVO can be fully vaporized in the upper mixed preheat section avoiding coke formation. So even a naphtha-only designed heater can be used to crack full HVO feed. This is rarely possible for a petroleum-based gas oil. Note that the TLE does not foul as rapidly when operating with HVO as compared to with gas oil, due to narrow boiling curve of the highly paraffinic HVO feed. In some rare cases where vaporization is still in the hot temperature zone, blending HVO with naphtha is sufficient to shift the vaporization. Depending the heater design, a suitable ratio of naphtha-HVO can be selected based on mixed feed dew point.

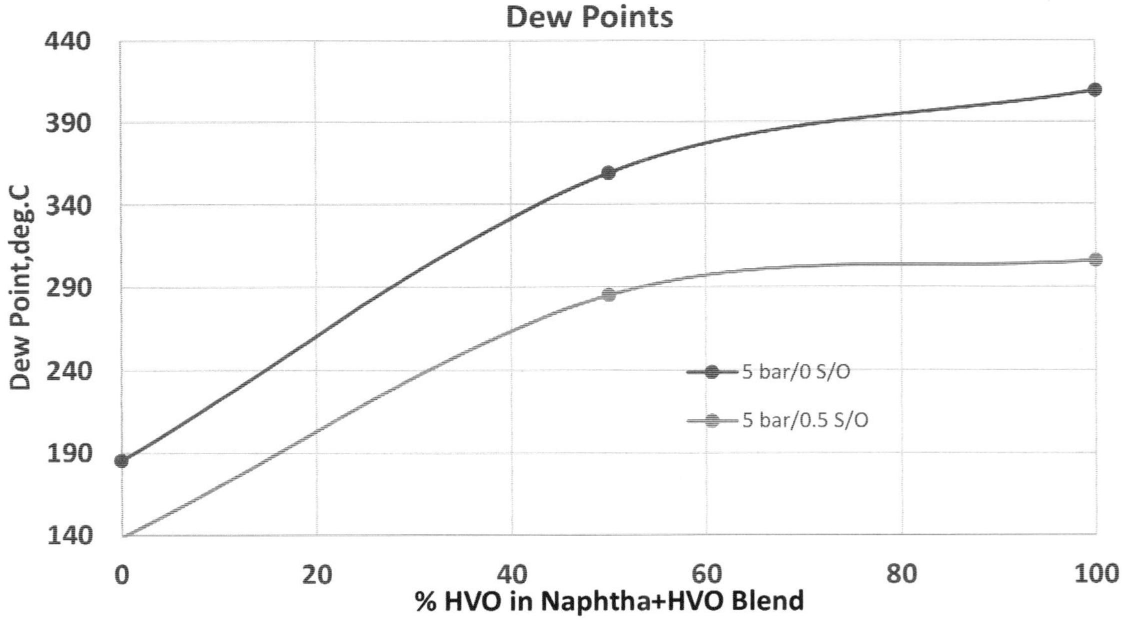

Figure 8: Dew point for different HVO-Naphtha Blends with and without dilution steam

Since there are no limitations in the heater at 100% ethylene production basis at design severity, the heater was rated for maximum severity producing the same once through ethylene production as the base naphtha case, which is shown as case 3. This reduced the feed rate further due to the high ethylene yield at maximum severity, and also reduced the fuel consumption significantly and reduced the SHP steam production. Since there are no heater limitations (burner, ID fan etc.) at the case 2 or case 3 conditions, the heater was rated for same feedrate as the naphtha case (Case 1) at high severity, which is shown as case 4. This produces about 16% more ethylene than the naphtha case. Though there are no limitations in the heater even at case 4 conditions, SOR TMT is slightly higher than the base naphtha case and a small reduction in run length is expected. Even with shorter run length, case 4 will run more than 45 days. Hence, HVO feed is more attractive than naphtha feeds and for a given heater, a better overall performance is expected. For case 2 operation, about 13,200 BPSD of vegetable oil is required, which is well within the current quantities available in many parts of the world.

As discussed earlier, biodiesel is the main focus of vegetable oil in many countries. Hence some refiners want flexibility either to produce biodiesel or cracker feed. Bio-diesel production uses isomerization and hence the product is rich in isoparaffins and still cyclic compounds are low in isomerized product. Figure 9 shows the distillation curve after isomerization of the same feed.

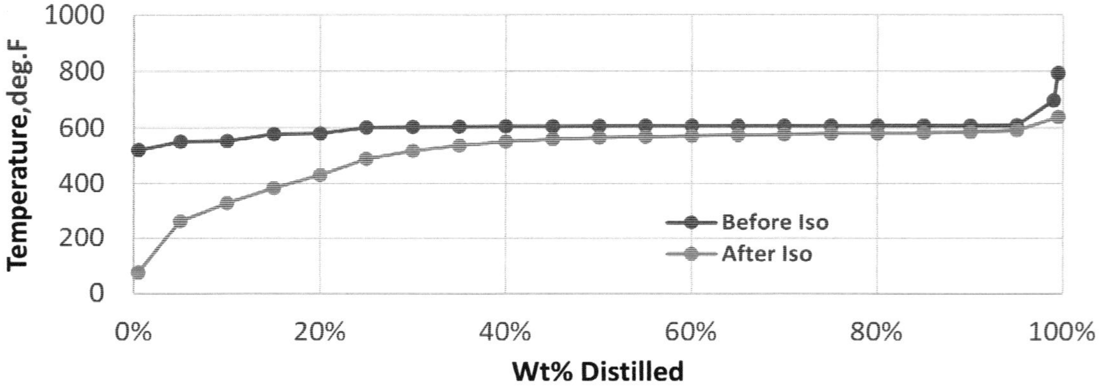

Figure 9: Distillation curve of HVO feed before and after isomerization (Soybean)

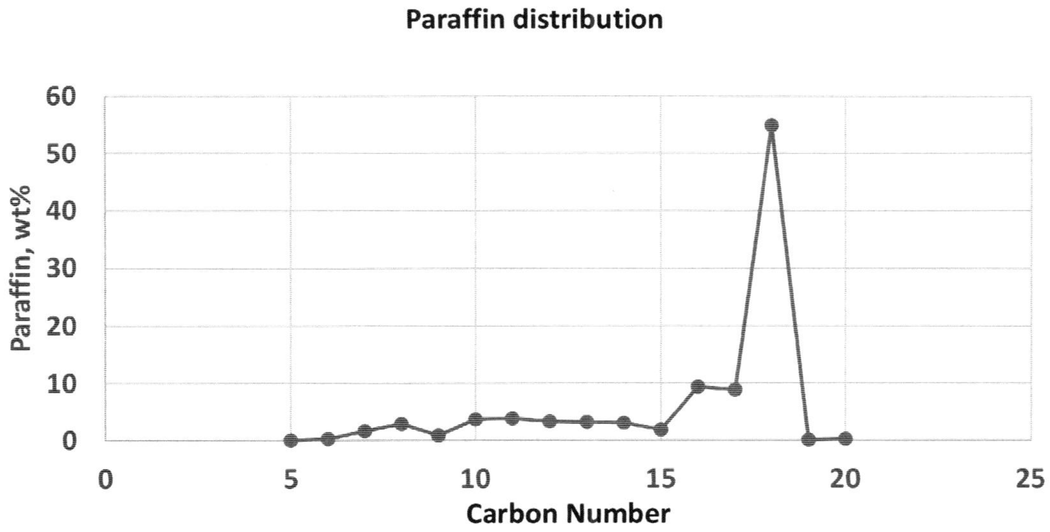

Figure 10: Total Paraffin distribution as a function of carbon number after Isomerization

Table 11: Properties of HVO feed after Isomerization

Product After Isomerization API	51.4
Sulfur,wppm	<5
Oxygen,wppm	<200
P-Total,wt%	98.26
N-Total	1.52
A-Total	0.16
I/N Ratio of paraf	19.5

Figure 10 shows the total paraffin distribution as a function of carbon number and the PIONA values are shown in Table 11. Isomerization almost preserves the total paraffins and produces only a small amount of cyclic compounds. During isomerization, some cracking also occurs and some lower molecular weight compounds are formed giving a low initial boiling point. As seen in Table 11, I/N ratio of paraffins is extremely high and thus it is not as good of a feed as the normal paraffins for olefin production but is an excellent product for fuel. Nevertheless, even this high isoparaffins can be cracked in the same heater when the diesel can not be sold. As mentioned, there are no issues in vaporizing the isomerized diesel. Calculated cracker yields are shown before and after isomerization at high severity in Table 12.

Table 12: Cracker Yields at high severity before and after isomerization

Typical yields before and after isomerization at high severity		
	BEFORE	**AFTER**
Ethylene, wt%	34.0	24.3
Propylene, wt%	14.0	14.1
Butadiene, wt%	6.8	6.2
C5-400F, wt%	12.8	19.0
PFO, wt%	7.8	9.3
Methane, wt%	16.0	16.2
P/E ratio, w/w	0.41	0.58

As seen from Table 12, even the isomerized diesel gives reasonable olefin yields. Notice the diesel contains mainly C18 isomers and hence on decomposition it can still produce normal hydrocarbon radicals and produce significant amount of ethylene unlike low molecular weight paraffins (i.C4 & i.C5). So the advantage of taking the cut before isomerization is obvious. The ISOTERRA plant can be designed with flexibility to meet both olefin feedstock and diesel product requirements by simply adjusting the product routing or operating conditions around the reactors. By designing for only olefin plant feed. the CAPEX is reduced since isomerization and fractionation steps are not required.

Summary and Conclusion

Currently petroleum-based ethane, propane, naphtha and gas oil are used to make ethylene. For sustainability and to conserve our limited resources plastic recycles, fuel oil and bio feeds such as vegetable oils and animal fats are gaining importance. Most feeds contain contaminants and hence are not suitable as cracker feeds. However, they can be cleaned up to be suitable for cracker. In Lummus' Thermal Crude to Chemicals (TC2C™) technology, pyrolysis fuel oil (PFO) is upgraded through CLG's hydroprocessing technology to produce olefins. Plastic is converted to Plastic PyOil and cracked to produce olefins. In this paper, vegetable oil and animal fats are hydroprocessed to produce feeds suitable for a cracker using CLG's ISOTERRA process. The ISOTERRA process is currently mainly used to produce biodiesels from vegetable oil. However, during hydroprocessing of the oil, it produces only a normal paraffinic product in the first reactor. As a clean product with free or low levels of contaminants, it is a good feed for a cracker. Since this is mainly C16-C20 rich normal paraffins, ethylene and olefin yields are high and much better than from naphtha and diesel feeds. It has a low coking tendency and hence existing naphtha crackers can be used to crack these feeds. Also, all the products, ethylene, propylene, butadiene and benzene are bio products and qualify for government credit in many countries. In most countries, used cooking oil and other non-food oils and wasted animal fats are disposed as waste and can be economically converted to valuable chemicals. HVO feed reduces the fuel consumption for the same amount of ethylene and propylene production compared with naphtha and gas oil crackers and hence CO_2 emission is also reduced by 5 to 10%. Though not discussed here, our in-house study shows HVO cracking is economical, and the payout is less than 2 years in most locations. It is economical to crack HVO and most of the existing crackers can be used without any modifications or only with minor modifications.

References

1 Statista, https://www.statista.com/statistics/1067372/global-ethylene-production-capacity/

2) K.M. Sundaram, U.K. Mukherjee, P.M. Santos and R.M. Venner, "Thermal Crude to Chemicals," paper presented at Ethylene Producer's Conference, Houston, March 13-16, 2023.

3) K.M. Sundaram, U.K. Mukherjee, P.M. Santos and R.M. Venner, "Crude to Chemicals," PTQ Part 1, Q4, 2023; PTQ Part 2, Q1, 2024.

4) https://cleanfuels.org/wp-content/uploads/GlobalData_UCO-Supply-Outlook_Sep2023.pdf

5) "CLG commissions unit at El Segundo Refinery," Hydrocarbon Eng., page 5, Dec. 2023.

6) E.M.A. Dauqan, H.A. Sami, A. Abdullah, Z.M. Kashim, "Fatty acid composition of four different vegetable oils by gas chromatography," 2nd Int. Conf. on Chem. And Chem. Eng., Vol. 24, 2011.

7) K.M. Van Geem, R. Abhari, S.P. Pyl, M.F. Reyniers & G.B. Marin, "From Biomass to ethylene; Steam Cracking of bio-synfined Naphtha," paper presented at EPC meeting held at St. Antonio, TX, March 21-25, 2010.

AIChE Paper Number: 190b

Maximizing Olefin Yields for Steam Cracking of Plastic Waste Pyrolysis Oils: An Experimental and Modeling Study

Tamás Buzogány
Drs. ir.
Ghent University, Ghent, Belgium

Ismaël Amghizar
ir.
AVGI, Ghent, Belgium

Marvin Kusenberg
Dr. ir.
Ghent University, Ghent, Belgium

Lin Chen
ir.
AVGI, Ghent, Belgium

David J. Brown
Dr. ir.
AVGI, Ghent, Belgium

Guy B. Marin
em. Prof. Dr. ir.
AVGI, Ghent, Belgium

Geraldine J. Heynderickx
Prof. Dr. ir.
Ghent University, Ghent, Belgium

Kevin M. Van Geem
Prof. Dr. ir.
Ghent University, Ghent, Belgium

Prepared for Presentation at the 2024 Spring National Meeting
New Orleans, LA, March 24 – March 28, 2024

AIChE Paper Number: 190b

Maximizing Olefin Yields for Steam Cracking of Plastic Waste Pyrolysis Oils: An Experimental and Modeling Study

Tamás Buzogány
Drs. ir.
Ghent University, Ghent, Belgium

Ismaël Amghizar
ir.
AVGI, Ghent, Belgium

Marvin Kusenberg
Dr. ir.
Ghent University, Ghent, Belgium

Lin Chen
ir.
AVGI, Ghent, Belgium

David J. Brown
Dr. ir.
AVGI, Ghent, Belgium

Guy B. Marin
em. Prof. Dr. ir.
AVGI, Ghent, Belgium

Geraldine J. Heynderickx
Prof. Dr. ir.
Ghent University, Ghent, Belgium

Kevin M. Van Geem
Prof. Dr. ir.
Ghent University, Ghent, Belgium

Abstract

The mismanagement of end-of-life plastics is no secret as the main part is being incinerated, discarded into the environment, exported, or landfilled [1, 2]. This leads to a linear economy for a large stream of products, which is not only unsustainable but also leads to the loss of a potentially valuable carbon source. To reduce society's dependence on fossil fuels and its burden on the environment, improved recycling of mixed plastic waste streams is necessary, both mechanically and chemically. Within chemical, or advanced recycling, thermal pyrolysis of plastic waste followed by the steam cracking of the produced oils is one of the most promising technologies. It is expected to grow exponentially in the coming years. This method is more robust compared to mechanical recycling in terms of the allowable degree of plastic waste mixing, since mechanical recycling typically requires pure waste polymer streams to close the material loop. However, chemical recycling also faces challenges related to impurities present in plastic-waste-derived feeds, particularly halogenated- and hetero-atom-containing components, and the high concentration of olefinic components [3-5]. As their effect on steam cracking in terms of yields and operating conditions is unknown, olefin producers are hesitant to feed the new feedstock into their existing steam crackers.

To shed more light on the effect of plastic waste pyrolysis oils on the steam cracking performance, a kinetic model, CRACKSIM, is developed for the steam cracking of plastic waste-derived feeds, enabling fast and reliable simulation of product yields using COILSIM1D [6]. Specifically, next to olefinic components, this model will also incorporate hetero-atom-containing components in the future. In this work, the method to develop kinetic schemes for the steam cracking of (hetero-atom-containing) hydrocarbons is presented. This framework allows capturing the essential chemistry for components commonly found in plastic-waste-derived feedstocks. Comparing the results of this methodology with pilot plant data gives valuable insights into pyrolysis oils' behavior during steam cracking.

Introduction

Chemical recycling of plastic waste is expected to grow exponentially in the coming decades [1]. Global legislation pushes for increasing the circularity of plastics [7], and mechanical recycling has limitations in handling highly mixed plastic waste streams, leading to recyclates of lower quality, i.e., so-called downcycling. Therefore, chemical or advanced recycling is the favored route, in particular, plastic waste pyrolysis. This process produces as the main product an oil that can be used to (partially) substitute fossil-based feedstocks to produce light olefins, with steam cracking being the most important process. The primary concern of utilizing pyrolysis oils in existing steam crackers is the substantially different composition of those synthetic oils and fossil-based feedstocks for which steam crackers were originally designed. Specifically, the high concentration of heteroatom-containing contaminants and up to 60% of unsaturated hydrocarbons [3-5] in typical pyrolysis oils will affect the (long-term) operation and product yields of industrial steam crackers. This is the reason why steam cracker operators are setting strict specifications on the degree of contamination for steam cracker feedstocks, particularly related to technical issues and off-spec products. Recent studies demonstrate the viability of using these oils under steam cracking conditions [3-5], however, several scientific challenges remain. New reaction possibilities and the formation of unwanted products (methanol, HCl, NO, etc.) need to be accurately described to conclusively evaluate the steam cracking feasibility at a large scale. Furthermore, olefin producers are uncertain about the optimal conditions for cracking these plastic waste oils due to their largely different composition in terms of unsaturated components. Therefore, a kinetic model,

CRACKSIM, is developed for the steam cracking of plastic waste-derived feeds, enabling fast and reliable simulation of product yields using the established simulation software COILSIM1D [6]. More specifically, within this work, the focus is on the branching structure of the plastic-waste derived feeds' iso-olefin and iso-paraffin content. For fossil-based feedstocks, the isomer distribution of paraffinic species is quasi-independent of its source [6]. However, for plastic-waste derived feeds this is no longer true. This means that kinetic schemes must be flexible enough to capture these structural differences and incorporate the necessary details to cope with the diversified product spectrum that elutes.

Modeling strategy

Kinetic model CRACKSIM

CRACKSIM forms the core of COILSIM1D, housing the necessary reaction network for accurate reactor simulations. The network is a single-event microkinetic model that has been developed at the Laboratory for Chemical Technology (LCT) of Ghent University for fossil-based feedstocks with a carbon numbers up to C_{50} [8, 9]. However, recent trends require the model to expand toward new feeds, including plastic-waste derived oils.

Several key components of this network are vital for its high performance. First, designated reaction families determine the decomposition pathways of all feedstock components, impacting final product yields. Second, specific assumptions simplify the network's generation process and reduce its complexity. The assumptions include the μ-hypothesis and the pseudo-steady-state assumption (PSSA) for μ-radical concentrations. Third, after generating reaction schemes, assigning kinetic and thermodynamic parameters to all elementary reaction steps is essential. Lastly, a lumping approach can be introduced to reduce the final models' size using exact mathematical calculations. Subsequent paragraphs will elaborate on these aspects further.

Reaction families

During steam cracking, a multitude of reactions take place. Fortunately, these numerous reactions can be categorized into just a few reaction families. This facilitates the integration of user-defined reactions into CRACKSIM. The model considers various reaction families, as outlined in Table 1, encompassing both radical and molecular reactions. The significance of these reaction families varies depending on the feedstock composition. For instance, in feeds derived from plastic waste with a substantial olefin content, retro-ene reactions play a significant role in determining final product yields, whereas their contribution is less pronounced in fossil-based feedstocks which contain mainly saturated hydrocarbons.

Table 1: Reaction families considered in CRACKSIM. Within the radical reactions a distinction can be made between: green = reactions initializing the decomposition of molecules; red = decomposition reactions for μ-radicals; blue = isomerization reactions.

Reaction family	Example	Ref.
	Radical reactions	
Intermolecular hydrogen abstraction	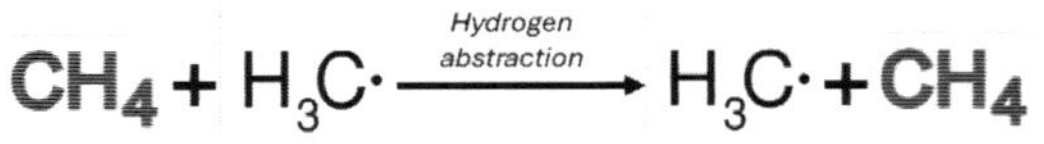	[10, 11]
Bond scission	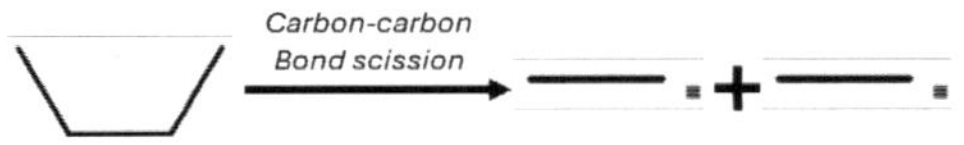	[12]
Addition	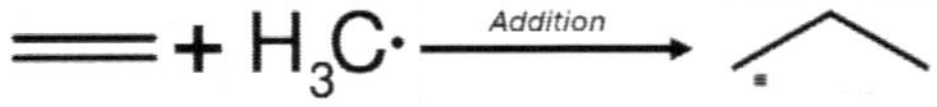	[13, 14]
Ipso-addition	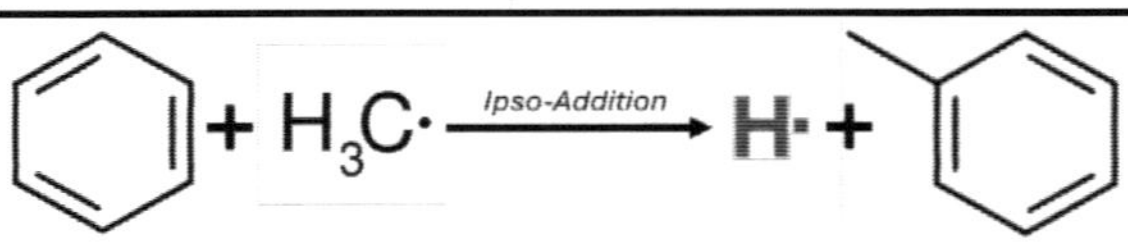	[15]
β-scissions [XX and XY]	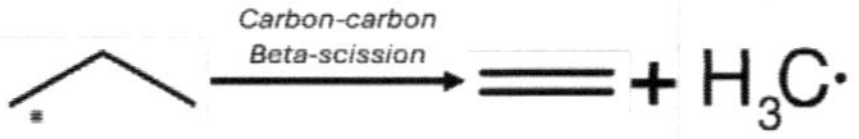	[13, 14]
Hydrogen shifts [1.2 to 1.6]	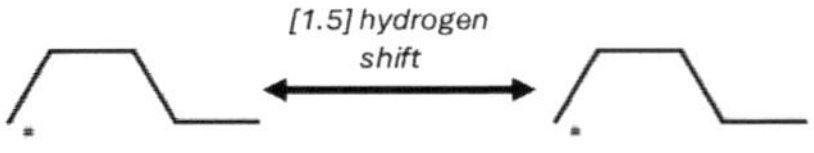	[16, 17]
Endo- and exo cyclisation and ring opening	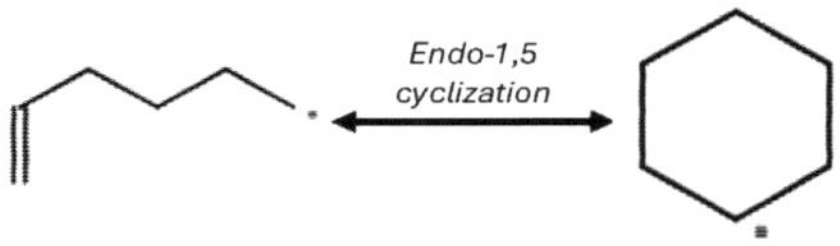	[17]
Mesomerization	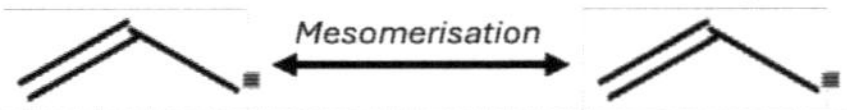	N.A.
	Molecular reactions	
Retro-ene	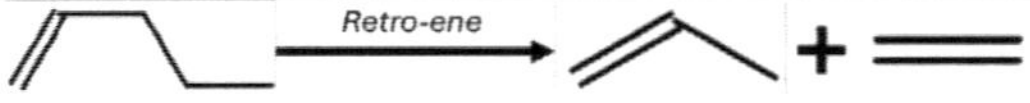	[18]
Retro Diels-Alder	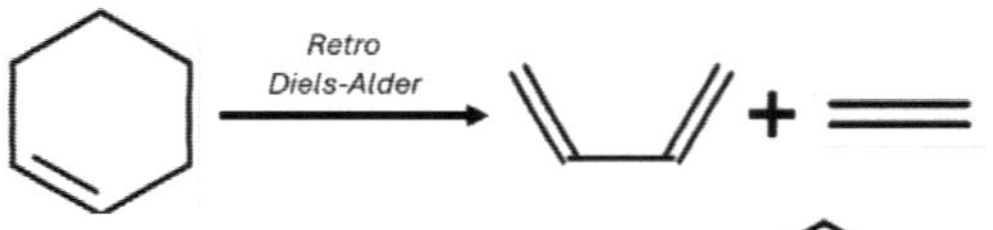	[19]
Electrocyclization	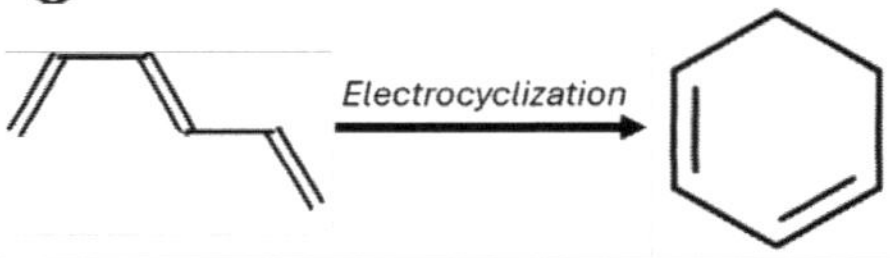	[6]

μ-hypothesis

Due to the extensive number of reactions and intermediates that necessitate consideration, the kinetic model could potentially become too unwieldy for swift simulations using COILSIM1D. However, the μ-hypothesis, which has been validated [6], offers a solution by dividing the network into two distinct parts: the μ-network and the β-network. It does so by assuming that C_{6+} molecules undergo only monomolecular reactions, thus splitting the full network.

The β-network encompasses the decomposition of C_{6-} molecules, which undergo both uni- and bimolecular reactions. The presence of the hydrocarbon matrix is crucial for these components, and simplifications in the reaction scheme are not permissible to ensure precise results. Within the β-network, all potential reactions and radicals are meticulously calculated and accounted for using the in-house developed reaction mechanism generator Genesys [20, 21].

On the contrary, the primary μ-network focuses on the decomposition pathway of C_{6+} radicals. For these components, it is assumed that bimolecular reactions are negligible compared to monomolecular reactions. Consequently, only a limited number of reaction families are considered within this network to decompose these components to the level of the β-network, as indicated in Table 1 (highlighted in red) and Figure 1. The primary μ-network is generated using the in-house developed network generator PRIM. Thus, all μ-radicals decompose into molecules and β-radicals accounted for in the β-network, establishing a connection between the two.

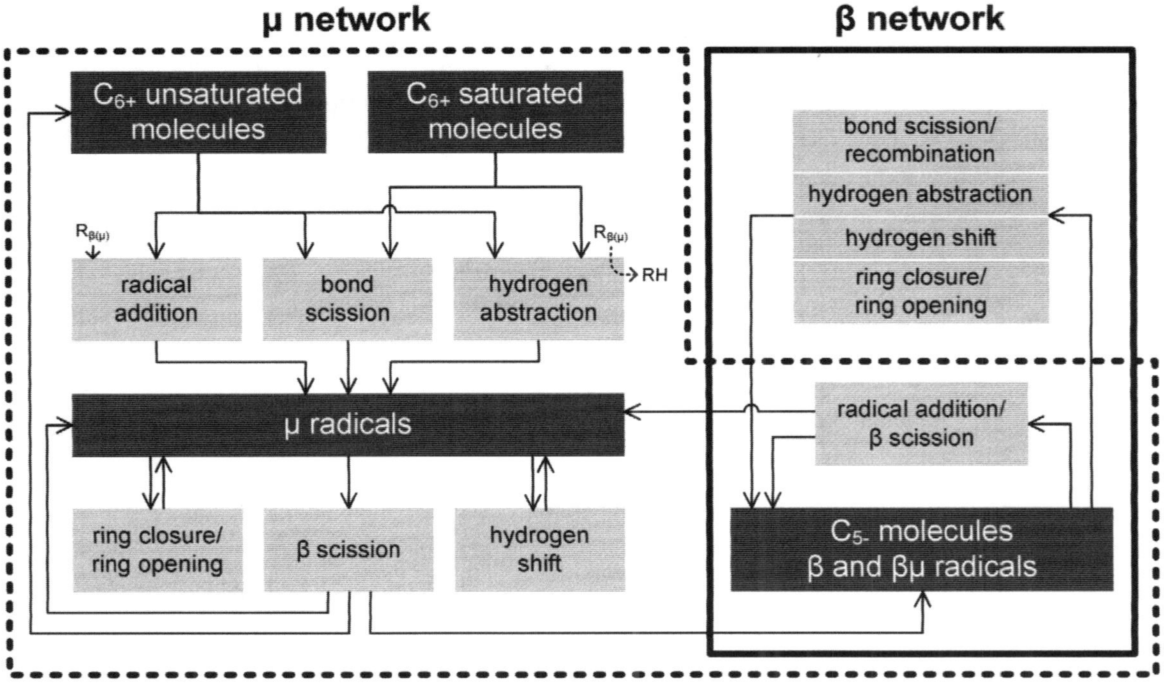

Figure 1: Overview of the construction of the single event microkinetic model. Interaction of the μ-network with the β-network. Figure adapted from Pyl [22].

Pseudo-steady-state approximation (PSSA) for the concentration of μ-radicals

The PSSA represents the second assumption within CRACKSIM, wherein the time-dependent concentration of μ-radicals is assumed to be zero. This simplification enables the removal of the μ-radicals' concentration from the continuity equations, leading to a significant reduction in system complexity and computational requirements. Additionally, because of the validated μ-hypothesis, it does so without any loss of accuracy for the final yields [6]. Equation (1) outlines the continuity equation governing these radicals.

$$\frac{d[\mu_i]}{dt} = S_i - \sum_{j=1}^{N_{dec}} k_{dec_{i,j}} [\mu_i] - \sum_{j=1}^{N_\mu} k_{i,j}[\mu_i] + \sum_{j=1}^{N_\mu} k_{j,i} [\mu_j] = 0 \qquad (1)$$

- t time [s]
- $[\mu_i]$ the concentration of the considered μ-radical [kmol m^{-3}]
- S_i the source term for the considered μ-radical from M_α [kmol m^{-3}]
- N_{dec} the total number of decomposition paths for the considered μ-radical [-]
- N_μ the number of μ-radicals in the considered decomposition scheme [-]
- $k_{dec_{i,j}}$ the rate coefficient for forming a β-radical from μ_i via decomposition path j [s^{-1}]
- $k_{i,j}$ the rate coefficient for forming radical μ_j from μ_i [s^{-1}]

The abovementioned equation can be expressed in the form as shown in Equation (2), where A contains all the kinetic rate coefficients related to the isomerization and decomposition of the considered μ-radical. This can be solved to get to the μ-radicals PSSA concentration as shown in Equation (3). To obtain the volumetric production rate of a product X, the μ-radicals' concentration from which it is formed is multiplied by the respective decomposition rate coefficient for all paths that form X. Considering that $[\mu_i]$ is linearly dependent on the initial molecules' concentration, and all other terms are rate coefficients, the volumetric production rate can be written as the product of a pseudo-rate coefficient $k^{pseudo}(X)$ and the molecules' concentration as in Equation (4).

$$\frac{d[\mu_i]}{dt} = [S] + [A][\mu_i] = \mathbf{0} \qquad (2)$$

$$[\mu_i] = -[A]^{-1}[S] \qquad (3)$$

$$R_v(X) = \sum_{i=1}^{N_\mu} \sum_{j=1}^{N_{dec}} \delta(X,i,j)\, k_{dec_{i,j}} [\mu_i] = k^{pseudo}(X)[M_\alpha] \qquad (4)$$

- X β-radical or molecule
- $R_v(X)$ volumetric rate of formation of X [kmol s^{-1}]
- $\delta(X,i,j)$ Dirac function equaling 1 if product X is formed via the jth decomposition of μ_i [-]

- $k_{dec_{i,j}}$ rate coefficient for the decomposition of μ_i via its jth decomposition [s^{-1}]
- $k^{pseudo}(X)$ the pseudo rate coefficient for product X [s^{-1}]
- $[M_\alpha]$ the concentration of the molecule from which the decomposition starts [kmol m^{-3}]

This approach allows reducing the decomposition path of a molecule to an equivalent global reaction as shown in Equation (5), with M and β molecules and β-radicals that can be connected to the β-network, respectively.

$$M_\alpha \xrightarrow{k^{pseudo}(M_\alpha)} \sum_{j=1}^{n_M} \frac{k^{pseudo}(M_j)}{k^{pseudo}(M_\alpha)} M_j + \sum_{j=1}^{n_\beta} \frac{k^{pseudo}(\beta_j)}{k^{pseudo}(M_\alpha)} \beta_j \qquad (5)$$

- n_M the number of molecules formed by the decomposition of M_α [-]
- n_β the number of β-radicals formed through the decomposition of M_α [-]
- $k^{pseudo}(mol_j)$: the pseudo rate coefficient as defined in Equation (3) [s^{-1} for monomolecular reactions | mol m^{-3} s^{-1} for bimolecular reactions]

CRACKSIM contains these equivalent global reactions for 21 temperatures ranging from 770 K to 1200 K. In a next step, thermokinetic properties must be assigned for all elementary reactions and intermediates, which will be explained in one of the following paragraphs. Finally, CRACKSIM interfaces with a fundamental reactor model, COILSIM1D, which contains all balance equations. This facilitates reactor simulations across different coil geometries, including the most important industrial reactor geometries. An example illustrates these principles next.

Decomposition example

To illustrate the derivations discussed earlier, the decomposition of 2-hexene through hydrogen abstraction proceeds as depicted in Figure 2. Please note that the color of each reaction arrow corresponds to the colors specified in Table 1.

In this scenario, the C_6 molecule can undergo a μ-radical forming reaction, such as intermolecular hydrogen abstraction by another β-radical. Each μ-radical formed through hydrogen abstraction (μ_i) may undergo isomerization reactions to another μ-radical within the system if permitted by the reaction families defined in the reaction network generator. Conversely, each of these μ-radicals can decompose via decomposition reactions. In this example, each of the decomposition products consists of β-radicals and molecules. Determining the stoichiometric coefficients of these species is essential for establishing the global equivalent reaction and to connect to the β-network, as outlined in Equation (4).

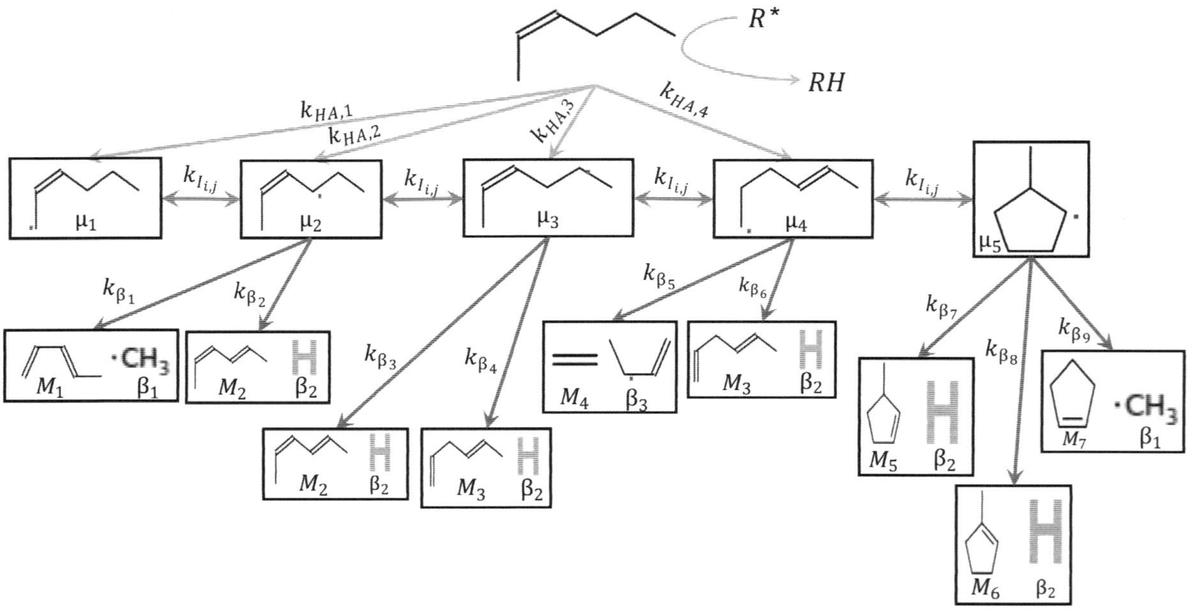

Figure 2: Decomposition of 2-hexene by intermolecular hydrogen abstraction. Mesomerization reactions are not included in the figure to keep the derivation compact.

To determine the stoichiometric coefficients of the β-radicals and molecules, the PSSA concentration of the μ-radicals must be determined first. Substituting the example of Figure 2 into Equation (2) gives for the S and A, respectively:

$$S = \begin{bmatrix} k_{HA,1} \\ k_{HA,2} \\ k_{HA,3} \\ k_{HA,4} \\ 0 \end{bmatrix} \cdot [C_6H_{12}][R^*] \tag{6}$$

$$A = \begin{bmatrix} -k_{I1,2} - k_{I1,3} - k_{I1,4} & k_{I2,1} & k_{I3,1} & k_{I4,1} & 0 \\ k_{I1,2} & \begin{matrix} -(k_{\beta_1} + k_{\beta_2}) \\ -k_{I2,1} - k_{I2,3} - k_{I2,4} \end{matrix} & k_{I3,2} & k_{I4,2} & 0 \\ k_{I1,3} & k_{I2,3} & \begin{matrix} -(k_{\beta_3} + k_{\beta_4}) \\ -k_{I3,1} - k_{I3,2} - k_{I3,4} \end{matrix} & k_{I4,3} & 0 \\ k_{I1,4} & k_{I2,4} & k_{I3,4} & \begin{matrix} -(k_{\beta_5} + k_{\beta_6}) \\ -k_{I4,1} - k_{I4,2} - k_{I4,3} \\ -k_{I4,5} \end{matrix} & k_{I5,4} \\ 0 & 0 & 0 & k_{I4,5} & \begin{matrix} -(k_{\beta_7} + k_{\beta_8} + k_{\beta_9}) \\ -k_{I5,4} \end{matrix} \end{bmatrix} \tag{7}$$

Please note that $k_{I4,5}$ and $k_{I5,4}$ are the rate coefficients for endo-1,5 cyclization and endo-1,5 ring-opening reactions, respectively. By substituting Equations (6) and (7) into Equation (3), the PSSA concentration of the μ-radicals for this decomposition results.

To determine the volumetric formation rate of all products, it is necessary to multiply the concentration of each μ-radical by their respective decomposition rate coefficient. For ethylene, this calculation yields:

$$R_v(C_2H_4) = k_{\beta_5}[\mu_4] = -k^{pseudo}(C_2H_4)[C_6H_{12}][R^*] \tag{8}$$

Finally, the stoichiometric coefficient for the global equivalent reaction is presented by Equation (9), which is of the same form as Equation (5).

$$C_6H_{12} + R^* \xrightarrow{k^{pseudo}(C_6H_{12})} \sum_{i=1}^{7} \frac{k^{pseudo}(M_i)}{k^{pseudo}(C_6H_{12})} M_i + \sum_{i=1}^{3} \frac{k^{pseudo}(\beta_l)}{k^{pseudo}(C_6H_{12})} \beta_l + RH \tag{9}$$

Thermokinetic properties

Assigning kinetic and thermodynamic properties to molecules and intermediate species is critical for acquiring accurate simulated product yields. The required data is obtained from the LCT internal database that comprises both experimental and ab initio calculated parameters at the CBS-QB3 level of theory, leading to a systematic and coherent database. In case no such data is available, group additive methods are applied to determine the desired parameters [13, 14, 23-25].

Lumping approach

Employing the aforementioned methodology, a comprehensive library of reaction schemes has been constructed for components ranging up to carbon number C_{50}. The complete μ-network encompasses approximately 2800 molecules, resulting in roughly 17000 equivalent global reactions. Furthermore, the β-network incorporates 117 radicals and 96 molecules, involving approximately 4300 elementary reactions. Taken together, this constitutes one of the most extensive networks ever developed.

To streamline the integration process with COILSIM1D, a lumping approach has been introduced to condense the final models' size [26]. This method allows for a further reduction in the number of continuity equations to be solved, leading to computational times of less than one second by introducing pseudo-components. The equivalent global reaction of these pseudo-components is derived from a weighted average of the abundance of each component within the lump as present in the considered feed. Importantly, this procedure does not compromise accuracy due to precise mathematical calculations.

A prerequisite for implementing this approach is the knowledge of the weight distribution of components within a lump in a specific feed. This technique has been successfully applied to iso-paraffins derived from fossil-based feedstocks, as their internal distribution is well-established and relatively independent of their source [6]. Through this method, a reduced model for fossil-based feeds has been developed, comprising approximately 1600 (pseudo-)components and around 10800 equivalent global reactions.

The detailed analysis of the feedstock followed by defining new components in CRACKSIM's library that represent the ones in plastic waste oils according to the analytical results summarizes the goal of this work. To increase the level of detail in CRACKSIM for the decomposition of plastic-waste-derived feeds, the number of components within a lump of iso-(di)olefins has been increased. This way, additional branching structures and double-bond configurations could be implemented, thus increasing the product spectrum and the accuracy of the yields.
In the subsequent sections, a detailed compositional analysis of two plastic-waste-derived feedstocks, namely Mixed Plastic Oil (MPO) and PE film-derived oil, will be presented. These feedstocks will be compared against a reference naphtha. To ensure the integrity of the equipment, the plastic waste oils have been blended with the reference naphtha in a [3:1] ratio. The composition of these blends serves as input for COILSIM1D to simulate their behavior under steam cracking conditions. The comparison between experimental and model results is elaborated upon in the final section of the paper, where the similarities and differences are discussed.

Analysis of plastic waste oils

Detailed analysis of the composition of plastic waste pyrolysis oils has been performed by Kusenberg et al. for both a reference naphtha fraction and an untreated plastic-waste derived naphtha cut using the analytical equipment of LCT. The analysis results are shown in

Table 2 [5]. Clearly, pyrolysis oils' carbon numbers range up to C_{50}, which is in line with CRACKSIM's library of decomposition schemes.
Next, the analyzed samples were cracked in an in-house bench scale steam cracker, which yielded the results presented in Figure 3. More information on the analytical equipment can be found in the work of Kusenberg et al. [5].

Table 2: Detailed composition of the feedstocks used in this study. Data obtained from Kusenberg et al. [5].

Feedstock	Reference naphtha	MPO pyrolysis oil	PE-film pyrolysis oil
IBP [°C]	10	14	14
50 % BP [°C]	61	334	340
FBP [°C]	136	603	592
Carbon number range	C_4-C_9	C_5-C_{52}	C_5-C_{50}
Paraffins [wt. %]	34.9	32.2	41.9
Iso-paraffins [wt. %]	44.1	3.7	3.5
α-Olefins [wt. %]	0.0	23.5	33.0
Iso-olefins [wt. %]	0.0	20.0	10.6

Diolefins [wt. %]	0.0	5.3	4.5
Naphthenes [wt. %]	17.0	12.4	5.5
Monoaromatics [wt. %]	4.0	2.7	1.0
Naphthenoaromatics [wt. %]	0.0	0.2	0.1
Diaromatics [wt. %]	0.0	0.1	0.0

As can be seen in

Table 2, the plastic-waste derived feedstocks have a significantly higher olefin content than the reference naphtha, while this is offset by a higher iso-paraffins content in the reference naphtha. These structural differences are crucial in determining the final product yields, as the position of the double bonds and the location of the branches affects the decomposition pathways and hence the product distribution strongly. Thus, the lumping approach introduced in the previous paragraph must be revisited to allow reducing the final models' size.

Reactor simulations using COILSIM1D

The compositional analysis of both the reference naphtha and the blend with plastic waste oil provided input data for COILSIM1D, a fundamental reactor model capable of conducting highly accurate steam cracking simulations by incorporating the developed kinetic schemes from CRACKSIM. These integrated models are compatible with various tubular reactor configurations, making them suitable for a wide range of applications, from laboratory-scale setups to full-scale crackers. Utilizing a specialized algorithm, simulations can be resolved in less than one second, offering users the opportunity to explore a diverse range of conditions and feedstocks within a limited timeframe. A comprehensive description of COILSIM1D and its functionalities is available in the existing literature [27].

All of these feedstocks have been modeled with the same composition as was performed by Kusenberg et al., with the simulation results shown in the parity plots of Figure 3.

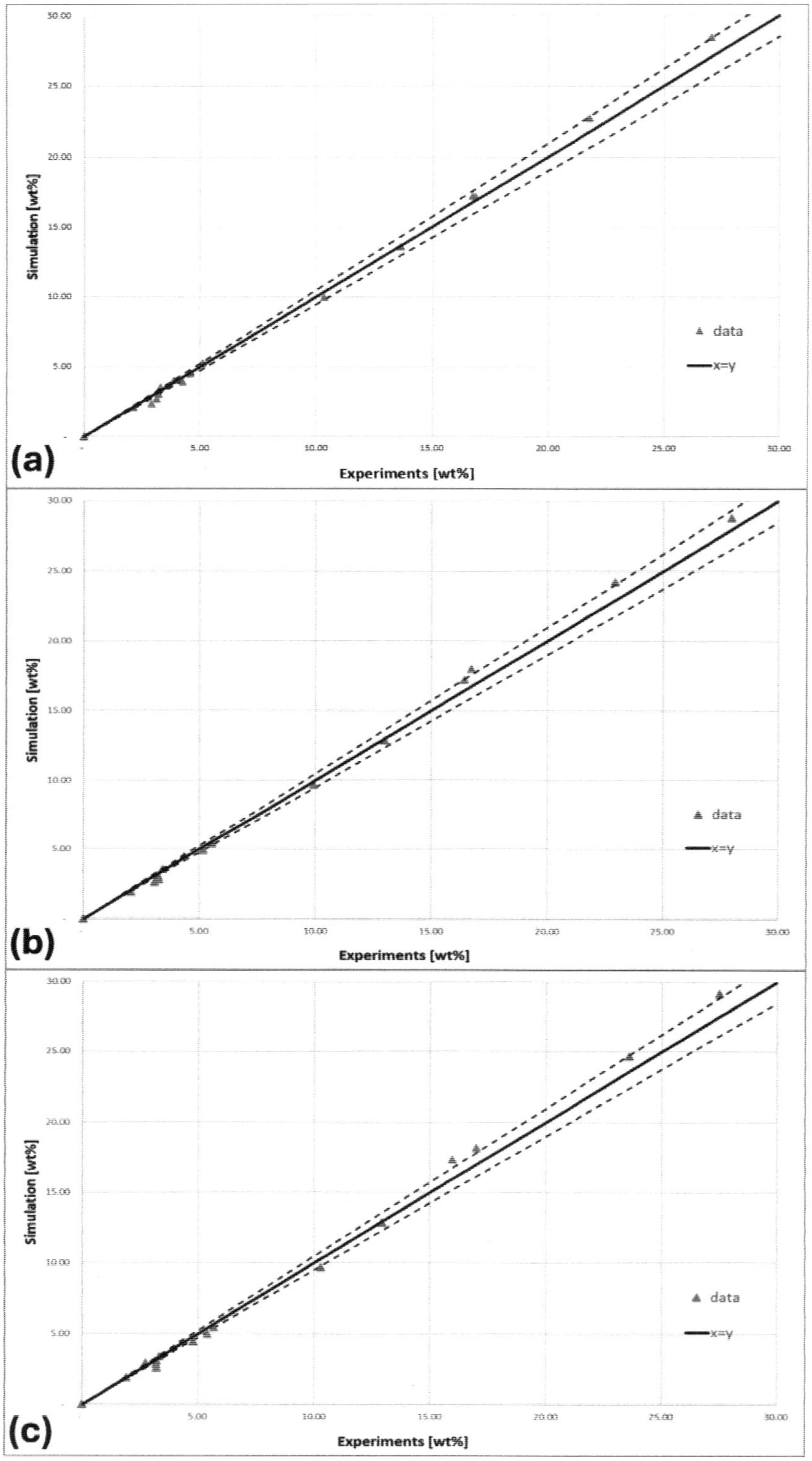

Figure 3: Parity plot of simulated yields for cracking various feedstocks under steam cracking conditions. (a) Reference naphtha; (b) Naphtha/MPO blend in 3:1 ratio; (c) Naphtha/PE oil blend in 3:1 ratio.

The simulation outcomes indicate a strong alignment between the model and experimental results, with deviations of less than 5% observed for the key decomposition products, achieved without the need for additional fitting. Consequently, by customizing kinetic schemes tailored to specific feeds, CRACKSIM demonstrates the capability to accurately predict steam cracking product yields. Looking ahead, the model will expand to incorporate hetero-atom-containing species, offering complete flexibility concerning plastic-waste-derived feeds. This advancement is poised to empower steam crackers to integrate renewable feedstocks into their processes with confidence.

Conclusions

A single-event microkinetic network known as CRACKSIM is currently under development for the decomposition of feedstocks derived from plastic waste pyrolysis. Within this work, additional details for iso-(di)olefins has been added to the library of decomposition schemes, increasing the models' capability to simulate the steam cracking of plastic waste pyrolysis oils. To evaluate the model's performance, a reference naphtha and plastic waste oil were analyzed using a comprehensive two-dimensional gas chromatography (GC×GC) setup to assess their compositional disparities. Subsequently, both oils were subjected to cracking in an in-house developed bench-scale steam cracker, with the reactor effluents analyzed using online analysis techniques. Meanwhile, the analytical composition of the feedstocks served as input for the kinetic model CRACKSIM, which, in conjunction with a fundamental reactor model, COILSIM1D, enabled the modeling of steam cracking for both feedstocks.

The findings demonstrated good agreement between experimental and model predictions. Overall, COILSIM1D exhibits promising results in predicting steam cracking yields for novel feedstocks. By incorporating comprehensive details of olefinic species, branching structure, and halogenated species, COILSIM1D has the potential to play a fundamental role for plant operators in transitioning and enhancing chemical recycling efforts.

Acknowledgements

We gratefully acknowledge the financial support of the Flemish Government and Flanders Innovation & Entrepreneurship (VLAIO) HBC.2023.0159.

References

1. SYSTEMIQ, *ReShaping Plastics: Pathways to a Circular, Climate Neutral Plastics System in Europe.* 2022.
2. Damayanti, D., et al., *Current Prospects for Plastic Waste Treatment.* Polymers (Basel), 2022. **14**(15).
3. Kusenberg, M., et al., *Towards high-quality petrochemical feedstocks from mixed plastic packaging waste via advanced recycling: The past, present and future.* Fuel Processing Technology, 2022. **238**.

4. Kusenberg, M., et al., *Opportunities and challenges for the application of post-consumer plastic waste pyrolysis oils as steam cracker feedstocks: To decontaminate or not to decontaminate?* Waste Manag, 2022. **138**: p. 83-115.
5. Kusenberg, M., et al., *Assessing the feasibility of chemical recycling via steam cracking of untreated plastic waste pyrolysis oils: Feedstock impurities, product yields and coke formation.* Waste Manag, 2022. **141**: p. 104-114.
6. Van Geem, K. and G.B.p. Marin, *Single event microkinetic model for steam cracking of hydrocarbons*. 2006, 2006.
7. Commission, E., *Proposal for a REGULATION OF THE EUROPEAN PARLIAMENT AND OF THE COUNCIL on packaging and packaging waste, amending Regulation (EU) 2019/1020 and Directive (EU) 2019/904, and repealing Directive 94/62/EC*, D.-G.f. Environment, Editor. 2022.
8. Hillewaert, L.P., J.L. Dierickx, and G.F. Froment, *Computer-Generation of Reaction Schemes and Rate-Equations for Thermal-Cracking.* Aiche Journal, 1988. **34**(1): p. 17-24.
9. Van Geem, K., M.F. Reyniers, and G.B. Marin, *First principles based reaction network for steam cracking*, in *AIChE Annual Meeting 2004*. 2004: Austin, TX, USA.
10. Paraskevas, P.D., et al., *Kinetic modeling of alpha-hydrogen abstractions from unsaturated and saturated oxygenate compounds by carbon-centered radicals.* Chemphyschem, 2014. **15**(9): p. 1849-66.
11. Paraskevas, P.D., et al., *Kinetic modeling of alpha-hydrogen abstractions from unsaturated and saturated oxygenate compounds by hydrogen atoms.* J Phys Chem A, 2014. **118**(40): p. 9296-309.
12. Mehl, M., et al., *Oxidation and combustion of the n-hexene isomers: A wide range kinetic modeling study.* Combustion and Flame, 2008. **155**(4): p. 756-772.
13. Sabbe, M.K., et al., *Carbon-centered radical addition and beta-scission reactions: modeling of activation energies and pre-exponential factors.* Chemphyschem, 2008. **9**(1): p. 124-40.
14. Sabbe, M.K., et al., *Hydrogen radical additions to unsaturated hydrocarbons and the reverse beta-scission reactions: modeling of activation energies and pre-exponential factors.* Chemphyschem, 2010. **11**(1): p. 195-210.
15. Baulch, D.L., et al., *Evaluated Kinetic Data for Combustion Modeling. Supplement I.* Journal of Physical and Chemical Reference Data, 1994. **23**(6): p. 847-848.
16. Wang, K., S.M. Villano, and A.M. Dean, *The impact of resonance stabilization on the intramolecular hydrogen-atom shift reactions of hydrocarbon radicals.* Chemphyschem, 2015. **16**(12): p. 2635-45.
17. Wang, K., S.M. Villano, and A.M. Dean, *Reactivity-Structure-Based Rate Estimation Rules for Alkyl Radical H Atom Shift and Alkenyl Radical Cycloaddition Reactions.* J Phys Chem A, 2015. **119**(28): p. 7205-21.
18. Sabbe, M.K., et al., *First principle-based simulation of ethane steam cracking.* AIChE Journal, 2011. **57**(2): p. 482-496.
19. Wang, Z., et al., *Experimental and kinetic modeling study on methylcyclohexane pyrolysis and combustion.* Combustion and Flame, 2014. **161**(1): p. 84-100.

20. Khandavilli, M.V., et al., *Group additive modeling of cyclopentane pyrolysis.* Journal of Analytical and Applied Pyrolysis, 2017. **128**: p. 437-450.
21. Vandewiele, N.M., et al., *Genesys: Kinetic model construction using chemo-informatics.* Chemical Engineering Journal, 2012. **207**: p. 526-538.
22. Pyl, S., *Duurzame productie van lichte alkenen: van fossiele naar hernieuwbare grondstoffen*, in *Department of Materials, Textiles and Chemical Engineering*. 2012, Ghent University: Ghent, Belgium. p. 212.
23. Sabbe, M.K., et al., *First principles based group additive values for the gas phase standard entropy and heat capacity of hydrocarbons and hydrocarbon radicals.* J Phys Chem A, 2008. **112**(47): p. 12235-51.
24. Sabbe, M.K., et al., *Group additive values for the gas phase standard enthalpy of formation of hydrocarbons and hydrocarbon radicals.* J Phys Chem A, 2005. **109**(33): p. 7466-80.
25. Pappijn, C.A.R., et al., *Bond additivity corrections for CBS-QB3 calculated standard enthalpies of formation of H, C, O, N, and S containing species.* International Journal of Chemical Kinetics, 2020. **53**(3): p. 345-355.
26. Ranzi, E., et al., *Initial product distributions from pyrolysis of normal and branched paraffins.* Industrial & Engineering Chemistry Fundamentals, 2002. **22**(1): p. 132-139.
27. Van Geem, K.M., M.F. Reyniers, and G.B. Marin, *Challenges of Modeling Steam Cracking of Heavy Feedstocks.* Oil & Gas Science and Technology - Revue de l'IFP, 2008. **63**(1): p. 79-94.

SPRING24
+20TH GCPS
A Joint AIChE and CCPS Meeting

AIChE Paper #197c

DEVELOPING THE CIRCULAR ECONOMY

Advances in Advanced Recycling and Pyrolysis Oil Characterization

Dave Smith
Senior Marketing Manager – CPI
Nalco Water, an Ecolab Company
mdsmith@ecolab.com

Kuldeep Wadhwa, PhD
Senior Staff Scientist – CPI RD&E
Nalco Water, an Ecolab Company
kuldeep.wadhwa@ecolab.com

Kameswara Vyakaranam, PhD
Staff Scientist – CPI RD&E
Kameswara.vyakaranam@ecolab.com

Prepared for Presentation at
American Institute of Chemical Engineers
2024 Spring Meeting and 20th Global Congress on Process Safety
New Orleans, LA
March 24-28, 2024

DEVELOPING THE CIRCULAR ECONOMY

Advances in Advanced Recycling and Pyrolysis Oil Characterization

Dave Smith
Senior Marketing Manager – CPI
Nalco Water, an Ecolab Company
mdsmith@ecolab.com

Kuldeep Wadhwa, PhD
Senior Staff Scientist – CPI RD&E
Nalco Water, an Ecolab Company
kuldeep.wadhwa@ecolab.com

Kameswara Vyakaranam, PhD
Staff Scientist – CPI RD&E
Kameswara.vyakaranam@ecolab.com

Keywords: recycling, advanced recycling, pyrolysis oil, pyoil, pyrolysate circularity

Abstract

As Advanced Recycling continues to advance, the feedstocks produced from the various chemical recycling processes can provide challenges for recyclers, chemical manufacturers, and refineries to produce and process. The purpose of this paper is as follows:

1. Present the results of extensive laboratory testing of various recycle plastic feedstocks.
2. Detail some of the potential challenges faced in producing and processing feedstocks derived from recycled plastic.
3. Discuss solutions that address these challenges to improve feedstock quality and reliability and reduce the potential negative impacts of producing and processing recycle feedstock.

1 Plastics and the Circular Economy

1.1 Introduction

Plastics are known to provide benefits to nearly every aspect of life. As the global economy and population expand, plastics demand is expected to grow from 350 million tons in 2020 to 950 million tons in 2050[1].

Historically, the plastics value chain was based on a "once-through", or "use-and-dispose" model: chemical manufacturers produced the building blocks used to make polymers; these polymers were used to make consumer products; the products were used and then discarded. As only approximately 10-12% of that plastic is recycled, millions of tons of plastic are incinerated, disposed of in landfills, or end up collecting in oceans and other waterways[2].

As public opinion has moved toward a greater emphasis on sustainability concerns, the response to plastics has turned towards increasing the level of circularity. Governments have moved towards taxing, regulating, or even banning the use and disposal of certain plastics, especially single-use products and packaging. Moreover, market studies show that consumers are demanding more ecologically friendly products[3]. Corporations, under pressure from investors focused on ESG (environment, social, governance) concerns, consumers, and regulatory agencies have responded: 72% of Fortune 300 companies have announced circularity goals[4].

1.2 Achieving Circularity

Mechanical methods traditionally have been employed to produce recycled plastic; typically, the plastic is ground or shredded, then melted, extruded, and cut into pellets or flakes. In this process, the chemical structure of the material is not changed; however, mechanical recycling has several challenges. For one, the process is typically restricted to three main grades of plastic: polyethylene terephthalate (PET), high-density polyethylene (HDPE), and polypropylene (PP). Also, over multiple cycles, the polymer quality will degrade. Sorting out the inappropriate grades of plastic can be very labor-intensive[5].

To increase the pool of recyclable plastics, manufacturers are looking to advanced recycling methods, also known as chemical recycling. In advanced recycling, the chemical structure of the polymer is changed. There are two main methods for chemical recycling:

- **Depolymerization** (also known as **re-monomerization** or **monomer recycling)**, where the polymer is broken down into its monomer building blocks. An example of this is the process developed by Agilyx ASA to recycle polystyrene material back to styrene monomer[6].
- **Plastics-to-Feedstock**, which uses heat in an oxygen-free or low-oxygen environment to break down plastics into material commonly called pyrolysis

oil (also known as pyrolysate or pyoil), which is similar to naphtha and is used as a feedstock for ethylene furnaces or refinery FCCs or cokers [7].

There are other variations, such as the Mura Technology process, which uses supercritical water to produce a feedstock similar to pyoil[8], and various catalytic, oxidative, or microbiological processes[9].

2 Pyrolysis Oils as an Ethylene Feedstock

As chemical manufacturers have started looking towards pyoils and similar feedstocks produced from recycled plastics, there has been considerable interest in the impact these feeds would have on production and processing within the ethylene plant. A non-exhaustive search of the Proceedings of the Ethylene Producers Conferences (EPC) returned seven different papers touching on some aspect of recycle feedstock impact[10].

2.1 Characterizing the Pyrolysis Oils

Because these pyoils are still new to the ethylene industry and not much is known about their properties, the Nalco Water Research, Development, and Engineering (RD&E) group began analyzing pyoils from various chemical recyclers and developed a battery of tests, using both ASTM methods and in-house proprietary tests. A list of some of the tests performed are shown below. At the time of this paper, over 200 pyoil samples have been tested. ASTM has established a working group to develop an industry standard battery of pyoil test methods.

Problem or Concern Area	Test	Method
Cold Flow	Pour point	ASTM D5950
Corrosion	NACE	TM0172
	TAN	ASTM D664
	Organic chloride content	ASTM D4929
Gum Formation and Fouling	Thermo-oxidative stability	proprietary Nalco Water method
	Total sediment	ASTM D4870
	Fouling potential	proprietary Nalco Water method

Table 1. Pyoil Test Methods

2.2 Challenges of Pyoil Feedstock

As a result of this testing, several fundamental issues were identified in the production, storage, transfer, and use of pyrolysis oils[11].

2.2.1 Product Stability

Pyrolysis oils can be highly olefinic in nature, breaking down to form films or gums. This is often, but not always, due to oxidation. Film formation can cause plugging of equipment, fouling of heat exchangers, and build-up of solids in storage equipment. Fouling in the convection section of an ethylene furnace is a particular concern.

Using a proprietary oxidative stability test, researchers found that treating the pyoil with an antioxidant can increase the stability and reduce or eliminate film formation. Table 2 shows the dose-response of pyoil stability using antioxidant, and Figure 1 shows an example of film formation of untreated versus treated pyoil.

Antioxidant Treat Rate (ppm)	Induction Period (h)
0 (blank)	2.28
250	3.91
500	6.60
1000	12.60

Table 2. Oxidative stability of a pyoil sample at varied antioxidant dosages

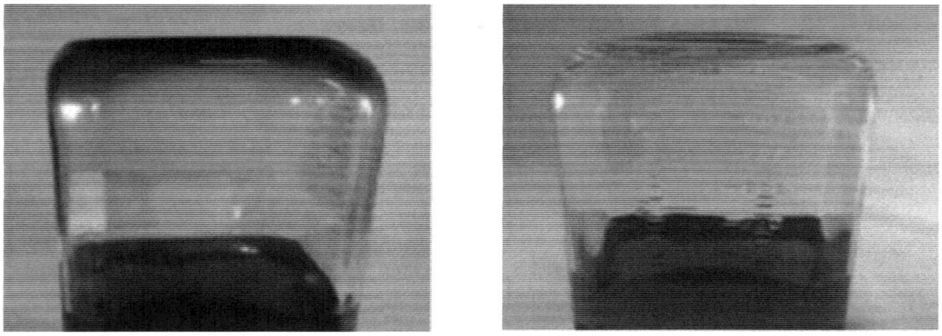

Figure 1. Film formation in untreated (left) vs. treated (right) samples

2.2.2 Flow Properties

Many of the pyoil samples exhibit undesirable cold flow properties. This can lead to difficult pumpability and require expensive capital investment on heated storage and transportation facilities, especially in colder climates. Use of a pour point depressant (PPD) chemistry can reduce expenditure on heated facilities and prevent constriction in pipes, valves, and other process sections. Table 3 shows the dose-response of pyoil using a pour point chemistry.

Pour Point Chemistry Treat Rate (ppm)	Pour Point (°C)
0 (blank)	6
200	-3
400	-12
600	-21

Table 3. Response of pyoil at various pour point depressant dosages

2.2.3 Heavy Components

Various of the pyrolysis processes produce heavy components ranging from a coke-like “char” to a tar-like heavy oil to a wax. These components can cause issues in the pyrolysis processes. In some conditions, higher-value, lighter pyoil may be downgraded to use as a “cutter” for the heavy material. In those cases, lab testing has shown that the use of a pour point chemistry can allow manufacturers to use less cutter stock, thus producing more higher value pyoil.

2.2.4 Variability

Analyses of pyoil samples have shown that they tend to vary widely from producer to producer, but even within the same producer, there can be wide variability from batch to batch. As chemical recyclers expand in scale from bench to demonstration to commercial operation, an increasing volume of waste feedstock is required. Sorting out known contaminant sources such as PET or polyvinyl chloride (PVC) is often done manually, making it both labor-intensive and subject to human error. Also, each pyrolysis and recovery section is unique.

Pyoil composition variability can impact ethylene manufacturers in multiple ways. Variable contaminant levels can cause process reliability issues, as discussed in more detail below. Presence of reactive components and their varying concentration can impact fouling risk. And varying amounts of saturates can impact ethylene yield across the furnaces. In benchmarking testing, researchers found that pyoils derived from catalytic processes had significantly higher concentrations of aromatics than thermal processes, as shown below.

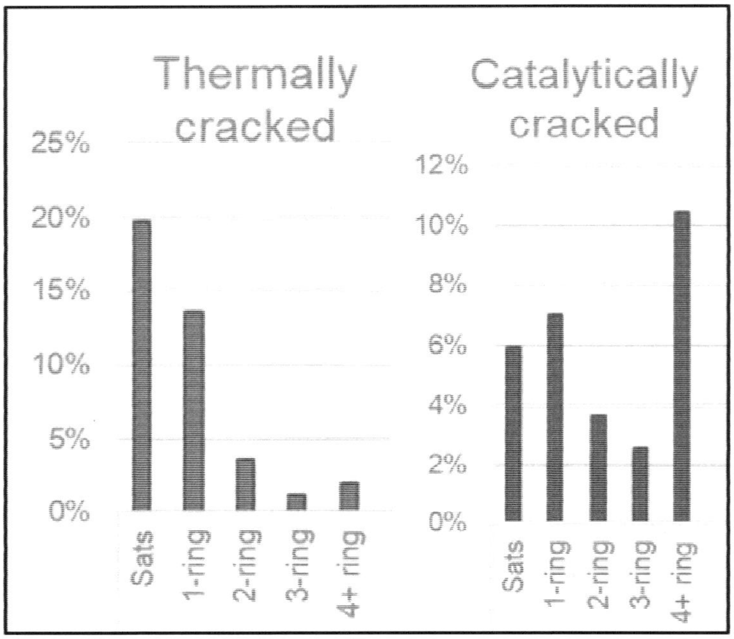

Figure 2. Comparison of pyoil composition from thermal vs catalytic processes

Variability tends to impact in three principal areas:

- Film formation (oxidative and non-oxidative)
- Corrosivity
- Feedstock contaminants

In one example, a chemical recycler sent two different sets of pyoil samples: one from March 2021, and the other from November 2022. Figure 3 shows the visual difference and the film formation in the two samples. Both samples were treated with stabilizer, so the film was not formed via an oxidative mechanism. Benchmarking shows that film formation and fouling propensity were observed for samples contain a high concentration of N,O-containing compounds such as caprolactam, imidazolines, and others, as shown below. Additionally, it was observed that when the film was filtered out, it re-formed in the samples. Testing showed that dispersants were effective at keeping the film solubilized in the pyoil and was even able to disperse deposits already formed.

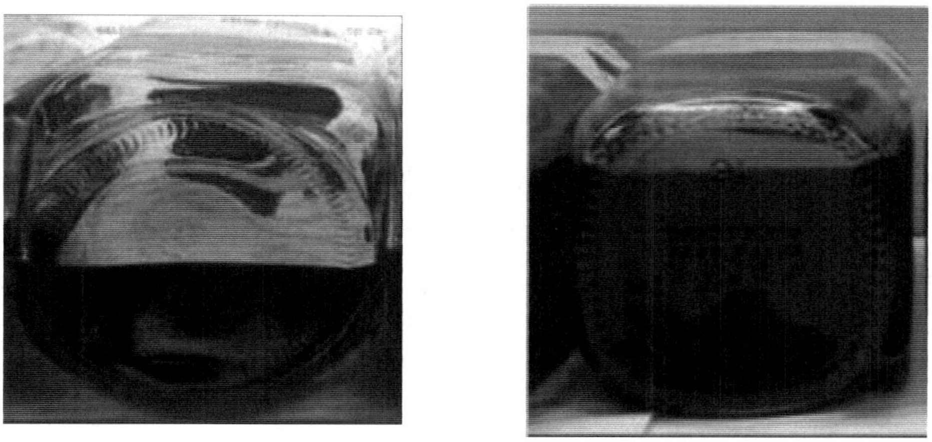

Figure 3. March 2021 sample (left) and November 2022 sample (right)

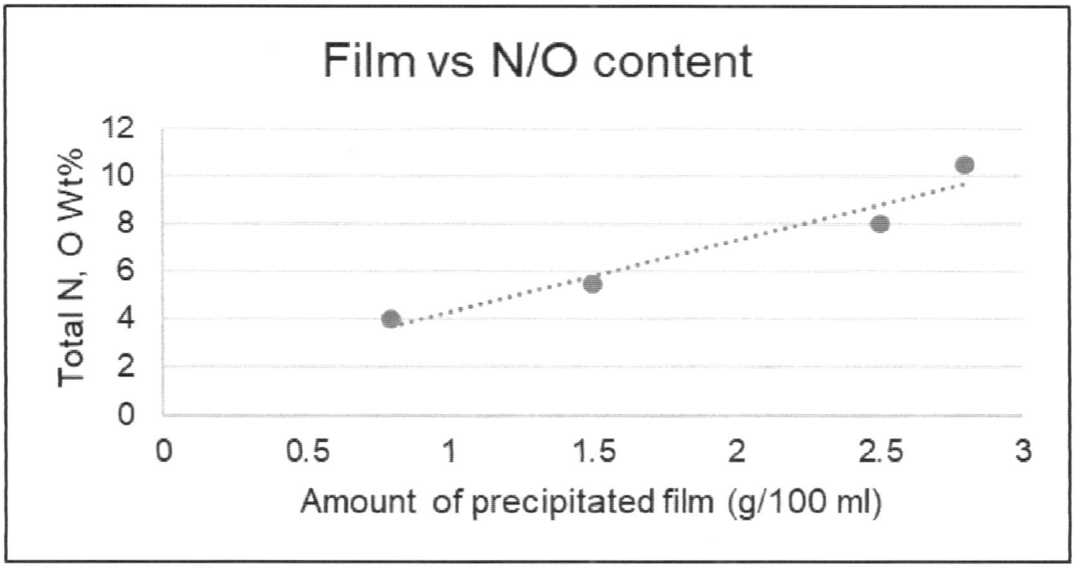

Figure 4. Film formation and N,O species

Another example of variability is observed from sets of samples from the same chemical recycler mentioned above, this between the March 2021 sample and a set from March 2023. In this case, there is again a noticeable difference in appearance, but testing also showed a difference in the corrosivity of the samples, as shown in Table 4 and Figure 5. Note that the March 2023 sample had a lower total acid number (TAN), presumably lower chlorides (a direct comparison could not be made as the researcher had transitioned to a different test method), but a more severe NACE rating. In benchmarking testing, results showed that the presence of a higher concentration of dioxygenates, such as benzoic acid, increased corrosion risk. This is consistent with what Nalco has observed operationally in some recycling facilities.

Test	March 2021	March 2023
TAN (mg KOH/g)	0.12	<0.05
NACE Rating	A	B+
Organic Cl (mg/kg)	39	Not Performed
Extractable Cl (lbs/1000 bbl)	Not Performed	0.91

Table 4. Comparison of sample corrosion test results

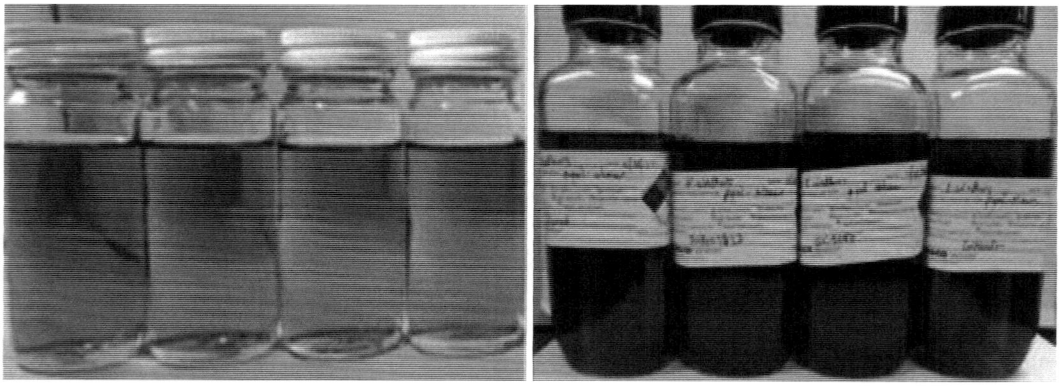

Figure 5. March 2021 sample (left) and March 23 sample (right)

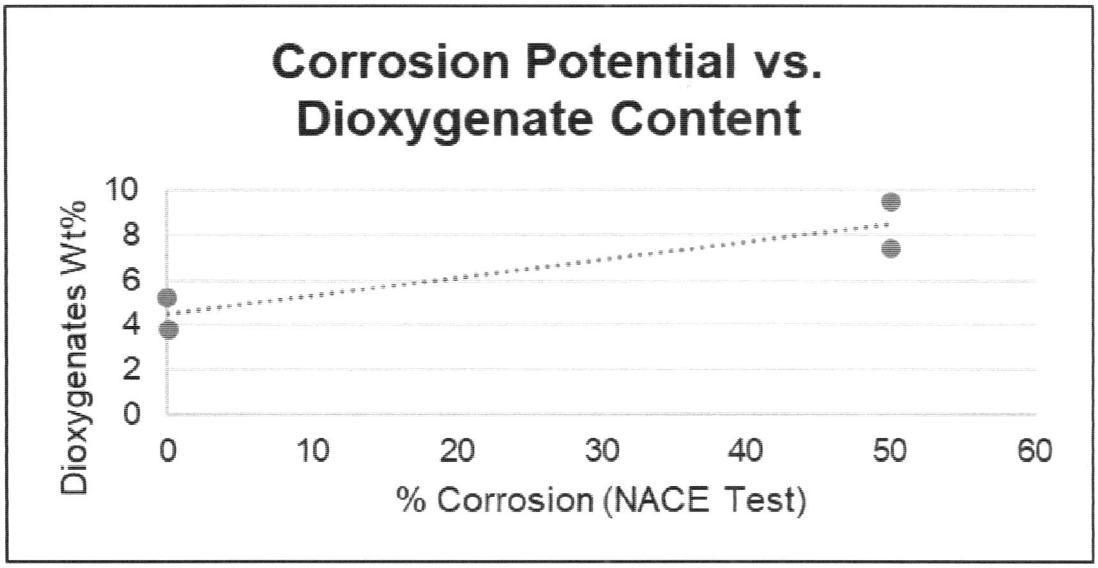

Figure 6. Corrosion potential and dioxygenate concentration

2.2.5 Contaminants

Feedstock contaminants are a significant concern to ethylene manufacturers, as evidenced by the fact that the Feedstock Contaminants Subcommittee of the EPC has featured multiple papers over the past few years dealing with contaminants in pyrolysate feedstocks, and multiple technologies have been introduced or repurposed to deal with contaminants.

Known potential contaminants in pyoil and their potential impact in the ethylene plant include[12]:

- **Oxygenates** can produce organic acids in the ethylene furnace increasing corrosive potential, or peroxides and carbonyls that increase fouling severity and downgrade crude C_4 value. The most common source of oxygenates is PET.
- **Chlorides** damage cracking furnaces and form HCl. The most common source of chlorides is PVC.
- **Nitrogen** can form ammonia in the furnace, and any N-O bonds present can form NO_x. Nitrogen can come from nylons, caprolactam, imidazolines, and other compounds.
- **Phosphorous** can be a catalyst poison.
- **Metals** can damage furnace tubes, cause risk of exposure at turnarounds, increase coking, and poison catalysts.
- **Silicon** can poison catalysts.

- **Other plastic additives** are still being studied for impact on the ethylene plant.

To reduce the potential negative impacts of contaminants in the ethylene production process, manufacturers have been diluting pyrolysis oil with other liquid feed. However, a longer-term, more cost-effective solution would be to remove contaminants and upgrade the quality of the pyrolysate feedstock. To this end, several approaches are being proposed in the marketplace, including adsorbents[13], hydroprocessing[14], and the use of extraction solvents.

In *Process Solutions for Improving the Stability and Handling Properties of Pyrolysis Oils*[15], Dreyer et. al. detailed the effectiveness of using extraction to not only remove non-oxidative film, but also inhibit future film formation, as shown in Figure 7.

Figure 7. Untreated (left) and treated (right) samples of pyrolysis oil using an extraction solvent

Since that publication, additional work using extraction solvents has been undertaken, for the purpose of removing contaminants, specifically chlorides. Initial testing was performed using traditional extraction techniques and an array of solvents, including the solvent referenced above. Additional testing was performed with a Nalco-developed proprietary extraction method. The results show that various Nalco chemistries were successful at removing approximately 30-40% of the chlorides. Using the proprietary extraction method, greater than 70% of the chlorides were removed.

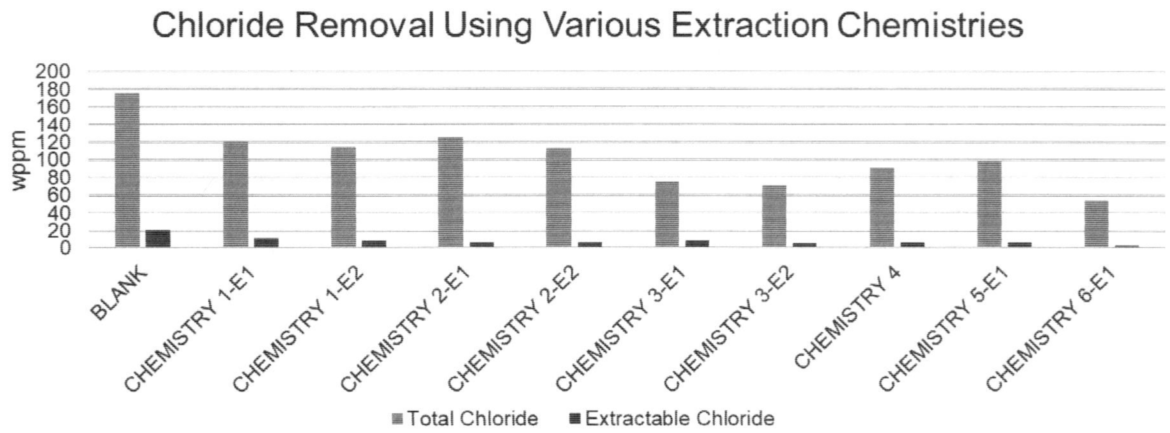

Figure 8. Results of chloride extraction using various extraction chemistries

3 Conclusions

Advanced recycling is an emerging technology that can complement traditional recycling to help promote plastics circularity. Billions of dollars have already been invested in this industry, with aggressive targets being set by nearly all major chemical and plastics manufacturers. There are, however, significant challenges to manufacturers seeking to use pyrolysis oil as feedstocks in their processes. Solvent stability, flow properties, heavy material, variability, and contaminants are among the challenges faced with the pyoils themselves, as well as additional challenges like to be discovered as the impacts of cracking this material on ethylene unit operations are documented.

Through extensive research of a variety of pyoils, Nalco Water research has shown that various chemical additives can address these challenges, providing pyoil that is more stable, less prone to fouling, and lower in contaminants.

4 References

[1] Rubel H, Jung U, Follette C, Meyer zum Felde A, Appathurai. A circular solution to plastic waste. *Boston Consulting Group,* 2019.

[2] Ryberg MW, Laurent A, Hauscgild. Mapping of global plastics value chain and plastics losses to the environment. *United Nations Environment Programme*, 2018.

[3] Smith D. Expanding the circular economy: sustainability gains through advanced recycling. *Society of Plastics Engineers International Polyolefins Conference,* 2023.

[4] Gold J. Accelerating plastic circularity with pyrolysis-based advanced recycling. Society of Plastics Engineers International Polyolefins Conference, 2024.

[5] Smith D, Pickett T, Bernard A, Lally J. Plastic waste to cracking feedstocks and the circular economy. *American Institute of Chemical Engineers Spring National Meeting, Ethylene Producers Conference,* 2020. Paper 159c.

[6] Agilyx. TruStyrenyx™: All-in-one technology platform for the chemical recycling of polystyrene. https://www.agilyx.com/trustyrenyx/ . Accessed December 12, 2023.

[7] Smith, 2023.

[8] Mura Technology. The next generation of advanced plastic recycling. https://muratechnology.com/ . Accessed December 12, 2023.

[9] Cooper T. Advances in physical and chemical plastics recycling. *Society of Plastics Engineers International Polyolefins Conference,* 2024.

[10] American Institute of Chemical Engineers. Proceedings of the Ethylene Producers Conference, 2023. https://epc.omnibooksonline.com/2023/ . Accessed December 12, 2023.

[11] Dreyer D, Vyakaranam K, Wadhwa K, Smith D. Process solutions for improving the stability and handling properties of pyrolysis oil. *American Institute of Chemical Engineers Spring National Meeting,* 2022. Paper 147a.

[12] Bernard A. Feedstock contaminants database presentation. *Ethylene Producers Conference*, 2019. Session 35.

[13] Shah N. Finding the right upgrading partner to make pyrolysis oils suitable for crackers. *Society of Plastics Engineers International Polyolefins Conference,* 2024.

[14] Haldor Topsoe. Give low-grade waste a high-grade purpose. https://www.topsoe.com/circular-plastic . Accessed December 13, 2023.

[15] Dreyer, et.al., 2022.

Unconventional and Sustainable Production Routes to Olefins

Christopher Dziedziak, John Murphy

The focus of recent R&D and commercial developments for novel processes and catalysts for olefins production goes well beyond traditional thermal steam cracking, fluid catalytic cracking (FCC) and propane dehydrogenation (PDH) routes, to include numerous unconventional, green, and circular approaches. These routes include, but are not limited to, utilization of biomass or waste plastic, renewably produced methane and syngas, direct CO_2 conversion, and the electrification of reactors. All these approaches must address certain critical factors affecting technology viability, notably the CO_2 footprint, lifecycle analysis, and overall sustainability in a move towards NetZero2050, and of course favorable economic conditions.

This presentation first establishes the state-of-the-art of the conventional olefins landscape, discussing any notable improvements in steam cracking and the growing utilization of on-purpose propylene technologies. The core of the presentation discusses the improvements and discoveries made to unconventional catalytic olefins production methods including direct methane/C_1 conversion, biomass and recycled waste, electrification, and CO_2 utilization, as well as concepts related to GHG/CO_2 reduction and increased sustainability and circular economy efforts in the petrochemical and polymer industries.

AIChE Paper Number: 190D

NUSCALE POWER CORPORATION SMALL MODULAR REACTOR (SMR) TECHNOLOGY

Kaleena Fisher
Process Director
Fluor Corporation

Prepared for Presentation at the Year (2024) Spring National Meeting
New Orleans, Louisiana, (March 28, 2024)

FLUOR®

NUSCALE POWER MODULE

Kaleena Fisher
Process Engineering Manager
Fluor Corporation

Abstract:

NuScale Power Corporation is a small modular reactor (SMR) technology company established in 2007 of which Fluor Corporation became the lead investor in 2011. They have designed the fully factory fabricated NuScale Power Module (NPM), each with the capacity of generating 77 MWe of electricity (gross). The advantage of SMRs as compared to typical pressurized water reactors is that they can operate more safely and can passively get to a safe cooled down state using natural convection without using pumps, external power or external water. SMRs also provide scalable power production by providing options for 4, 6, or 12 NPMs. The NPMs are capable of black start and load following.

FLUOR®

NuScale Background

NuScale Power was formed in 2007 to complete the design and commercialization of the NuScale Power Module™, which is a small modular reactor (SMR). The initial concept originated as a project at the Oregon State University in 2000 funded by the U.S. Department of Energy under the Multi-Application Small Light-Water Reactor Program. In 2011, Fluor Corporation became NuScale's lead investor. In 2013, NuScale won a competitively bid, cost-share award from the U.S. DOE. NuScale received final design approval for their NuScale Power Module from the Nuclear Regulatory Commission (NRC) in 2020. In 2023, the U.S. NRC accepted the Standard Design Approval (SDA) Application for the NuScale VOYGR-6 plant design for the 77 MWe small modular reactor. NuScale is the only company that has achieved this acceptance for a small modular reactor design.

NuScale Power Module (TM)

The NuScale Power Module features a fully factory fabricated SMR that is small enough to be shipped by truck, rail or barge. The NuScale Power Module has an integral nuclear steam supply system in which the reactor core, the steam generators, and the pressurizer are all contained in a single vessel. The advantage to this simple design is that because it uses natural convection it eliminates the need for reactor coolant pumps, large bore piping and other systems and components that are typical for large conventional reactors. Each module produces up to 77 MWe. A facility can have up to 12 modules for a total of 924 MWe total output. Each NuScale Power Module vessel is close in size to a typical large Pressurized Water Reactor (PWR) steam generator. Each module feeds one turbine generator train, eliminating single-shaft risk. Each turbine generator train has 100% turbine bypass capability. The design has demonstrated resiliency for every configuration (black-start, island mode, seismically robust, cyber security, etc.).

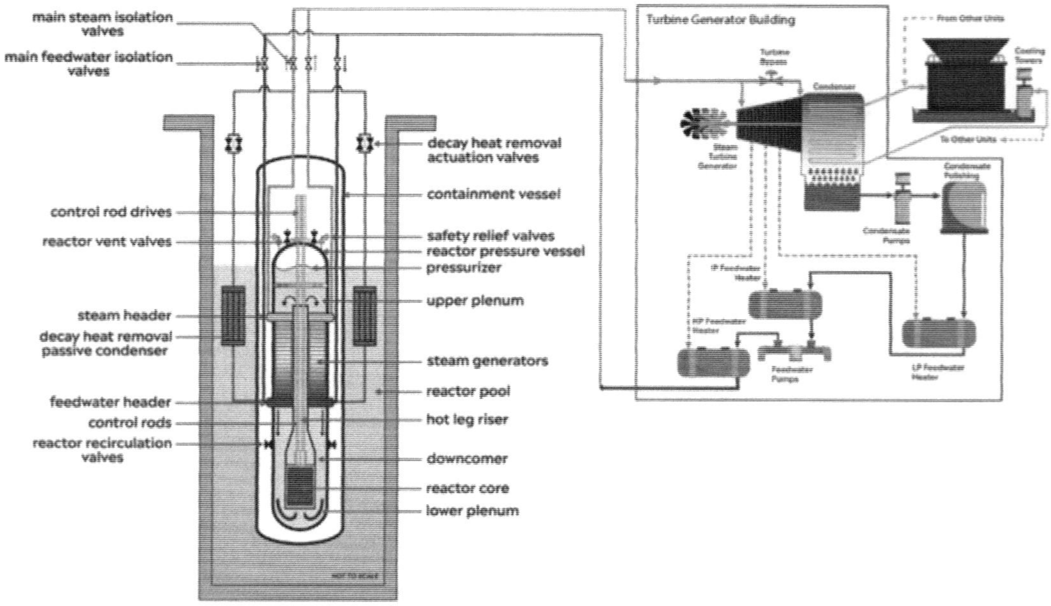

Figure 1: NuScale Power Train, Copyright © 2021 NuScale Power, LLC

FLUOR®

Safety Advantages

The simplicity and size of the NuScale Power Module allows for the following safety advantages:

1. Cooling water circulates through the nuclear core by natural convection eliminating the need for pumps and emergency generators.
2. The NuScale Power Module vessel is submerged in a below-grade pool of water in a building that is resistant to:
    a. earthquakes impact
    b. aircraft impact
    c. hurricane impact
    d. tornado impact, and
    e. floods.
3. The SMR reactor core is 1/20th the size of a typical large reactor core. The small size and the integrated reactor design eliminates the risk of large-break loss-of coolant accidents.
4. Module and plant protection systems are non-microprocessor based using field programmable gate arrays that do not use software and are therefore not vulnerable to internet cyber-attack.
5. NuScale design has resilience to solar-induced geomagnetic disturbances (GMDs) and electromagnetic pulse (EMP) events beyond the current nuclear fleet.
6. A single module can be black-started and can power the entire plant in case of loss of the grid with no external source of AC/DC power or additional water required to keep the reactors safe.
7. The plant has Island Mode capability, meaning it can be disconnected from the transmission system and using electricity from a single module to power the facility.

NuScale's design ensures that on loss of all electricity to the power plant, each NuScale Power Module will safely shut down and self-cool, indefinitely, with no need for operator or computer action, AC or DC power, or additional water. This capability is a first for commercial nuclear power.

Typical PWR reactors have a probability of 1 core damage event per 10,000 and 1,000,000 years. The probability of a NuScale reactor equipment failure is 1 event per module every ~3 billion years.

Other Advantages

The NuScale reactor building includes space in the stainless steel lined concrete pool for at least 10 years of used fuel storage. The NuScale standard facility design includes an area for dry storage cask storage of all the spent fuel produced during the 60-year life of the plant. The NuScale Power Module design allows for load following, which allows more flexibility than a typical large conventional reactor.

AICHE PAPER NUMBER: 190E

INCREMENTAL APPROACH TO DECARBONIZATION FOR STEAM CRACKERS

Melanie O'Sullivan
Technology Manager for Carbon Capture & Energy Storage
Lummus Technology

Galip Guvelioglu
Technology Director for Gas Processing
Lummus Technology

Dr. Baozhong Zhao
Director of Engineering for Thermal Engineering
Lummus Technology

Peter Kuzma
Sr. Technology Director for Ethylene
Lummus Technology

Daniel McKenzie
Technology Manager for Ethylene
Lummus Technology

Prepared for Presentation at the 2024 Spring National Meeting
New Orleans, LA, March 24-28, 2024

INCREMENTAL APPROACH TO DECARBONIZATION FOR STEAM CRACKERS

Melanie O'Sullivan
Technology Manager for Carbon Capture & Energy Storage
Lummus Technology
Galip Guvelioglu
Technology Director for Gas Processing
Lummus Technology
Dr. Baozhong Zhao
Director of Engineering for Thermal Engineering
Lummus Technology
Peter Kuzma
Sr. Technology Director for Ethylene
Lummus Technology
Daniel McKenzie
Technology Manager for Ethylene
Lummus Technology

Abstract:

Ethylene production generates more than 260 million metric tons of CO_2 each year. As global ethylene demand and capacity continues to grow, decarbonization of this process will become critical to achieving emissions goals. While making this change is imperative, it may be a challenge to implement today based on the uncertainty or nonexistence of carbon tax credits. This presentation will review Lummus' Carbon Capture Ready Design, which can be applied to a grassroots cracker today for decarbonization of the future. This concept considers certain pre-investment steps to minimize CO_2 generation during current operation by minimizing fuel firing and maximizing electrification by means such as combustion air preheating and Lummus' Hybrid and Electric Heaters (SRT-h/SRT-e). The Carbon Capture Ready Design also considers pre-investment to easily facilitate the implementation of carbon capture in the future. This paper will review the economic and emissions impact of the Carbon Capture Ready Design on current operation without carbon capture and future operation with carbon capture.

Introduction

Decarbonization is a complicated challenge facing new and existing petrochemical and refining facilities. Particularly for new facilities, it is difficult to determine the best way to decarbonize an asset with unclear or unknown carbon tax credits during project development. Relatively small initial investments can be considered to help easily facilitate carbon capture in the future as well as ensuring it is possible to implement carbon capture down the line. These investments will vary depending on the type of carbon capture being considered. An additional investment step may be taken in the initial plant design to reduce the size of the carbon capture unit by reducing fuel firing via combustion air preheating or heater electrification. This paper will explore these different levels of pre-investment for a 1200 KTA mixed feed ethylene plant for three types of carbon capture: Post Combustion, Oxyfuel Combustion, and Pre-Combustion.

Description of Types of Carbon Capture

Post Combustion Carbon Capture

In Lummus' Post Combustion Technology, the flue gas from the fired heater is compressed with a blower (if needed) and cooled against circulating water in a Direct Contact Cooler before feeding the CO_2 Absorber. In the CO_2 Absorber, an amine-based solvent is circulated through the tower to capture CO_2 from the flue gas. A water wash section is provided at the top of the tower to minimize solvent losses before the flue gas is vented to atmosphere. The CO_2-rich solvent from the CO_2 Absorber is heated and sent to the Solvent Stripper where CO_2 is stripped from the solvent with low-pressure steam. The regenerated, CO_2-lean solvent is then cooled and returned to the CO_2 Absorber. The net overhead of the Solvent Stripper is the wet CO_2 product, which is compressed to an intermediate pressure, dried with triethylene glycol (TEG), and compressed to final battery limit conditions.

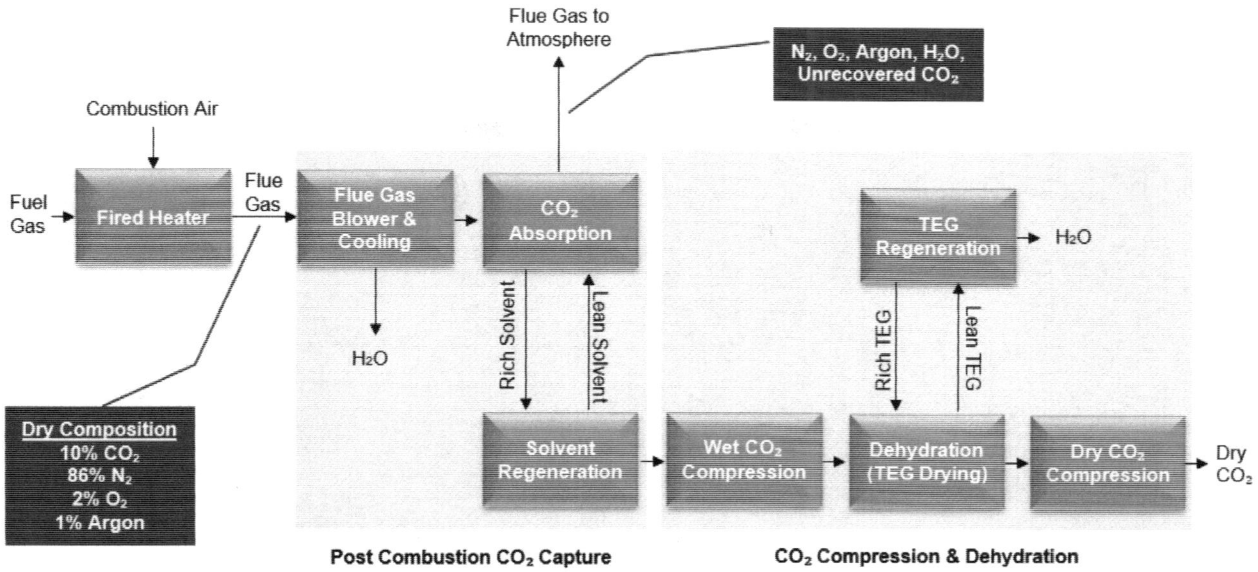

Figure 1: Block Flow Diagram of Post Combustion Carbon Capture

Post Combustion Carbon Capture is an energy-intensive process and requires a significant amount of steam to regenerate the solvent. Second generation solvents may be considered to reduce the specific energy of the process. Lummus has a master collaboration agreement with Toshiba Energy Systems & Solutions Corporation for advanced carbon capture solutions. Toshiba's post combustion solvent will be considered for this analysis. This solvent boasts approximately 30% reduction in specific energy compared to first generation solvents as well as reduced solvent degradation and emissions. The optimal CO_2 recovery for this carbon capture technology is 90%.

Oxyfuel Combustion with CO_2 Purification

In Oxyfuel Combustion, fuel is combusted with oxygen from an Air Separation Unit (ASU) rather than air to significantly reduce inert nitrogen in the resultant flue gas. The gross flue gas from the fired heater is cooled against circulating water in a Direct Contact Cooler and compressed with a blower. A portion of the cooled flue gas is recycled back to the fired heater to reduce the oxygen concentration in the combustion zone, minimize NOx emission, and obtain combustion and heat transfer characteristics similar to conventional combustion air to minimize the impact on furnace design and/or operation.

The elimination of inert nitrogen from the flue gas, which is present when air is used for combustion, reduces the amount of net flue gas that must be processed in the Purification Unit by an order of magnitude and the CO_2 concentration in the net flue gas is nearly an order of magnitude higher compared to Post Combustion. The net flue gas is compressed, dehydrated using a molecular sieve dryer, and purified via a cryogenic distillation process using a refrigeration system. A reject light stream is vented to atmosphere from the distillation tower overhead, which consists of excess oxygen required for efficient combustion, trace inert nitrogen and argon from the ASU, and unrecovered CO_2. A liquid CO_2 stream is generated, which may be pumped and heated to final battery limit conditions. The optimal CO_2 recovery for this carbon capture technology is 95%.

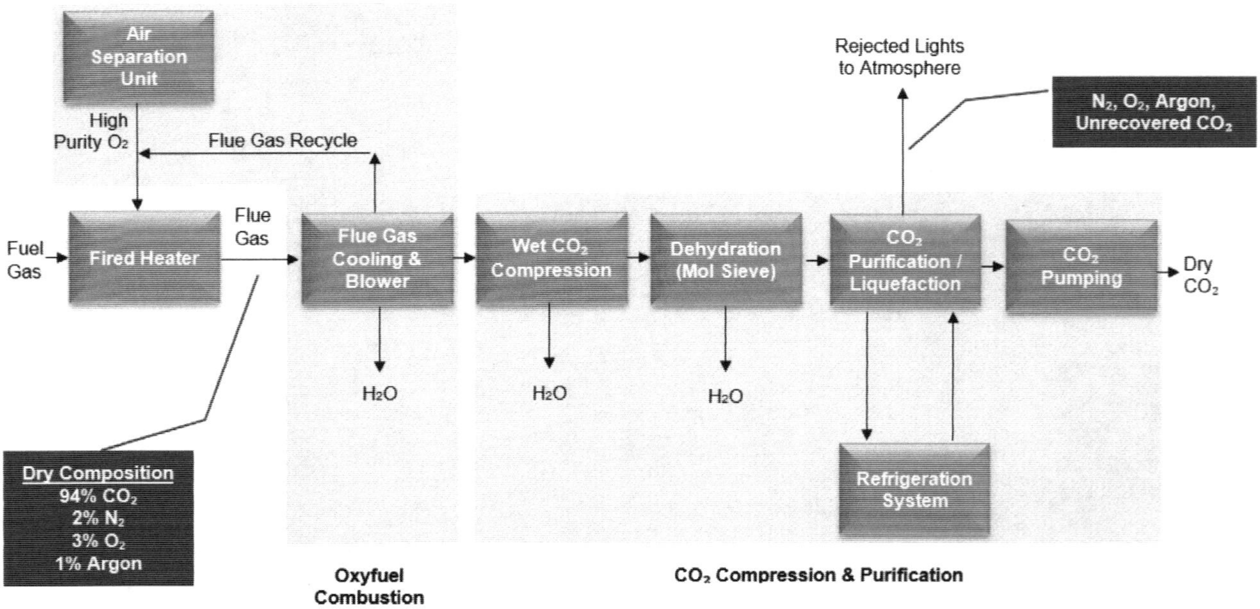

Figure 2: Block Flow Diagram of Oxyfuel Combustion with CO_2 Purification

Pre-Combustion Carbon Capture

Pre-Combustion Carbon Capture generates blue hydrogen, which is used as fuel gas and results in negligible CO_2 emissions from the fired heater. Process offgas streams, which are traditionally used for fuel firing, are instead compressed, combined with supplemental natural gas, and sent to a Feed Fired Heater. The Feed Fired Heater is also fired by internally-generated blue hydrogen, resulting in negligible emissions from its flue gas. The effluent from the Feed Fired Heater is then sent to an Auto-Thermal Reformer (ATR) with oxygen from an ASU to convert hydrocarbons to carbon monoxide and hydrogen. The ATR effluent is sent to a Water Gas Shift (WGS) Reactor to convert carbon monoxide to carbon dioxide. The effluent from the Water Gas Shift Reactor is then sent to an amine-based carbon capture unit. This unit is the same configuration as the Post Combustion Unit and typically uses a first generation amine-based solvent due to the higher partial pressure of CO_2 feeding the unit. The vent gas from the CO_2 Absorber overhead is the blue hydrogen product. An acceptable hydrogen purity for fuel firing can be generated without a Pressure Swing Adsorption unit. The captured CO_2 from the Solvent Stripper overhead is compressed and dried, similar to a Post Combustion Unit. Steam is generated as a part of the heat recovery from the reactor effluents, which is used for solvent regeneration in the carbon capture unit. As a result, no external steam is required for this process. Pre-combustion carbon capture can achieve up to 98% overall scope 1 CO_2 recovery.

A Steam Methane Reformer (SMR) with carbon capture may also be considered to generate blue hydrogen, however, this configuration will require two carbon capture units to achieve a similar CO_2 recovery: one on the process stream (pre-combustion) and another on the reformer flue gas (post combustion). For this reason, an ATR was selected as the hydrogen generation technology for this study.

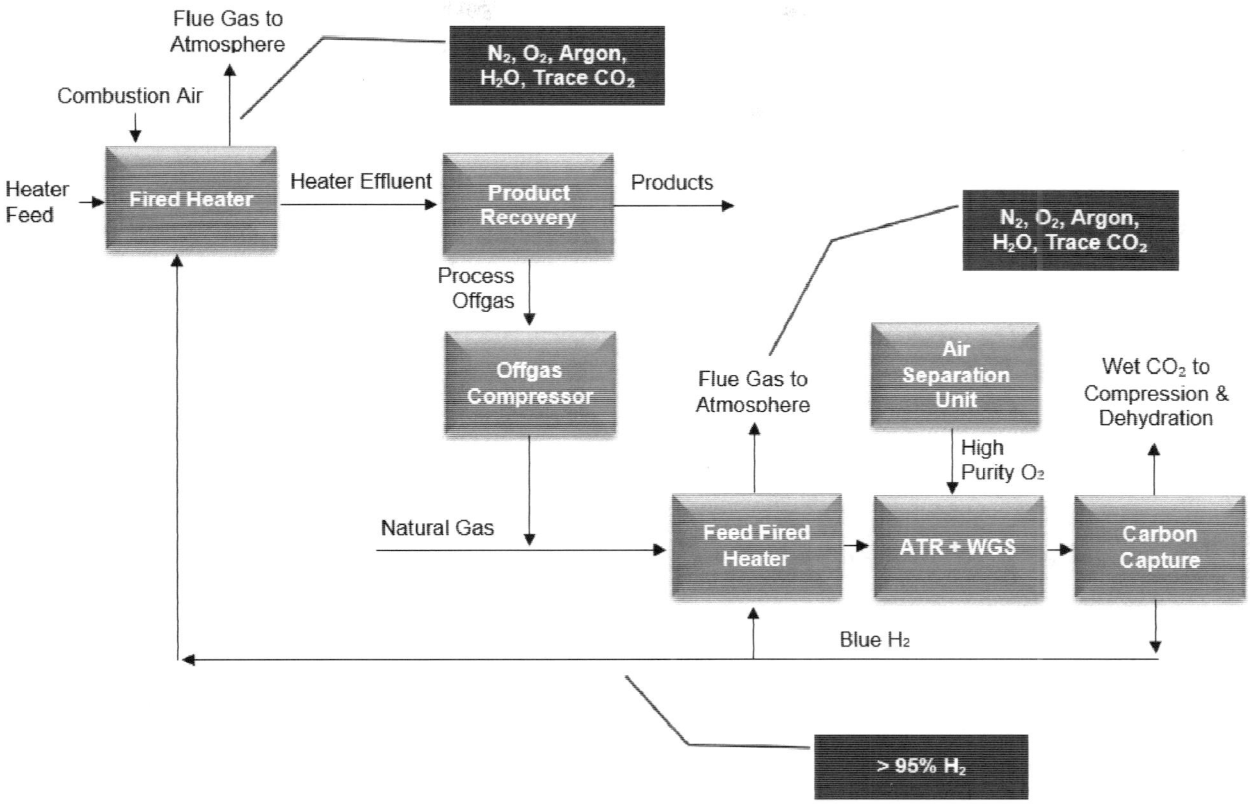

Figure 3: Block Flow Diagram of Pre-Combustion Carbon Capture

Carbon Capture Ready Considerations

Different levels of pre-investment may be considered for the Carbon Capture Ready design. Level 0 is a conventional design and includes no pre-investment for the future implementation of carbon capture, while Level 1 includes certain pre-investments for a future carbon capture unit. Level 1 considerations will differ depending on the carbon capture technology being applied.

Level 1 Pre-Investment

Level 1 considerations for Post Combustion will require tie-ins and accommodations for future ducting to direct the heater flue gas to the capture unit. The heater structure should be pre-invested to handle the future load of this additional ducting. Solvents used in Post Combustion experience increased degradation when exposed to NOx above certain concentrations. Therefore, space should be left in the heater convection section for a future selective catalytic reduction bed (SCR) to convert NOx to nitrogen and water, if selective catalytic reduction is not already required for permitting purposes. The other significant pre-investment consideration for Post Combustion is plot space around the heaters. The largest plot consideration is the towers that process the low-pressure flue gas. The Direct Contact Cooler and CO_2 Absorber towers have large cross-sectional areas and should be located near the existing heaters. The Solvent Stripper could be located further away because it processes liquid streams that are pumped to/from the CO_2 Absorber. The lowest cost design option will be for all equipment within the CO_2 capture and CO_2 compression/dehydration unit to be collocated near the existing heaters. Considerations may also be given to the induced draft (ID) fans in the heater stack for future carbon capture operation. Depending on the pressure head requirement, the ID fan may be pre-invested for future carbon capture operation. If the head requirement is too high to accommodate both pre-carbon capture operation and carbon capture operation, the heater structure should be pre-invested for the future ID fan required for carbon capture operation. Without this consideration, a blower will be required to provide sufficient driving force to send the flue gas through the ducting and packed beds in the capture unit. By pre-investing the ID fan for this higher head operation (or the heater for the future ID fan replacement), an additional piece of equipment, and the associated plot, may be eliminated.

Level 1 considerations for Oxyfuel Combustion will require similar tie-ins and accommodations for future ducting to direct the flue gas to the capture unit. Oxyfuel will likewise require some plot space near the existing heaters. The Direct Contact Cooler and Flue Gas Recycle Blower should be located near the existing heaters. The CO_2 Purification Unit may be located further away, but this will result in an energy penalty. Again, the optimal solution will be for all equipment within the Flue Gas Recycle and CO_2 Purification Unit to be collocated near the existing heaters. The Air Separation Unit may be located further away. The flue gas recycle stream typically accounts for between 60-70% of the heater gross flue gas for a steam cracker, so sufficient space will need to be left for these return lines as well. Burner flexibility to accommodate the high concentration of CO_2 in future operation may also be considered with the initial heater investment. Lastly, the CO_2 Purification Unit will use propylene refrigerant for the cryogenic separation. This is a relatively small percentage of the total propylene refrigerant capacity of a typical cracker; therefore, it is optimal to consider pre-investing the refrigeration system for a future higher capacity rather than considering a parallel, smaller refrigeration system for the CO_2 Purification Unit or modifying the equipment within the existing refrigerant system when implementing carbon capture.

Pre-combustion requires the least Level 1 pre-investment. Burner flexibility to accommodate 100% hydrogen firing in the future may be considered with the initial heater investment. Hydrogen firing will also result in higher NOx formation compared to methane firing due to the higher flame temperature. Space should be left in the heater convection section for a future selective catalytic reduction bed (SCR) to convert NOx to nitrogen and water, if selective catalytic reduction is not already required for permitting purposes.

Level 2 Pre-Investment

Level 2 pre-investment includes the same considerations as Level 1 while also reducing fuel firing in the cracking heaters. Level 2 pre-investment will result in a reduction of the amount of CO_2 generated in the heaters as well as a reduction in the capacity of the carbon capture unit. This paper considers three different approaches to Level 2 pre-investment.

The first approach (Level 2A) is to apply combustion air preheating using the Lummus External Air Preheat (LEAP™) technology. LEAP uses a thermal fluid to transfer heat from the cracking heater convection section to the combustion air at grade, resulting in a simplified structural configuration compared to conventional air preheat. This air preheat technology was selected for several reasons. When Post Combustion or Oxyfuel Combustion is applied, there will be quite a bit of additional ducting on the heater stack. To minimize crowding, it is preferred to consider air preheating at grade rather than above the convection section. Preheating at grade will also allow the design to easily switch from air preheating to flue gas recycle preheating for Oxyfuel Combustion. LEAP does not require any additional plot above a conventional heater design, as the heat exchanger is located within the heater structure. Conventional air preheat at grade will require substantial plot near the heaters for the additional heat exchanger. While Level 2 pre-investment is considered during the initial plant construction, and therefore, this additional plot could be accounted for, it will result in other hydraulically critical equipment being further away, so the optimal solution would be to consider air preheating without any additional plot requirements near the heaters.

The next two approaches replace fuel firing duty with electrification. The first approach (Level 2B) is to include electric heaters (SRT-e™) in place of conventional heaters. This approach will result in the highest electrical demand.

The last approach (Level 2C) is to consider hybrid electric heaters (SRT-h™) instead of conventional heaters. A hybrid cracking heater provides a portion of the heater duty via fuel firing and a portion via electricity. One significant advantage of the SRT-h for this application is that it can be designed to maintain the same steam generation as a conventional heater by operating with higher excess air and therefore, no changes to the recovery section of the plant are needed. The SRT-h can also be operated with typical excess air, which will reduce the amount of flue gas and capacity of the carbon capture unit. In this scenario, it would be similar to the air preheat case (Level 2A) and the steam production will be reduced, which will impact the recovery section. Therefore, for this paper, SRT-h operation with high excess air and constant steam production will be considered.

High oxygen concentration in the flue gas is known to increase solvent degradation for Post Combustion solvents. However, even when considering a plant consisting of only SRT-h heaters, the oxygen concentration in the flue gas is below this limit. Furthermore, Toshiba's post combustion solvent can tolerate significantly higher oxygen concentrations than first generation solvents. The CO_2 concentration in the flue gas will also be lower due to the higher excess air, further emphasizing the importance of using a high performing solvent for capture to limit the specific energy of the unit.

In Oxyfuel Combustion, a higher flue gas recycle rate will be required to maintain heater performance and steam generation for the SRT-h. Rather than 60-70% of the gross flue gas being recycled back to the heaters, more than 70% will be returned to the heaters. For both Post Combustion and Oxyfuel Combustion, there will be a negligible impact to the capacity of the capture unit (for Post Combustion) and the flue gas recycle unit (for Oxyfuel Combustion) compared to Level 0 and 1 because the flue gas rate will be relatively unchanged. However, the CO_2 compression unit and purification units will be lower capacity due to the lower CO_2 product rate.

Level 2 considerations are optimally applied on a grass-roots plant. If, for example, combustion air preheating is applied in the future, modifications must be made to compressor drivers to avoid significant increases in steam import, which will diminish the overall emissions reduction potential. Unless coupled with an expansion or unless this modification is required as a part of plant maintenance, this will not be economical. The same issue will be encountered with heater electrification. Unless an existing heater has reached end of life, or if coupled with an expansion, it will not be logical to replace a functioning heater. The hybrid heater design considered does not have the same steam shortfall issue as the other technologies and a conventional heater could be revamped to be a hybrid heater. However, the simplest approach would be to consider a hybrid heater for the initial design, which is what this analysis has considered.

Impact of Carbon Capture Ready Pre-Investment on Current Operation

Prior to implementing carbon capture, there will be some changes to operation compared to a Level 0 implementation depending on the level of pre-investment. Level 1 will see no change to operation from Level 0 while Level 2 will have some changes to utility consumption, emissions, and OPEX.

Table **1**: Impact of Carbon Capture Pre-Investment on Operation Prior to Carbon Capture Implementation

Pre-Investment Level	**0**	**1**	**2A**	**2B**	**2C**
Fuel Reduction Method	**None** (Apply 6/6 Operating Heaters)	**None** (Apply 6/6 Operating Heaters)	**LEAP** (Apply 6/6 Operating Heaters)	**SRT-e** (Apply 1/6 Operating Heaters)	**SRT-h** (Apply 3/6 Operating Heaters)
Carbon Capture in Operation	**No**	**No**	**No**	**No**	**No**
Utilities					
Heater Fuel Firing	BASE	BASE	-20%	-15%	-10%
Fuel Gas Import / Export	Import	Import	Export	Export	Export
SHP Steam Production	BASE	BASE	-45%	-16%	BASE
HP Steam Import	BASE	BASE	x 3.7	x 0.0	BASE
% Total Compression Power via Motor Drive	0%	0%	59%	24%	0%
Electrical Consumption	BASE	BASE	x 4.6	x 9.3	x 5.4
Emissions					
Scope 1 CO_2-e Generated	BASE	BASE	-20%	-15%	-9%
Scope 2 CO_2-e Generated (2020) (1)	BASE	BASE	x 4.4	x 7.3	x 4.5
Scope 1 + 2 CO_2-Emitted (2020)	BASE	BASE	-3%	+14%	+7%
Total Scope 2 CO_2-e Generated (2020 - 2050)	BASE	BASE	x 4.3	x 6.3	x 4.0
Total Scope 1 + 2 CO_2-e Emitted (2020 - 2050)	BASE	BASE	-9%	+2%	+0%

1) Scope 2 emissions consider HP steam import based on an 80% boiler efficiency burning natural gas and predicted electrical emissions curve taken at 5 year intervals. [1]

Impact to Fuel Gas Balance

Level 0 and Level 1 require a small fuel gas import from another unit in the facility for the specific plant used for this study. All Level 2 balances will export a portion of the fuel gas generated in the cracker recovery section due to the reduction in heater fuel firing duty.

An argument could be made that by reducing fuel firing in the cracking heaters and increasing the fuel gas export, the ethylene plant's scope 1 emissions are reducing, but the fuel gas export is likely still being burned elsewhere as a part of another facility's scope 1 emissions. This is a fair critique, but it's important to note that this fuel gas is being generated as a byproduct of the ethylene process and can displace natural gas production, which has its own emissions related to production. The emissions associated with natural gas production are difficult to quantify, but there will be an overall net reduction in emissions when the reduction in on-purpose natural gas production is taken into account compared to the byproduct natural gas production in the ethylene plant.

Alternately, Lummus' Oxidative Coupling of Methane (OCM) technology may be considered to consume the fuel gas export. OCM is a catalytic process that transforms methane into commodity chemicals in an efficient, cost-effective, scalable manner using processes that can be seamlessly integrated into existing industry infrastructure. At the core of this process is the Lummus proprietary Gemini™ OCM reactor and catalyst, which enables the reaction that converts methane to ethylene in an adiabatic catalytic fixed bed and the subsequent cracking reaction of ethane and/or propane to ethylene and/or propylene in an adiabatic zone. The results presented below consider increased fuel gas export, but an OCM unit is another potential design option.

Impact to Steam Balance

Level 0 and Level 1 steam balances require a small HP steam import stream. Considering air preheating via LEAP (Level 2A) with all heaters in operation, two major compressors (the Propylene Refrigerant and Binary Refrigerant Compressors) will need to be converted from steam turbine drives to motor drives to achieve a similarly low HP steam import as Level 0 and Level 1 (< 40 T/h). Considering one SRT-e heater (Level 2B), one large compressor (Propylene Refrigerant Compressor) will require a dual drive (with a portion of the compression power supplied via turbine and the remainder supplied by electricity) to achieve steam balance without import/export steam. Considering half the operating heaters are SRT-h heaters (Level 2C), the steam production is the same as Level 0 and Level 1, so there is no change to steam production or HP steam import compared to Level 0 and 1.

Impact to Electrical Demand

All Level 2 cases will require additional electrical consumption to account for the displaced energy provided by fuel firing. Considering LEAP (Level 2A), this additional power will be for the electrification of compressor drives. Considering SRT-e (Level 2B), the additional power will be from electrification of compressor drives plus the electric cracking heater. Considering SRT-h

(Level 2C), this additional power will only be from the partial electrification of the cracking heaters.

However, electrical generation has its own associated emissions factor. It can be seen in Table **1** that considering a US grid electrical emissions factor from 2020, the total Scope 1 + 2 CO_2-e emitted for Level 2B & 2C is higher than that of Level 0 & 1. However, there are commitments in the US to decarbonize the electrical grid. When these predictions are taken into account, and considering carbon capture will be implemented 30 years in the future, it can be seen that the total Scope 1 + 2 CO_2-e emitted between 2020 and 2050 is either lower than Level 0 & 1 or approximately the same for all Level 2 cases. [1]

Impact of Carbon Capture Ready on Future Operation

In the following sections, the impacts of pre-investment are investigated for each carbon capture technology.

Post Combustion Carbon Capture

The impact of pre-investment will first be examined for Post Combustion. No pre-investment (Level 0) runs the risk of carbon capture being impossible to implement if there is no plot available around the cracking heaters. For the particular test case of this paper, there are several locations where the carbon capture unit may be located, as shown in Figure **4**. The unit could be located south of the future heaters (red hatch), south of the hot section process equipment (blue hatch), or across the road east of the heaters (purple hatch). The depiction in Figure **4** only shows the potential general location of the unit, but does not show the plot area to scale, as this will depend on the specific case and capacity.

In order to locate the unit south of the hot section process equipment or south of the future heaters, space must be available in the pipe rack for the large flue gas ducting. For Level 0 pre-investment, this would not have been considered and is likely not possible. Therefore, the only location available is east of the heaters (purple hatch). This location will also need to ensure access for heater maintenance is not compromised by the carbon capture unit. The heater structure will need to be reinforced in order to support the load of the ducting or a new structure and foundation may be needed to support the ducts. It is a very realistic possibility that there may not be sufficient space to execute this. For the purpose of this exercise, it is assumed the structure and foundation were able to be reinforced.

Because space in the heater convection section was not saved for a future SCR, a Lummus Compact SCR design will need to be considered. A separate blower must be included, as the ID fan/heater structure did not consider the higher pressure drop of the carbon capture unit. Furthermore, a significant HP steam import will be required. This will impact the overall emissions reduction possible for the system.

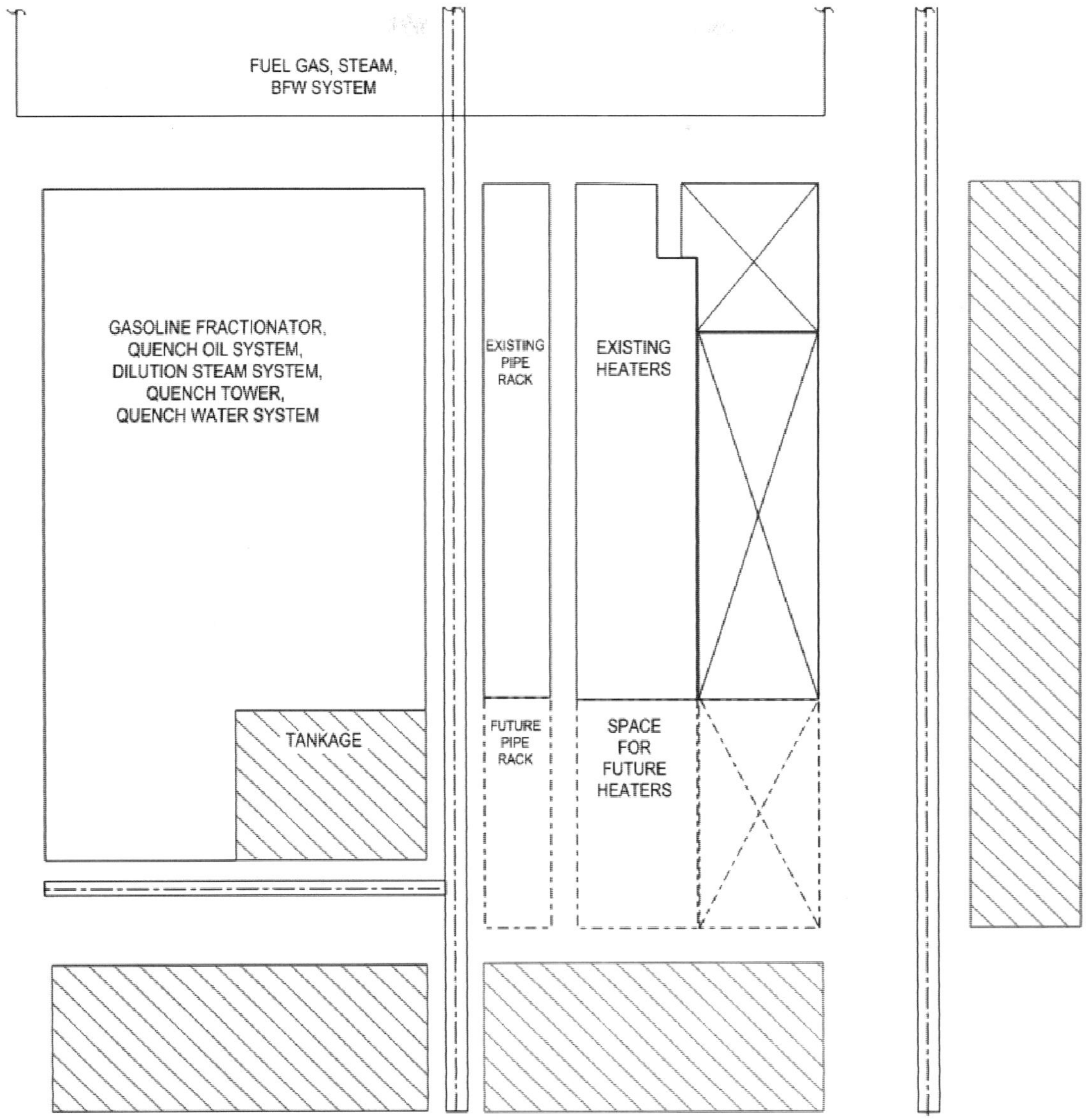

Figure 4: Plot Plan of Potential Carbon Capture Locations

In Level 1, the plot considered future carbon capture unit during construction. The tanks in the blue hatch area were relocated, as they are pumped systems and not hydraulically critical. The pipe rack also saved space for the future ducting. The Direct Contact Cooler / CO_2 Absorber may be located in the area occupied by these tanks (blue hatch in Figure **4**). A single carbon capture unit may be considered for this plant. Furthermore, the Direct Contact Cooler and Absorber may be located in a single tower shell to reduce the plot requirements further while staying within acceptable L/D limitations. The Solvent Stripper and CO_2 Compression / Dehydration Systems may be located across the road, south of the Direct Contact Cooler / CO_2 Absorber (blue hatch in Figure **4**). Ducting will need to be added to the heater stack at the available tie-in point and the SCR will need to be installed. The heater structure already considered the load of these ducts, so no reinforcement is needed. During the initial plant construction, the Propylene Refrigerant Compressor (PRC) was designed with dual-drive capabilities. Before carbon capture is implemented, the compressor will operate in full turbine-mode (same as Level 0 as shown in Table **1**). When carbon capture is implemented, a portion of the total compressor power will be supplied electrically. This results in the HP steam import reducing compared to Level 0 (as shown in Table **2**). This driver flexibility has a marginal impact to the compressor cost.

In Level 2, the fuel fired heaters will have the same considerations as Level 1. However, the size of the Post Combustion unit will be reduced due to the fuel firing reduction measures. Level 2B also has the flue gas, and all associated ducting, from one entire heater completely eliminated. The SRT-e should be optimally located on the north end of the heater island (furthest from the carbon capture unit).

During the initial plant construction for Level 2A, the Charge Gas Compressor (CGC) will also be designed with dual-drive capabilities. Before carbon capture is implemented, this compressor will operate in full turbine-mode. When carbon capture is implemented, a portion of the total compressor power will be supplied electrically. This results in the HP steam import reducing compared to Level 0. Level 2B will require dual-drive capabilities for the Propylene Refrigerant Compressor, similar to Level 1, but the motor capacity will account for a larger portion of the total compression power.

The Level 2 options will require a significant increase in electrical demand compared to Level 0 & 1, which will correspond to higher OPEX. For locations where the cost of electricity is high and/or carbon capture will not be implemented for an extended timeframe, the additional OPEX spent before carbon capture implementation will be likewise high and negatively impact the payback period of the unit. Therefore, for Level 2 pre-investment, a shorter timeline for carbon capture implementation is considered (5 years instead of 30 years) as well as a lower electrical cost ($0.04/kWh instead of $0.07/kWh). This is considered for all carbon capture technologies presented.

As shown in Table **2**, Level 2A results in the largest reduction in heater fuel firing. Level 1 & 2C result in the least impact to SHP steam production and lowest HP steam import. Level 1 requires the lowest increase in electrical consumption, as no heater duty is supplied via electricity and the additional electrical consumption is only for compressor electrification.

Level 2A will yield the lowest scope 1 CO_2 generation and emissions, as this directly correlates with fuel firing duty. Level 1 and Level 2C represent the lowest scope 2 CO_2 generation, as these cases have the lowest steam import. Level 2A will have the lowest scope 1+2 CO_2 generation, mostly due to the lower scope 1 generation, while Level 1 will have the lowest scope 1+2 CO_2 emissions, mostly due to the lower scope 2 generation.

Level 2A will have the smallest plot requirement, as it processes the lowest amount of flue gas and CO_2 product. Therefore, Level 2A may be considered optimal if the available plot for the unit is limiting.

The total investment cost at year 0 (or when the plant is being built) includes any additional capital spent compared to a Level 0 design, which spends no additional capital for carbon capture prior to year X (or when carbon capture is implemented). This includes additional cost for heater pre-investment as discussed above, additional cost for change in heater type (i.e. LEAP, SRT-e, and SRT-h relative to conventional heater), and additional cost for change in compressor drives compared to Level 0. The total investment cost at year X will be the cost of implementing carbon capture. This will include any heater modifications needed at year X (mostly required for Level 0), the cost of the carbon capture unit, and additional OPEX spent by year X compared to Level 0. The additional OPEX will pertain to Level 2 pre-investment, where there is increased electrical demand in place of utilizing internally-generated fuel gas to supply the required heater duty. Lastly, the total investment from year 0 to year X is the sum of the previous two investment values.

Level 1 can be achieved with very minimal additional cost at year 0. Level 2A will have the lowest investment at year X. This is mostly due to the fact that it will have the highest fuel gas export (or lowest fuel firing duty) to outweigh the higher electrical cost compared to Level 0. This will result in a lower OPEX compared to Level 0 & 1 before carbon capture is implemented. Level 2A will also have the lowest capacity (and CAPEX) carbon capture unit. It can be seen that Level 1, 2A, and 2C all have a lower total investment cost compared to Level 0.

The final result shows that Level 2A will have the lowest payback period, although other limiting factors may favor other cases. For example, Level 2C may be preferred if steam import is limiting, Level 1 may be preferred if minimal upfront investment is favorable, and Level 2B may be preferred if electrification is incentivized.

Table **2**: Impact of Pre-Investment Levels when Implementing Post Combustion Carbon Capture

Pre-Investment Level	**0**	**1**	**2A**	**2B**	**2C**
Fuel Reduction Method	**None** (Apply 6/6 Operating Heaters)	**None** (Apply 6/6 Operating Heaters)	**LEAP** (Apply 6/6 Operating Heaters)	**SRT-e** (Apply 1/6 Operating Heaters)	**SRT-h** (Apply 3/6 Operating Heaters)
Carbon Capture in Operation	**Yes**	**Yes**	**Yes**	**Yes**	**Yes**
Utilities (Year X)					
Heater Fuel Firing	BASE	BASE	**-20%**	-15%	-10%
SHP Steam Production	**BASE**	**BASE**	-45%	-16%	**BASE**
HP Steam Import	BASE	**-41%**	-23%	-13%	-39%
% Total Compression Power via Motor Drive	0%	15%	63%	25%	0%
Electrical Consumption	**BASE**	x 1.2	x 2.0	x 3.8	x 2.5
Emissions (Year X)					
Scope 1 CO_2-e Generated	BASE	BASE	**-20%**	-15%	-9%
CO_2 Recovery			90%		
Scope 1 CO_2-e Emitted	BASE	BASE	**-18%**	-16%	-4%
Scope 2 CO_2-e Generated (Year X)	BASE	**-38%**	-17%	+1%	-30%
Scope 1 + 2 CO_2-e Generated (Year X)	BASE	-7%	**-19%**	-12%	-13%

Pre-Investment Level	0	1	2A	2B	2C
Fuel Reduction Method	**None** (Apply 6/6 Operating Heaters)	**None** (Apply 6/6 Operating Heaters)	**LEAP** (Apply 6/6 Operating Heaters)	**SRT-e** (Apply 1/6 Operating Heaters)	**SRT-h** (Apply 3/6 Operating Heaters)
Carbon Capture in Operation	**Yes**	**Yes**	**Yes**	**Yes**	**Yes**
Scope 1 + 2 CO_2-e Emitted (Year X)	BASE	-27%	-17%	-4%	-22%
Cost					
Total Investment at Year 0	BASE	+1%	+10%	+4%	+2%
Total Investment at Year X	BASE	-8%	-31%	+4%	-7%
Total Investment Year 0 ~ X	BASE	-6%	-22%	+4%	-5%
OPEX Year X+	BASE	-22%	-77%	+7%	-36%
Payback Period, years	**8.8**	**6.9**	5.1	**13.6**	**7.4**

1) Scope 2 emissions consider HP steam import based on an 80% boiler efficiency burning natural gas. Electrical emissions consider 2050 estimation of 40 kg CO_2-e/MWh
2) Year X considered to be 30 years for Level 0~1. Year X considered to be 5 years for Level 2A~2C.
3) OPEX considers the following utility costs:
   Operation & Maintenance: 3% CAPEX at year X
   Electricity: $0.07/kWh (Level 0~1); $0.04/kWh (Level 2A~2C)
   Steam: $7.50/MMBTU
   Natural Gas: $6.00/MMBTU
   CO_2 Product: $85/ton
4) Optimal value (either min/max) highlighted in red font.

Oxyfuel Combustion

Oxyfuel Combustion will have similar plot concerns as Post Combustion. Level 0 pre-investment will also require the unit be located east of the heaters (purple hatch in Figure **4**). Oxyfuel Combustion will have the additional complication of requiring flue gas recycle to return to the heaters. Similar to Post Combustion, this location will also need to ensure access for heater maintenance is not compromised by the carbon capture unit. The heater structure will need to be reinforced in order to support the weight of the ducting. The heater burners will also need to be replaced. Additionally, in order for Oxyfuel Combustion to retain the compelling characteristics of the technology, air ingress into the system must be minimized. Depending on the state of the existing heaters when Oxyfuel Combustion is implemented, additional heater modifications may be required for resealing.

The revamp of the refrigerant system will be more involved without pre-investment. The total compressor power will increase by approximately 9% when Oxyfuel Combustion is incorporated. The largest individual stage power increase will be on the 3rd stage, which will see an increase of approximately 14%. The compressor and driver should be able to accommodate this higher capacity by operating at a higher speed without modification to the equipment. The total system capacity will increase by approximately 11%. An additional shell or shell replacement/modification may be required for the refrigerant condenser to accommodate the higher duty. The higher capacity of the refrigerant system will also increase the flare load. Additional flare mitigation may be required to stay within the design flare capacity. Some of the vapor-controlled suction drums will be revamped by replacing internals, while some of the liquid-controlled suction drums will need to operate with lower holdup than design.

In Level 1, the carbon capture unit will be located in the blue hatch area (Figure **4**) with the tanks relocated and the Direct Contact Cooler and Flue Gas Recycle Blower located in this area. Because Oxyfuel Combustion requires both a flue gas feed line and flue gas recycle line going to/from the heater area, appropriately locating the equipment relative to the heaters is critical. The CO_2 Purification Unit can be located south of the hot section equipment. This location can also easily access propylene refrigeration from the cracker system for the CO_2 Purification Unit, as the compressor house is located just west of the end of the depiction shown in Figure **4**. The ASU has no specific location requirements relative to the carbon capture unit. Ducting will need to be added to the heater stack to direct the flue gas to the capture unit and at the heater inlet for the flue gas recycle at the available tie-ins. The heater structure already considered the load of the ducts. No changes are required for the burners, as they already considered flexibility for Oxyfuel operation. The refrigerant system was pre-invested so slightly larger drums were initially installed (relative to Level 0) and no modifications or mist eliminator replacement will be needed.

In Level 2, the fuel fired heaters will have the same considerations as Level 1, however, the size of the carbon capture unit will be reduced due to the fuel firing reduction measures. Level 2A will also require some additional modifications. The combustion air that is preheated prior to carbon capture implementation will need to be replaced with the flue gas recycle. The original design would have already considered this future operation and no changes are required except for routing. Level 2B also has the flue gas, and all associated ducting, from one entire heater completely eliminated. The SRT-e should be optimally located on the north end of the heater island (furthest from the carbon capture unit).

As shown in Table **3**, there are many similar trends between Post Combustion Capture technology results and Oxyfuel Combustion technology results. Level 2A results in the largest reduction in heater fuel firing. Level 1 & 2C results in the least impact to SHP steam production. There is no additional steam demand for Oxyfuel Combustion, with the exception of the higher Propylene Refrigeration Compressor power, so similar trends are seen for the steam demand as prior to carbon capture implementation, with Level 2B requiring the lowest HP steam import. Level 1 requires the lowest increase in electrical consumption, as no heater duty is supplied via electricity and no major compressors have motor drives.

Level 2A will yield the lowest scope 1 CO_2 generation, as this directly correlates with fuel firing duty. Level 1 represent the lowest scope 2 CO_2 generation due to the low electrical demand. Level 2A will have the lowest scope 1+2 CO_2 generation, mostly due to the lower scope 1 generation, while Level 2B will have the lowest scope 1+2 CO_2 emissions due to the elimination of the steam import and lower scope 1 emissions. For Oxyfuel Combustion it can be observed that Level 2A & 2C result in a higher scope 1 and 2 emitted due to the higher scope 2 emissions.

Level 2A will have the smallest plot requirement, as it processes the lowest amount of flue gas and CO_2 product. Therefore, Level 2A may be considered optimal if the available plot for the unit is limiting.

Level 1 can be achieved with very minimal additional cost at year 0. Level 2A will have the lowest investment at year X. This is mostly due to the fact that it will have the highest fuel gas export (or lowest fuel firing duty) to outweigh the higher electrical cost compared to Level 0. This will result in a lower OPEX compared to Level 0 prior to implementing carbon capture. Level 2A will also have the lowest capacity (and CAPEX) carbon capture unit. It can be seen that Level 1 & 2A have a lower total investment cost compared to Level 0.

The final result shows that Level 2A will have the lowest payback period, although other limiting factors may favor other cases with somewhat higher payback periods. For example, Level 1 or 2B may be preferred if reducing overall emissions is preferred.

Table **3**: Impact of Pre-Investment Levels when Implementing Oxyfuel Combustion

Pre-Investment Level	**0**	**1**	**2A**	**2B**	**2C**
Fuel Reduction Method	**None** (Apply 6/6 Operating Heaters)	**None** (Apply 6/6 Operating Heaters)	**LEAP** (Apply 6/6 Operating Heaters)	**SRT-e** (Apply 1/6 Operating Heaters)	**SRT-h** (Apply 3/6 Operating Heaters)
Carbon Capture in Operation	**Yes**	**Yes**	**Yes**	**Yes**	**Yes**
Utilities (Year X)					
Heater Fuel Firing	BASE	BASE	-20%	-15%	-10%
SHP Steam Production	BASE	BASE	-45%	-16%	BASE
HP Steam Import	BASE	BASE	+26%	-100%	-7%
% Total Compression Power via Motor Drive	0%	0%	60%	27%	0%
Electrical Consumption	BASE	BASE	x 1.7	x 2.7	x 1.9
Emissions (Year X)					
Scope 1 CO_2-e Generated	BASE	BASE	-20%	-15%	-10%
CO_2 Recovery	95%				
Scope 1 CO_2-e Emitted	BASE	BASE	-20%	-15%	-10%
Scope 2 CO_2-e Generated (Year X)	BASE	BASE	+42%	+11%	+32%
Scope 1 + 2 CO_2-e Generated (Year X)	BASE	BASE	-17%	-14%	-8%

Pre-Investment Level	0	1	2A	2B	2C
Fuel Reduction Method	**None** (Apply 6/6 Operating Heaters)	**None** (Apply 6/6 Operating Heaters)	**LEAP** (Apply 6/6 Operating Heaters)	**SRT-e** (Apply 1/6 Operating Heaters)	**SRT-h** (Apply 3/6 Operating Heaters)
Carbon Capture in Operation	**Yes**	**Yes**	**Yes**	**Yes**	**Yes**
Scope 1 + 2 CO_2-e Emitted (Year X)	BASE	BASE	+12%	-2%	+12%
Cost					
Total Investment at Year 0	BASE	+1%	+9%	+4%	+2%
Total Investment at Year X	BASE	-1%	-38%	+52%	+21%
Total Investment Year 0 ~ X	BASE	-0%	-17%	+31%	+13%
OPEX Year X+	BASE	BASE	-39%	+5%	-6%
Payback Period, years	**6.9**	**6.9**	3.4	**44.2**	**10.7**

1) Scope 2 emissions consider HP steam import based on an 80% boiler efficiency burning natural gas. Electrical emissions consider 2050 estimation of 40 kg CO_2-e/MWh
2) Year X considered to be 30 years for Level 0~1. Year X considered to be 5 years for Level 2A~2C.
3) OPEX considers the following utility costs:
   Operation & Maintenance: 3% CAPEX at year X
   Electricity: $0.07/kWh (Level 0~1); $0.04/kWh (Level 2A~2C)
   Steam: $7.50/MMBTU
   Natural Gas: $6.00/MMBTU
   High-Purity Oxygen: $40/ton
   CO_2 Product: $85/ton
4) Optimal value (either min/max) highlighted in red font.
5) OPEX and resulting scope 2 emissions include ASU electrical demand based on 190 kWh/ton O_2

Pre-Combustion

Pre-Combustion is generally the easiest technology to implement without pre-investment. The Hydrogen Unit and ASU have no specific location requirements relative to the heater. Level 0 pre-investment will result in the heater burners requiring replacement and a Lummus Compact SCR design. Level 1 pre-investment will only require the installation of the SCR and routing the hydrogen to the heaters and fuel gas to the hydrogen unit.

Again, similar trends are observed as the other carbon capture technologies. Level 2A pre-investment results in the largest reduction in heater fuel firing. Level 2C pre-investment results in the no impact to SHP steam production. There is no external steam demand for Pre-Combustion, so similar trends are seen as prior to carbon capture implementation, with Level 2B eliminating the HP steam import by increased compressor electrification. Level 1 pre-investment requires no increase in electrical consumption compared to Level 0, as no heater duty is supplied via electricity and no major compressors have motor drives.

Level 2A pre-investment will yield the lowest scope 1 CO_2 generation and emissions, as this directly correlates with the hydrogen unit capacity, which correlates to fuel firing duty. Level 1 represent the lowest scope 2 CO_2 generation. Level 2A will have the lowest scope 1+2 CO_2 generation, mostly due to the lower scope 1 generation, while Level 1 will have the lowest scope 1+2 CO_2 emissions due to the low scope 2 emissions.

Level 2A will have the smallest plot requirement, as it is the lowest capacity. Therefore, Level 2A may be considered optimal if the available plot for the unit is limiting.

Level 1 can be achieved with very minimal additional cost at year 0. Level 2A will have the lowest investment at year X. This is mostly due to the fact that it will have the highest fuel gas export (or lowest fuel firing duty) to outweigh the higher electrical cost compared to Level 0. This will result in a lower OPEX compared to Level 0 before implementing carbon capture. Level 2A will also have the lowest capacity (and CAPEX) carbon capture unit. It can be seen that Level 1 & 2A have a lower total investment cost compared to Level 0.

The final result shows that Level 2A will have the lowest payback period, although other limiting factors may favor other cases with somewhat higher payback periods. For example, Level 1 may be preferred if reducing overall emissions is preferred.

Table **4**: Impact of Pre-Investment Levels when Implementing Pre-Combustion Carbon Capture

Pre-Investment Level	**0**	**1**	**2A**	**2B**	**2C**
Fuel Reduction Method	**None** (Apply 6/6 Operating Heaters)	**None** (Apply 6/6 Operating Heaters)	**LEAP** (Apply 6/6 Operating Heaters)	**SRT-e** (Apply 1/6 Operating Heaters)	**SRT-h** (Apply 3/6 Operating Heaters)
Carbon Capture in Operation	**Yes**	**Yes**	**Yes**	**Yes**	**Yes**
Utilities (Year X)					
Heater Fuel Firing	BASE	BASE	**-20%**	-15%	-10%
SHP Steam Production	**BASE**	**BASE**	-45%	-16%	**BASE**
HP Steam Import	BASE	BASE	x 3.7	**x 0.0**	BASE
% Total Compression Power via Motor Drive	0%	0%	59%	24%	0%
Electrical Consumption	**BASE**	**BASE**	x 1.7	x 2.7	x 1.9
Emissions (Year X)					
Scope 1 CO_2-e Generated	BASE	BASE	**-21%**	-16%	-9%
Overall CO_2 Recovery			98%		
Scope 1 CO_2-e Emitted	BASE	BASE	**-21%**	-16%	-10%

Pre-Investment Level	**0**	**1**	**2A**	**2B**	**2C**
Fuel Reduction Method	**None** (Apply 6/6 Operating Heaters)	**None** (Apply 6/6 Operating Heaters)	**LEAP** (Apply 6/6 Operating Heaters)	**SRT-e** (Apply 1/6 Operating Heaters)	**SRT-h** (Apply 3/6 Operating Heaters)
Carbon Capture in Operation	**Yes**	**Yes**	**Yes**	**Yes**	**Yes**
Scope 2 CO_2-e Generated (Year X)	**BASE**	**BASE**	x 2.3	x 1.8	x 1.6
Scope 1 + 2 CO_2-e Generated (Year X)	BASE	BASE	**-17%**	-14%	-7%
Scope 1 + 2 CO_2-e Emitted (Year X)	**BASE**	**BASE**	+68%	+41%	+31%
Cost					
Total Investment at Year 0	**BASE**	+0%	+9%	+4%	+2%
Total Investment at Year X	BASE	+0%	**-20%**	+24%	+7%
Total Investment Year 0 ~ X	BASE	+0%	**-12%**	+18%	+5%
OPEX Year X+	BASE	BASE	**-33%**	-5%	-9%
Payback Period, years	**10.7**	**10.7**	**6.9**	**31.2**	**12.7**

1) Scope 2 emissions consider HP steam import based on an 80% boiler efficiency burning natural gas. Electrical emissions consider 2050 estimation of 40 kg CO_2-e/MWh
2) Year X considered to be 30 years for Level 0~1. Year X considered to be 5 years for Level 2A~2C.
3) OPEX considers the following utility costs:
   Operation & Maintenance: 3% CAPEX at year X
   Electricity: $0.07/kWh (Level 0~1); $0.04/kWh (Level 2A~2C)
   Steam: $7.50/MMBTU
   Natural Gas: $6.00/MMBTU
   High-Purity Oxygen: $40/ton
   CO_2 Product: $85/ton
4) Optimal value (either min/max) highlighted in red font.
5) OPEX and resulting scope 2 emissions include ASU electrical demand based on 170 kWh/ton O_2

Comparison of Carbon Capture Technologies

A similar payback period can be achieved for each carbon capture technology discussed above, but certain technologies may be more desirable depending on the specific plant and decarbonization goals. The optimal pre-investment level for each carbon capture technology is shown in Table **5**.

Pre-combustion and Oxyfuel require the least steam import, as there is no external steam consumed by the unit for either technology. Post Combustion swaps a larger portion of compressor drives to motors in order to reduce the steam import required for the carbon capture unit. Oxyfuel Combustion and Pre-Combustion have similarly high electrical consumption.

Pre-combustion has the highest scope 1 CO_2 generation, mostly from within the hydrogen unit, but also the largest CO_2 recovery, which results in the lowest overall CO_2 emissions. Oxyfuel, by comparison, generates the least CO_2. There are minimal CO_2 utilization technologies available today at the capacities required for an ethylene plant, so it may be preferred to minimize CO_2 generation while achieving similar CO_2 emissions, as shown with Oxyfuel Combustion. Pre-combustion requires the least investment prior to carbon capture implementation, while Oxyfuel requires the least total investment due to the lower CAPEX of the Oxyfuel unit. Pre-combustion requires the highest OPEX after carbon capture implementation. This is because it has the lowest fuel gas export, as additional feed to the hydrogen unit is required compared to fuel gas feed to the heaters for the other cases.

Each of these factors discussed above may alter which carbon capture technology is preferable depending on a specific site's access to utilities, these utilities' emissions factors, a site's decarbonization goals, footprint limitations, or rollout of investment commitments among others.

Table **5**: Comparison of Carbon Capture Technologies' Optimal Pre-Investment Level

Pre-Investment Level	**0**	**2A**	**2A**	**2A**
Fuel Reduction Method	**None** (Apply 6/6 Operating Heaters)	**LEAP** (Apply 6/6 Operating Heaters)	**LEAP** (Apply 6/6 Operating Heaters)	**LEAP** (Apply 6/6 Operating Heaters)
Carbon Capture in Operation	**No**	**Yes Post Combustion**	**Yes Oxyfuel**	**Yes Pre-Combustion**
Utilities (Year X)				
Fuel Firing	BASE	-20%	**-22%**	-12%
SHP Steam Production	**BASE**	-45%	-45%	-45%
HP Steam Import	**BASE**	x 16.7	x 3.7	x 3.7
% Total Compression Power via Motor Drive	0%	63%	60%	59%
Electrical Consumption	**BASE**	x 5.1	x 7.7	x 7.8
Emissions (Year X)				
Scope 1 CO_2-e Generated	BASE	-20%	**-23%**	+13%
Overall CO_2 Recovery	0%	90%	95%	**98%**
Scope 1 CO_2-e Emitted	BASE	-92%	-96%	**-98%**
Scope 2 CO_2-e Generated (Year X)	**BASE**	x 13.2	x 4.9	x 4.9
Scope 1 + 2 CO_2-e Generated (Year X)	BASE	-1%	**-16%**	+18%

Pre-Investment Level	0	2A	2A	2A
Fuel Reduction Method	**None** (Apply 6/6 Operating Heaters)	**LEAP** (Apply 6/6 Operating Heaters)	**LEAP** (Apply 6/6 Operating Heaters)	**LEAP** (Apply 6/6 Operating Heaters)
Carbon Capture in Operation	**No**	**Yes Post Combustion**	**Yes Oxyfuel**	**Yes Pre-Combustion**
Scope 1 + 2 CO_2-e Emitted (Year X)	BASE	-73%	-89%	-91%
Cost				
Total Investment at Year 0		BASE	+1%	-1%
Total Investment at Year X		BASE	-69%	-20%
Total Investment Year 0 ~ X		BASE	-48%	-14%
OPEX Year X+		BASE	x 4.5	x 7.9
Payback Period, years		**5.1**	**3.4**	**6.9**

1) Scope 2 emissions consider HP steam import based on an 80% boiler efficiency burning natural gas and predicted electrical emissions curve. Electrical emissions considers 2050 estimation of 40 kg CO_2-e/MWh
2) Year X considered to be 30 years for Level 0~1. Year X considered to be 5 years for Level 2A~2C.
3) OPEX considers the following utility costs:
   Operation & Maintenance: 3% CAPEX at year X
   Electricity: $0.07/kWh (Level 0~1); $0.04/kWh (Level 2A~2C)
   Steam: $7.50/MMBTU
   Natural Gas: $6.00/MMBTU
   High-Purity Oxygen: $40/ton
   CO_2 Product: $85/ton
4) Optimal value (either min/max) highlighted in red font.
5) OPEX and resulting scope 2 emissions include ASU electrical demand based on 190 kWh/ton O_2 for Oxyfuel Combustion and 170 kWh/ton O_2 for Pre-Combustion.

Conclusion

In conclusion, pre-investment during plant construction can ensure it is not only possible to implement carbon capture in the future, but that carbon capture is optimally applied while producing an attractive economic payback period. Furthermore, it should be noted that this paper only considers decarbonization incentives and not mandates. The observed trend to date has largely only considered decarbonization incentives by governments, but it is anticipated that in the future these incentives may increase in value or include mandates. Being prepared for this realistic possibility with minimal pre-investment can prevent significant difficulties down the road.

As shown, Level 1 CAPEX pre-investment is minimal at the initial plant construction. Therefore, Level 1 is preferrable if the future in uncertain and minimal upfront investment is preferred. Level 1 may also be optimal if carbon capture is expected to be implemented as a part of an expansion, in which case, carbon reduction measures like heater electrification may be considered as a part of that expansion. Level 2 will have a higher investment before implementing carbon capture, largely due to the higher OPEX relative to Level 0. Rather than consuming internally generated fuel gas or steam, these cases require energy sources be purchased externally, resulting in inevitably higher operating cost. Therefore, Level 2 may be implemented if there are (or expected to be) incentives for switching to electrification, the cost of electricity is low and the electrical source is clean, carbon capture is expected to be implemented in the short-term, or if minimizing scope 1 CO_2 generation is preferred.

A brief comparison of carbon capture technologies was also investigated. While Pre-combustion requires the least pre-investment, it also generates the most CO_2 that must either be stored or utilized in some way. Oxyfuel Combustion requires the least total investment and can achieve a similar overall emissions reduction to Pre-combustion. Post Combustion has the lowest OPEX after carbon capture implementation, as it does not consume purified oxygen.

In short, this paper only provided a brief summary of a handful of options, but there are many permutations and combinations of these cases that may be tailored to achieve an attractive solution based on a specific client, plant, location, and decarbonization goal.

References

[1] E. Grubert, "Emissions Projections for US Utilities Through 2050," *Environmental Research Letter,* 2021.

AIChE Paper Number: R08-R09

INTRODUCING ELECTRIC CRACKING TOWERS FOR NET-ZERO ETHYLENE PRODUCTION

Authors:

Edwin Rodriguez

Davidson School of Chemical Engineering, Purdue University, West Lafayette, IN, USA

Rakesh Agrawal

Davidson School of Chemical Engineering, Purdue University, West Lafayette, IN, USA

Prepared for Presentation at the 2024 Spring National Meeting

New Orleans, Louisiana, USA, March 24-28, 2024

INTRODUCING ELECTRIC CRACKING TOWERS FOR NET-ZERO ETHYLENE PRODUCTION

Authors:

Edwin Rodriguez

Rakesh Agrawal

Davidson School of Chemical Engineering, Purdue University, West Lafayette, IN, USA

Abstract

Ethylene is an essential building block for countless modern products and services. Global production surged to ~150 million tons in 2020 and is projected to grow to ~255 million tons by 2035. While Flame-heated Tubular Crackers (FTCs) have dominated ethylene production for the last century, they face challenges with scalability, thermal management, and carbon emissions, accounting for ~0.6% of global emissions. In response, electrically heated crackers have re-emerged as a more sustainable option aligned with net-zero goals. However, existing electric cracker designs encounter limitations akin to traditional FTCs, relying on external heating through the tube walls. This study introduces Electric Cracking Towers (E-CRATOS), a novel reactor configuration designed to overcome the constraints of both traditional FTCs and current electric crackers. E-CRATOS employs electric heating elements organized into stages along the reactor length, enabling superior thermal performance and scalability. As an electrically powered solution, E-CRATOS retains the benefits of decarbonization, reducing CO_2 emissions by up to ~95% compared to FTCs. The internal heating elements allow the use of larger diameter tubes, reducing construction metal and transportation weight. Additionally, strategic control of the heating elements facilitates temperature profile optimization to enhance ethylene yield while reducing hot spots. A case study under ideal heat transfer conditions is presented to demonstrate as an initial assessment under simplified conditions of E-CRATOS against current cracking technologies. The results reveal the potential for an increase in ethylene production by a factor of ~1.17 while substantially reducing the mean gas temperature across the reactor length. By fundamentally changing the reactor design, E-CRATOS represents a opportunity to achieve dramatic carbon emissions reductions while boosting production, bringing ethylene production one step closer to a net-zero future.

1. Introduction

Ethylene (C_2H_4) is a pillar of the global industrial economy. Its applications span everything from internet communication and food packaging to textiles and medical sterilization [1]–[3]. Its level of importance is such that in 2017 the American Association of Fuel and Petrochemical Manufacturers (AFPM) referred to it as the "World's Most Important Chemical" [4].

1.1 Flame-heated Tubular Crackers (FTCs)

Over the past century, Flame-heated Tubular Crackers (FTCs) have been the dominant force in ethylene production [5]–[7]. As illustrated in Figure 1, FTCs are mainly comprised of a series of long metallic tubes heated externally by the combustion of fossil fuels. Temperatures inside the reaction tubes reach ~850 °C, promoting the cracking of feedstocks like ethane or naphtha into ethylene [8]. FTCs have been known since at least 1908 [9] and gained widespread popularity after 1918 when George Curme employed them for industrial-scale ethylene production [5]–[7].

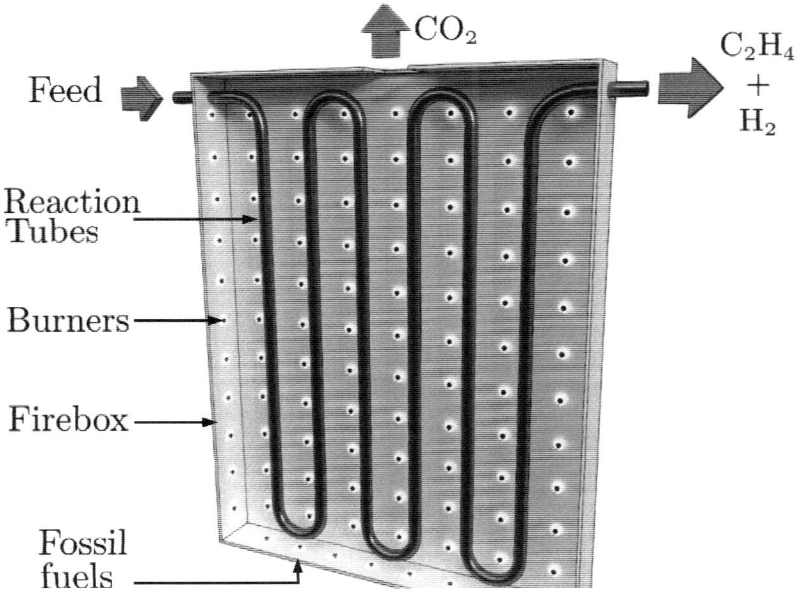

Figure 1. Illustration of a Flame-heated Tubular Cracker (FTC).
External burners heat the reaction tubes to facilitate cracking.
This configuration is commonly known as steam cracking when steam serves as the diluent.

Despite their longstanding dominance, FTCs still face persistent limitations that undermine their performance. These challenges arise primarily from external heating and include constrained heat transfer, pronounced thermal gradients, and restricted temperature control [10]. Additionally, being primarily fueled by the combustion of fossil fuels, FTCs constitute the second-largest CO_2-emitting unit in the chemical industry [11], [12].

1.2 FTCs and CO_2 Emissions

For every ton of ethylene produced, FTCs release ~1.3 tons of CO_2 [13]. In 2021 alone, with an ethylene market of ~150 million tons (Mt) [14], FTCs released ~200 Mt of CO_2 or ~0.6% of global emissions [15]. Ethylene demand is projected to reach ~255 Mt by 2035 [14]. In a business-as-usual scenario [16], annual emissions may surpass ~335 Mt, equivalent to the CO_2 output of ~73 million cars [17]. This trajectory directly conflicts with global decarbonization goals. For example, the United States aims to reduce greenhouse gases 50-52% below 2005 levels by 2030 and achieve net-zero emissions by 2050 [18]. Likewise, the European Union has committed to a reduction of at least 55% by 2030 and aims to attain climate neutrality by 2050 [19]. Decarbonizing FTCs constitutes one of the primary opportunities for the chemical industry to reduce CO_2 emissions substantially [20].

1.3 Electrically Heated Crackers

Electrically heated crackers present a promising pathway to decarbonize traditional FTCs [21], [22]. They have attracted renewed interest, especially when paired with low-carbon energy sources like nuclear, solar, or wind [23].

While electric cracking may seem like a novel concept, its origins trace back to the early 1800s when John Dalton and William Henry conducted the first known ethylene cracking experiment using a Spark Cracker, illustrated in Figure 2 [24], [25].

Figure 2. Illustration of a Spark Cracker.
Electrodes provide electric discharges to crack ethylene molecules.
This configuration is also known as a Volta Eudiometer which originates from the Greek words eúdios (clear weather) and metron (measure).

Transitioning to industrial scale, electrically heated crackers have been considered as early as the 1910s. George Curme himself acknowledged their potential, as evidenced in the following excerpt from his 1918 patent:

"The method of heating the ethane is not, in general, important... It is recognized, however, that some advantage is to be had in electrically heated reaction tubes on account of the ease afforded for temperature control, and the possibility of more thorough heat insulation." [7]

Despite this recognition, the historical prevalence of lower fossil fuel heating prices compared to electric heating favored the dominance of FTCs over their electric counterparts.

Today, the industry is actively exploring various electric cracking configurations. Notably, in September 2022, BASF, in collaboration with Linde and SABIC, began the construction of a demonstration unit to test both Impedance Tubular Crackers (ITCs) and External Resistance Tubular Crackers (ERTCs), illustrated in Figures 3 and 4 respectively [26]. However, despite their potential, it is crucial to note that both ITCs and ERTCs, relying on external heating from the tube walls, present intrinsic limitations akin to those observed in FTCs.

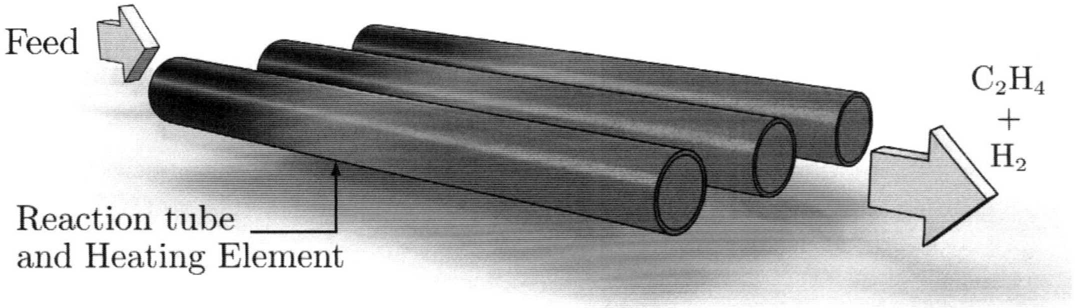

Figure 3. Illustration of an Impedance Tubular Cracker (ITC).
Electricity flows directly through the reaction tubes, generating heat.
This configuration is also known as direct heat cracking [27].

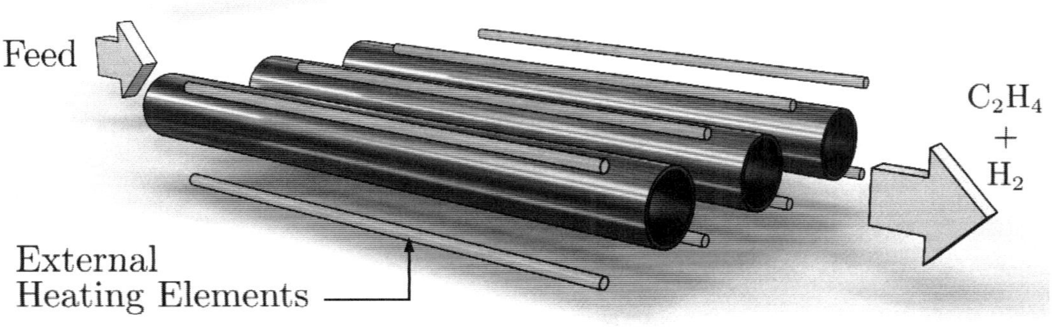

Figure 4. Illustration of an External Resistance Tubular Cracker (ERTC).
Electric heating elements outside the reaction tube provide heat to facilitate cracking.
This configuration is also known as indirect heat cracking [28].

1.4 Limitations of External Heating

Externally heated crackers, regardless of the heating method, face inherent design limitations that undermine their performance across multiple levels.

A major constraint stems from the intense radial thermal gradients within each tube, often reaching ~100°C from walls to center [10]. These uneven reaction conditions result in diminished conversion rates at the center and promote undesired reactions near the wall. To mitigate this effect, tube diameters are kept remarkably small, sometimes as narrow as 3 cm [29]. However, the reduced flow areas of these tubes require hundreds of parallel tubes to scale up production, consequently necessitating massive amounts of metal during construction. A single externally heated cracker may demand over 2500 tons of raw metal, representing about half a billion USD in costs [30], [31].

Another significant constraint is temperature control. With over 200 tubes with lengths ranging from 10 to 100 m, surrounded by 1100°C+ heating elements, continuously measuring, and regulating internal temperatures is challenging [29]. Even when areas demanding more heat are identified, the distribution of heating elements, as illustrated in Figures 1, 3, and 4, restricts localized delivery. Moreover, any heat produced in a specific zone outside the tubes tends to disperse across a larger area due to the conductivity of the metal.

The lack of temperature control contributes to the formation of hot spots, reducing reactor longevity and increasing coke formation. At typical reaction conditions, an increase of just 20°C in the metal temperature can halve the reactor lifespan [32]. With current metallurgy, a ~6-year service is typical [33]. Coke formation exacerbates this problem by decreasing flow area, elevating pressure drop, and reducing heat transfer [33]–[35]. Consequently, higher external temperatures become necessary, further intensifying hot spots. This cycle continues until temperatures reach a critical level and decoking is required, disrupting operations for days [36]. A naphtha cracker typically will be out of operation for 20-40 hours every 15-40 days [37].

Typical thermal efficiencies of externally heated crackers range between 38 and 48% [13]. This is due to the larger temperature difference between the environment and the interior of the tube. While adding a recovery section improves efficiency, it further increases complexity and costs.

These limitations of external heating reveal the need for innovative designs that enhance temperature control, reduce coking, minimize materials usage, improve efficiency, and increase flexibility. Novel approaches will be essential to drive ethylene production into a net-zero future.

2. Electric Cracking Towers (E-CRATOS)

To address the limitations of existing cracking reactor designs, this research introduces the concept of Electric Cracking Towers (E-CRATOS). As illustrated in Figure 5, E-CRATOS employ electric heating elements arrayed in stages along the reactor length. This innovative system enables substantially larger tube diameters, reducing construction metal demand and unit weight. By strategically changing the spacing and power to each heating element, E-CRATOS can precisely tailor the heat profile along the reactor length (US App. No 63/561,869).

E-CRATOS can incorporate turbulence-promoting elements for enhancing internal mixing and more homogeneous reaction conditions. This flexible design allows the allocation of more heating elements where more heat is needed, providing greater heating capacity without raising wall temperatures. These advantages translate to higher ethylene yields and lower coke formation. Moreover, when powered by low-carbon energy sources [23], E-CRATOS has the potential to reduce CO_2 emissions by up to ~95% compared to FTCs.

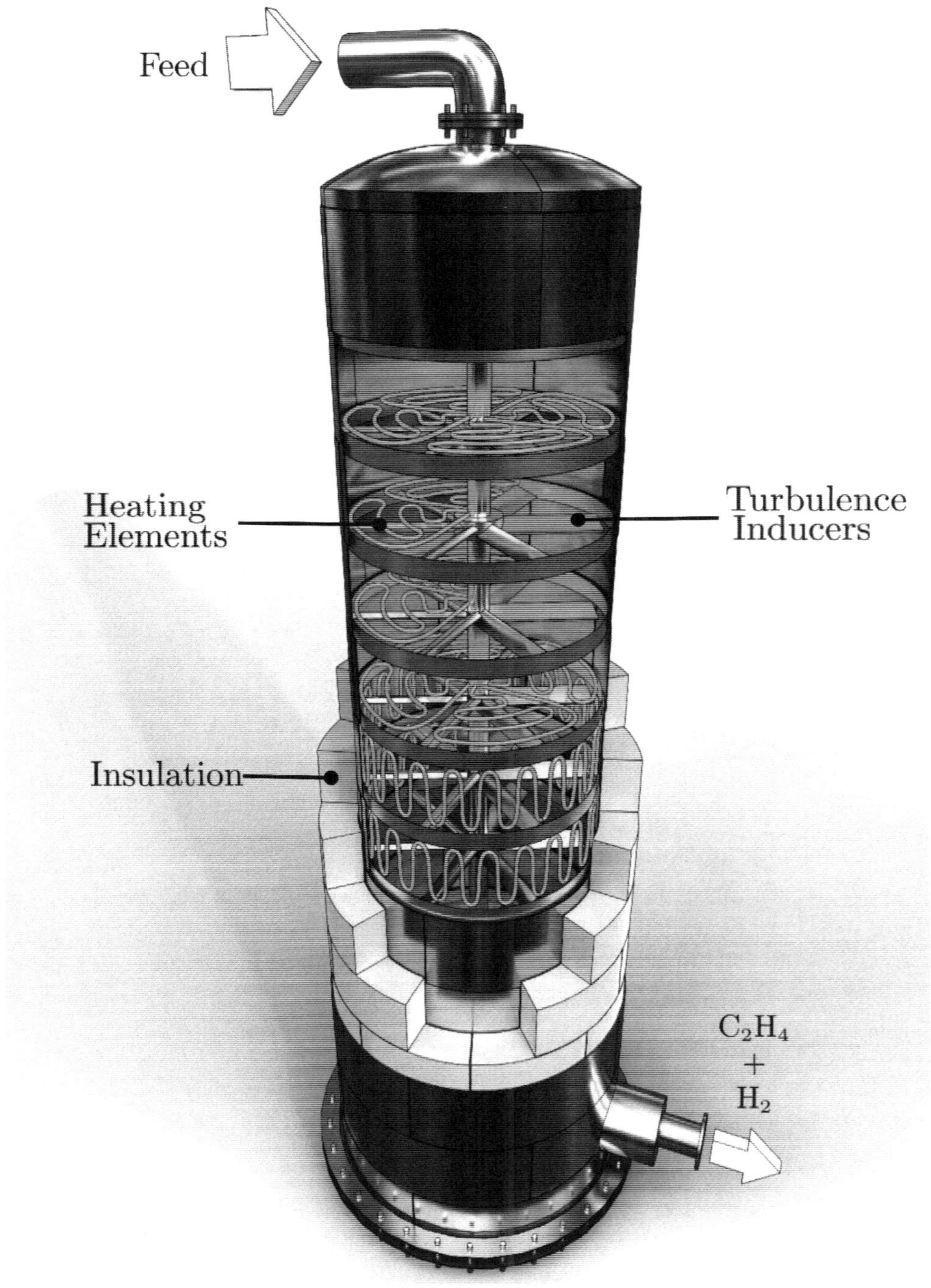

Figure 5. Illustration of an Electric Cracking Tower (E-CRATOS).

3. Case Study

While E-CRATOS holds promise over both traditional FTCs and current electric crackers, it is imperative to demonstrate its performance quantitatively through rigorous modeling and experimentation. This study sets the stage for such comparison by providing an initial assessment of E-CRATOS against FTCs and ITCs/ERTCs under simplified conditions through simulation. The subsequent sections delve into the specifics of one of the models used and the results obtained.

2.1 Simulation Methodology

The following simulations compare the internal behavior of three reactors: (i) ITCs/ERTCs, (ii) FTCs, and (iii) E-CRATOS. No distinction is made between ITCs and ERTCs, as their behavior is identical under the assumptions made.

To ensure a fair comparison, all cases employ the same feed conditions, reactor dimensions, and energy input. Ideal plug-flow and isobaric conditions are assumed to isolate configuration effects. No diluent is used to eliminate dilution influences. Material and transport effects are discarded through ideal heat transfer and constant conductivity assumptions. Tables 1 and 2 present the reactor dimensions and the feed conditions, respectively.

Table 1: Reactor dimensions

Parameter	Value	Units
Reactor Length	100	m
Reactor Diameter	0.124	m

Table 2: Feed conditions

Parameter	Value	Units
Inlet Temperature	500	°C
Inlet Pressure	3.5	bar
Feed Flow Rate	1	kg/s
Composition	Pure Ethane (C_2H_6)	

The kinetic constants employed in this study are sourced from Sundaram and Froment (1978) [38]. For modeling thermodynamic properties, the NASA 7-coefficient polynomial system is adopted with coefficients obtained from the Active Thermochemical Tables (ATcT) database [39], [40].

The resulting system of differential equations is solved using Ansys Chemkin-Pro. These equations encompass the energy balance, mass balance for each component, and pressure drop, represented by Equations (1) to (3) respectively [41], [42]. The variable definitions and the units employed are presented in Table 3.

$$\rho \overline{C_p} v_z \frac{dT}{dz} - \sum_{j=1}^{M} (-\Delta H_{rxn\,j}) r_j = -\frac{\dot{Q}_L}{A_S} \quad (1)$$

$$v_z \frac{dC_i}{dz} - \sum_{j=1}^{M} \alpha_{i,j} r_j = 0 \quad (2)$$

$$\frac{dP}{dz} = 0 \quad (3)$$

Table 3: Variable Definition

Variable	Symbol	Units
Average heat capacity	$\overline{C_p}$	J/kg-K
Mixture density	ρ	kg/m^3
Axial velocity	v_z	m/s
Gas temperature	T	K
Axial length	z	m
Enthalpy of reaction for equation j	$\Delta H_{rxn\,j}$	J/mol
Rate of reaction for equation j	r_j	mol/s-m^3
Number of chemical reactions	M	#
Heat per unit length	$\dot{Q}_L$	W/m
Surface area	A_S	m^2
Concentration of component i	C_i	mol
Pressure	P	Pa

2.2 Heat Profiles

Under the outlined assumptions, the sole difference in reactor performance stems from the heat profile, denoted as $\dot{Q}_L$, provided by each reactor. This heat profile is a function of the axial length, z, and its specific form depends on the distinct configuration of each unit. For example, as depicted in Figure 6, ITCs/ERTCs generate constant heat profiles. The energy consumption per unit length, set at 40 kW/m, corresponds to the optimal amount of energy obtained for this case, as illustrated in Figure 7. Below this threshold, insufficient energy is available to fully process the feed, while exceeding it leads to over-cracking and diminished performance.

The total amount of power is obtained by calculating the area under the curve, expressed by equation (4). In the case of Figure 6, this value corresponds to 4 MW.

$$W_T = \int_{z=0}^{Z=L} \dot{Q}_L \, dz \quad (4)$$

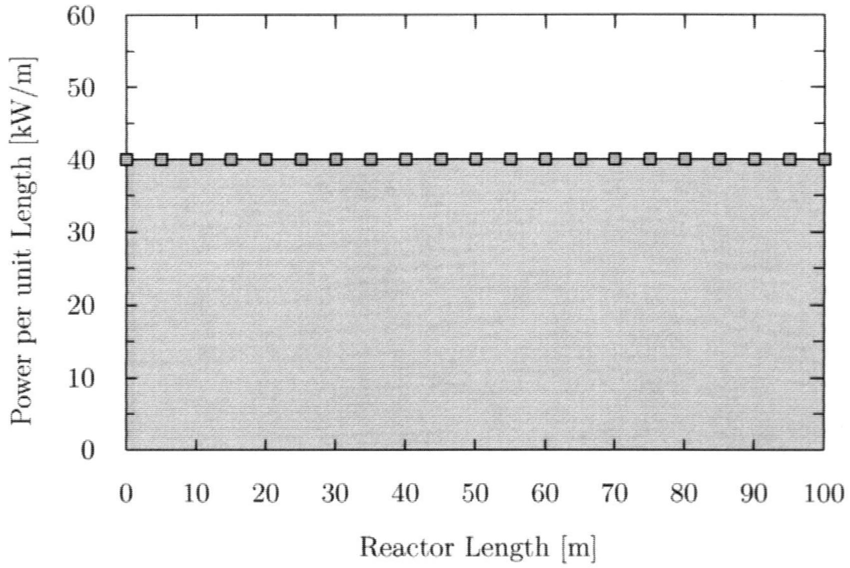

Figure 6. Heat profile for the ITC/ERTC unit

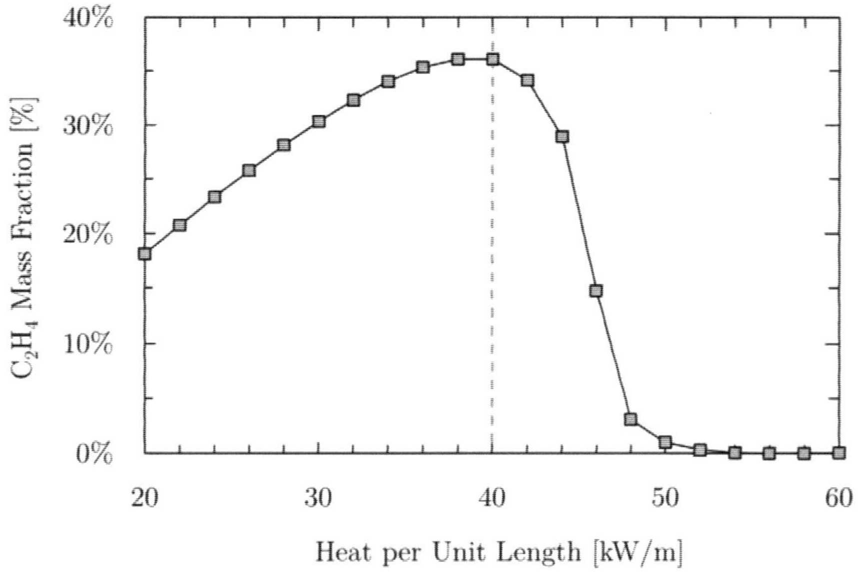

Figure 7. Heat profile for optimization for the ITC/ERTC unit

For FTCs, the heat profile is influenced by numerous variables such as the burner locations, tube spacing, and material emissivity [43], [44]. To streamline the calculations, this study adopts the form of the heat profile outlined by Van Geem et al. (2004) [10]. To ensure equitable comparison, the area under the curve is set at 4 MW while preserving its original shape. The final heat profile obtained is depicted in Figure 8.

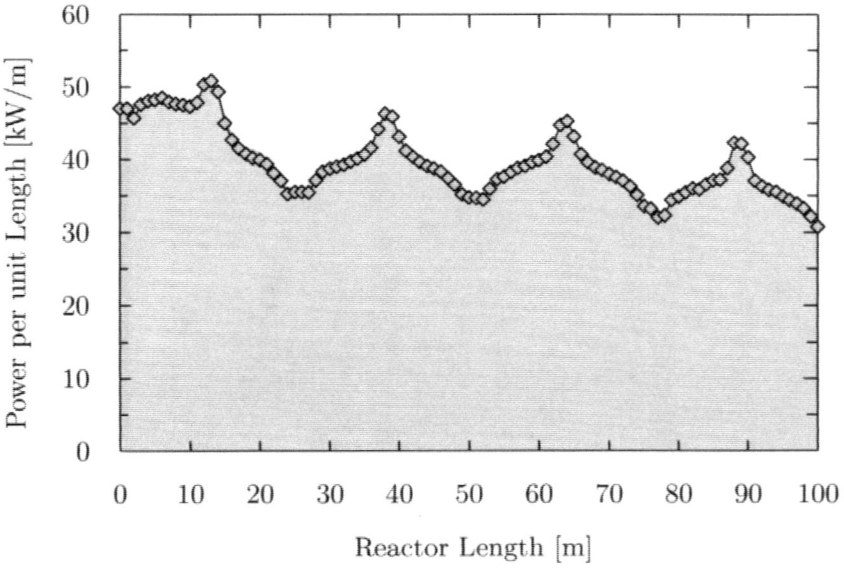

Figure 8. Heat profile for the FTC unit

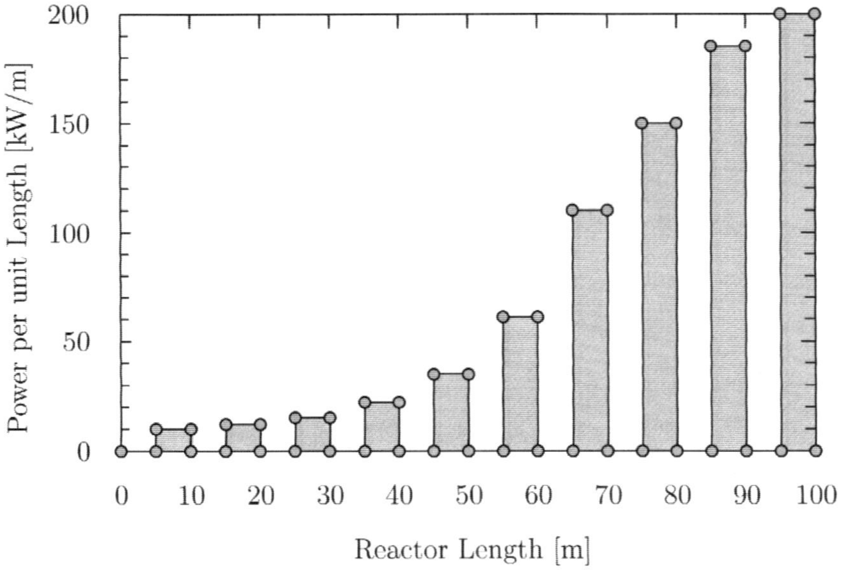

Figure 9. Heat profile for the E-CRATOS unit

E-CRATOS provides additional flexibility to tailor the heat profile. Parameters such as the spacing between heating stages, power distribution per stage, and their lengths can be adjusted. In this study, a simplified heat profile resembling a normal distribution is employed, with a constant stage length of 5 m and a space of 5 m between stages as well, as illustrated in Figure 9. Various methods, including Fourier optimization, variational calculus, and genetic algorithms, can be utilized to derive alternative heat profiles for E-CRATOS. For visual comparison, Figure 10 presents the heat profiles for all three cases. Despite the distinct shapes of each profile, all three units have the same integral, or area under the curve, meaning the three units transfer the same amount of heat to the gas.

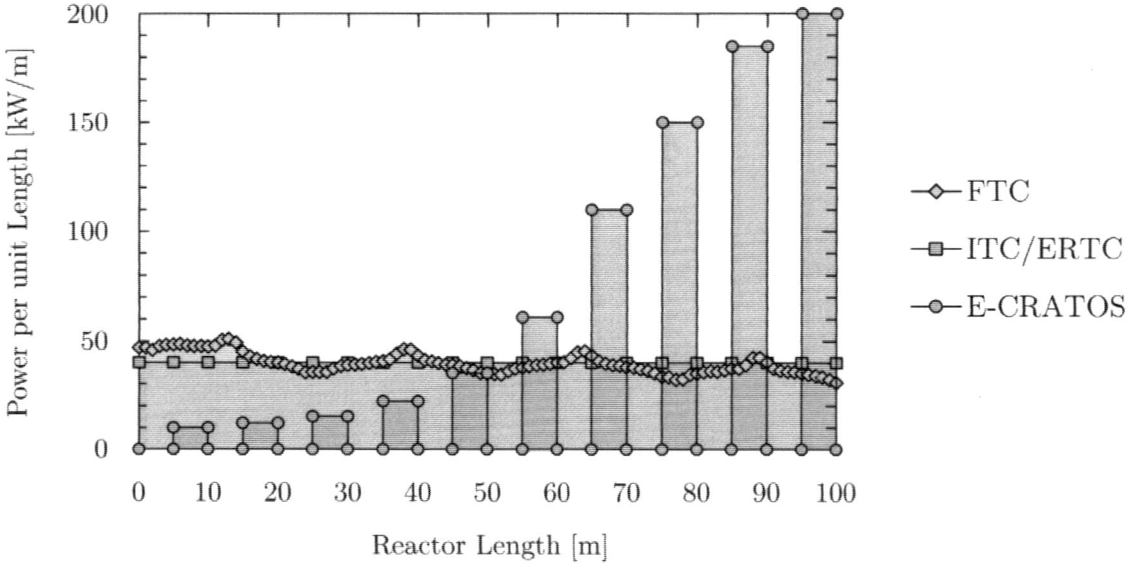

Figure 10. Comparison between heat profiles

4. Results

Figures 11 and 12 illustrate the gas temperature profile and ethylene composition obtained for the three reactors. Across all cases, there is a discernible delay in the increase of ethylene composition upon entering the reactor. This behavior is attributed to the gas requiring a certain length to reach a temperature at which the reaction becomes significant. Above ~720°C ethylene composition rises rapidly, and more energy is consumed by the endothermic reactions. While FTCs and ITC/ERTC present a similar behavior, the temperature profile of E-CRATOS exhibits a "serrated" shape, corresponding to the space between stages.

As shown in Table 4, E-CRATOS demonstrates superior performance by achieving a higher outlet ethylene composition by a factor of ~1.17 while maintaining a lower mean gas temperature and comparable outlet temperature to FTC and ITC/ERTC.

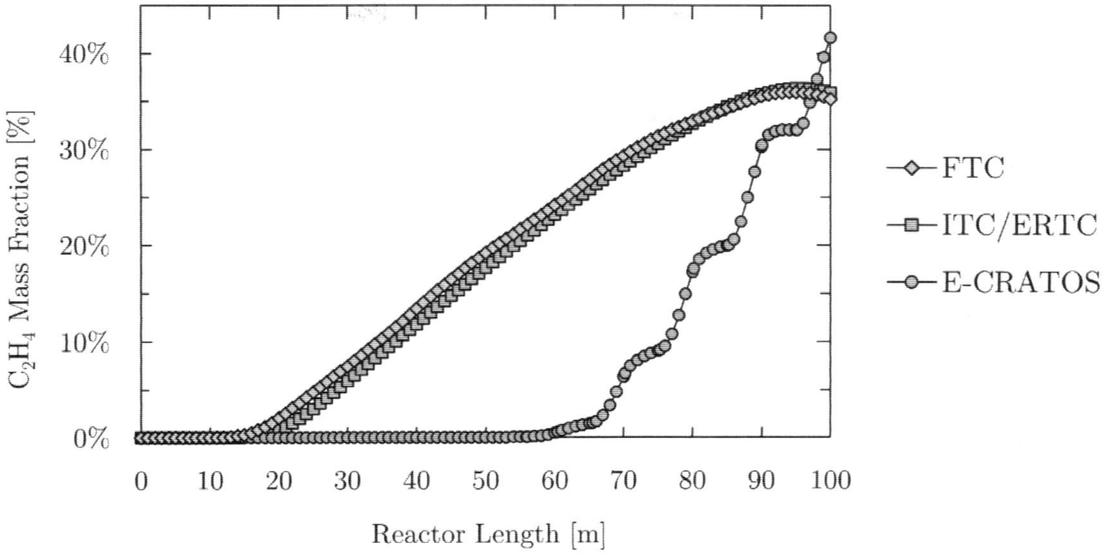

Figure 11. Gas temperature profile. Case study

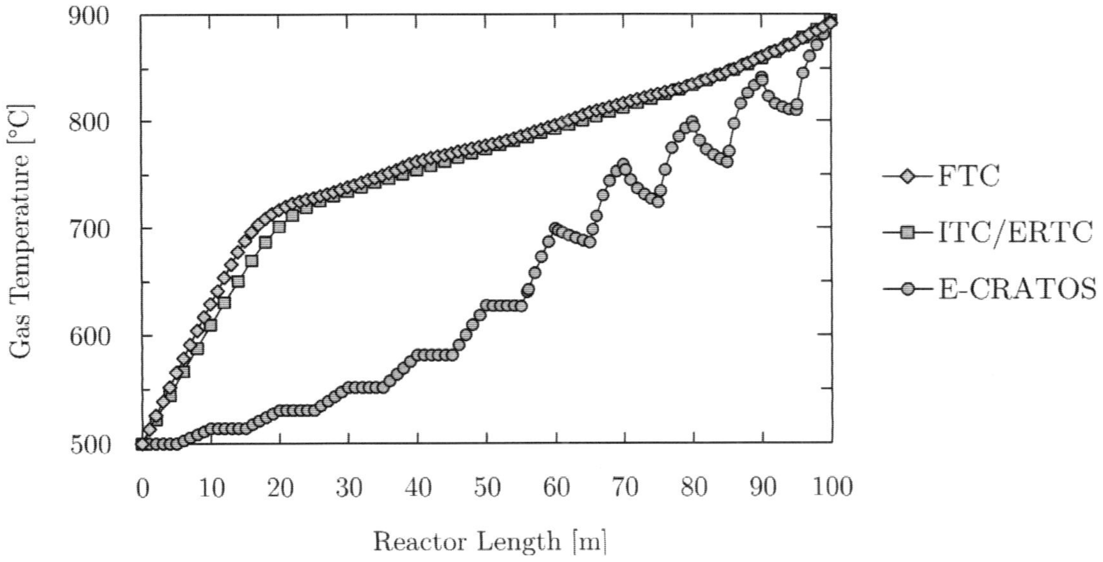

Figure 12. Ethylene mass fraction profile. Case study

Table 4: Reactor dimensions

Reactor	Outlet Temp. [°C]	Average Temp. [°C]	Median Temp. [°C]	Outlet C_2H_4 mass frac. [%]
FTC	891.5	762.7	776.6	35.2%
ITC/ERTC	894.5	756.6	773.2	35.9%
E-CRATOS	892.2	542.4	627.3	41.7%

5. Conclusions

This study introduces the concept of Electric Cracking Towers (E-CRATOS), a novel reactor design featuring electric heating elements organized into stages along the reactor length. A comparative analysis under simplified conditions is conducted to assess E-CRATOS against traditional FTCs, and ITCs/ERTCs. The results underscore the potential of E-CRATOS in reducing internal gas temperatures and enhancing ethylene production.

The findings emphasize the pivotal role of optimizing the heat profile along the reactor length for achieving superior performance. E-CRATOS emerges as the configuration offering the highest degree of flexibility in tailoring this profile. Notably, E-CRATOS exhibits a distinct advantage in maintaining lower average and median temperatures.

These encouraging results motivate further in-depth analysis to better assess the capabilities of E-CRATOS, opening a new path for the development of net-zero cracking reactors.

6. Acknowledgements

We would like to acknowledge the National Science Foundation under cooperative agreement no. EEC-1647722 for CISTAR ERC funding.

6. References

1. K. Yamaguchi, H. Kishi, and T. Nagasawa, "On the use of Polyethylene Insulation in the Manufacture of Submarine Cables," *IEEE Trans. Electr. Insul.*, no. 1, pp. 75–79, 1977.
2. "Ethylene Oxide (EtO)," *EPA*. Environmental Protection Agency. [Online]. Available: https://www.epa.gov/ingredients-used-pesticide-products/ethylene-oxide-eto
3. "Poly-ethylene-terephthalate," *American Chemical Society, 6(11), 951–952.*, vol. 2. 2019. [Online]. Available: https://www.acs.org/molecule-of-the-week/archive/p/poly-ethylene-terephthalate.html?cid=home_motw
4. "Ethylene: The 'World's most important chemical,'" *American Fuel & Petrochemical Manufacturers (AFPM)*. Aug. 2022. [Online]. Available: https://www.afpm.org/newsroom/blog/ethylene-worlds-most-important-chemical
5. "Landmark for the birthplace of the petrochemical industry," *American Chemical Society*. [Online]. Available: https://www.acs.org/content/acs/en/education/whatischemistry/landmarks/petrochemical-industry-birthplace.html
6. "Revisiting the first cracker, in West Virginia, as shale rekindles interest," *Home / Chemical Week*. [Online]. Available: https://chemweek.com/CW/Document/59713/Revisiting-the-first-cracker-in-West-Virginia-as-shale-rekindles-interest
7. H. Pierre and G. Curme, "PRODUCTION OF ETHYLENE," U.S. Patent 1,460,545 (1923)
8. Z. Chen, Y. Li, W. P. Oladipupo, E. A. R. Gil, G. Sawyer, and R. Agrawal, "Alternative ordering of process hierarchy for more efficient and cost-effective valorization of shale resources," *Cell Reports Phys. Sci.*, vol. 2, no. 10, 2021.
9. W. A. Bone and H. F. Coward, "The thermal decomposition of hydrocarbons. Part I.[Methane, ethane, ethylene, and acetylene.]," *J Chem Soc*, vol. 93, pp. 1197–1225, 1908.
10. K. M. Van Geem, G. J. Heynderickx, and G. B. Marin, "Effect of Radial Temperature Profiles on Yields in Steam Cracking," *AIChE J.*, vol. 50, no. 1, pp. 173–183, Jan. 2004, doi: 10.1002/aic.10016.
11. A. Isella and D. Manca, "GHG Emissions by (Petro)Chemical Processes and Decarbonization Priorities—A Review," *Energies*, vol. 15, no. 20, 2022, doi: 10.3390/en15207560.
12. "MIT decarbonization of ethylene production." [Online]. Available: https://greengroup.mit.edu/decarbonization-ethylene-production#:~:text=Its high demand in combination,emitter of the chemical industry.
13. H. Zimmermann and R. Walzl, "Ethylene," *Ullmann's Encycl. Ind. Chem.*, 2000.

14. "Global Ethylene Market Analysis," *Research and Markets*. [Online]. Available: https://www.researchandmarkets.com/reports/5743468/global-ethylene-market-analysis-plant-capacity?utm_source=GNOM&utm_medium=PressRelease&utm_code=z75q4z&utm_campaign=1828184+-+Global+Ethylene+Market+Analysis+Report+2023%3A+Plant+Capacity%2C+Production%2C

15. Z. Liu, Z. Deng, S. J. Davis, C. Giron, and P. Ciais, "Monitoring global carbon emissions in 2021," *Nature Reviews Earth and Environment*, vol. 3, no. 4. Springer Nature, pp. 217–219, Apr. 01, 2022. doi: 10.1038/s43017-022-00285-w.

16. J. Montgomery and M. Van Clieaf, *Net Zero Business Models: Winning in the Global Net Zero Economy*. John Wiley & Sons, 2023.

17. "Greenhouse Gas Emissions from a Typical Passenger Vehicle," *EPA*. Environmental Protection Agency. [Online]. Available: https://www.epa.gov/greenvehicles/greenhouse-gas-emissions-typical-passenger-vehicle#:~:text=A typical passenger vehicle emits,of carbon dioxide per year.

18. J. KERRY and G. MCCARTHY, *THE LONG-TERM STRATEGY OF THE UNITED STATES Pathways to Net-Zero Greenhouse Gas Emissions by 2050*. 2021. [Online]. Available: https://www.whitehouse.gov/wp-content/uploads/2021/10/US-Long-Term-Strategy.pdf

19. "Climate change: what the EU is doing." [Online]. Available: https://www.consilium.europa.eu/en/policies/climate-change/#:~:text=At least 55%25 fewer emissions by 2030,-Exactly a year&text=As an intermediate step towards,greenhouse gas emissions by 2030.&text=This goal is a major,of cutting emissions by 40%25.

20. "The search for greener ethylene." [Online]. Available: https://cen.acs.org/business/petrochemicals/search-greener-ethylene/99/i9

21. R. Agrawal, Z. Chen, and P. Oladipupo, "ELECTRICALLY HEATED DEHYDROGENATION PROCESS," U.S. Patent 11,578,019 B2 (2023)

22. G. G. der ploeg, "ELECTRICALLY HEATED REACTOR, A FURNACE COMPRISING SAID REACTOR AND A METHOD FOR GAS CONVERSION USING SAID REACTOR," WO 2021/130107 A1 (2020)

23. P. E. Hardisty, T. S. Clark, and R. G. Hynes, "Life cycle greenhouse gas emissions from electricity generation: a comparative analysis of Australian energy sources," *Energies*, vol. 5, no. 4, pp. 872–897, 2012.

24. J. Dalton, *A new system of chemical philosophy - Olefiant Gas*, vol. 2. Smithsonian Institution libraries, 1810.

25. W. Malisoff and E. Gustav, *Methane*, vol. 22. The Journal of Physical Chemistry, 1918.

26. "Joint News Release - BASF, Sabic and Linde start construction of the world's first demonstration plant for large-scale electrically heated steam cracker furnaces," *BASF*. [Online]. Available: https://www.basf.com/global/en/media/news-releases/2022/09/p-22-326.html

27. S. Sankaran and et al., "ELECTRICALLY POWERED FURNACES TO HEAT A FEED AND RELATED METHODS," WO 2023/006475 A1, Feb. 2023

28. S. Stevenson, "FURNACE INCLUDING ELECTRICALLY POWERED HEATING ELEMENTS ARRANGED FOR UNIFORM HEATING AND RELEVANT METHODS," WO 2023/016968 A1, Feb. 2023

29. S. M. Sadrameli, "Thermal/catalytic cracking of hydrocarbons for the production of olefins: A state-of-the-art review I: Thermal cracking review," *Fuel*, vol. 140. Elsevier Ltd, pp. 102–115, Jan. 15, 2015. doi: 10.1016/j.fuel.2014.09.034.

30. "US ethane cracker construction costs rise 1-2 percent year on year," *Reuters Events*. [Online]. Available: https://www.reutersevents.com/downstream/engineering-and-construction/infographic-us-ethane-cracker-construction-costs-rise-1-2-year-year

31. "Shale Gas Fuels U.S. Chemical Plant Construction." [Online]. Available: https://www.enr.com/articles/45920-shale-gas-fuels-us-chemical-plant-construction

32. A. I. Foster and B. J. Cromarty, "The theory and practice of steam reforming," in *4th European Technical Seminar on Hydrogen Plants, Lisbon*, 2003.

33. S. H. Symoens *et al.*, "State-of-the-art of Coke Formation during Steam Cracking: Anti-Coking Surface Technologies," *Ind. Eng. Chem. Res.*, vol. 57, no. 48, pp. 16117–16136, Dec. 2018, doi: 10.1021/acs.iecr.8b03221.

34. M. Geerts, S. H. Symoens, P. A. Reyniers, G. B. Marin, M. F. Reyniers, and K. M. Van Geem, "Steam Cracking Coke Properties and Their Influence on Furnace Run Length Predictions: Experimental and Modeling Study," *Ind. Eng. Chem. Res.*, vol. 59, no. 52, pp. 22460–22472, Dec. 2020, doi: 10.1021/acs.iecr.0c04727.

35. G. J. Heynderickx, E. M. Schools, and G. B. Marin, "Simulation of the decoking of an ethane cracker with a steam/air mixture," *Chem. Eng. Sci.*, vol. 61, no. 6, pp. 1779–1789, 2006.

36. "A novel approach to cleaning furnace coils." [Online]. Available: www.digitalrefining.com/article/1000854

37. H. Eerens and D. van Dam, "DECARBONISATION OPTIONS FOR LARGE VOLUME ORGANIC CHEMICALS PRODUCTION, DOW CHEMICAL TERNEUZEN," 2022.

38. K. M. Sundaram and G. F. Froment, "Modeling of thermal cracking kinetics. 3. Radical mechanisms for the pyrolysis of simple paraffins, olefins, and their mixtures," *Ind. Eng. Chem. Fundam.*, vol. 17, no. 3, pp. 174–182, 1978.

39. D. H. B. Branko Ruscic, "Active Thermochemical Tables," *Active Thermochemical Tables - Home*. [Online]. Available: https://atct.anl.gov

40. A. Burcat and B. Ruscic, "Third millenium ideal gas and condensed phase thermochemical database for combustion (with update from active thermochemical tables).," 2005.

41. K. M. Van Geem, R. Žajdlík, M. F. Reyniers, and G. B. Marin, "Dimensional analysis for scaling up and down steam cracking coils," *Chem. Eng. J.*, vol. 134, no. 1–3, pp. 3–10, Nov. 2007, doi: 10.1016/j.cej.2007.03.065.

42. R. Karimzadeh and M. Ghashghaee, "Design of a flexible pilot plant reactor for the steam cracking process," *Chem. Eng. Technol.*, vol. 31, no. 2, pp. 278–286, Feb. 2008, doi: 10.1002/ceat.200700326.

43. M. Rezaeimanesh, A. Asghar Ghoreyshi, S. M. Peyghambarzadeh, and S. Hassan Hashemabadi, "A coupled CFD simulation approach for investigating the pyrolysis process in industrial naphtha thermal cracking furnaces," *Chinese J. Chem. Eng.*, vol. 44, pp. 528–542, Apr. 2022, doi: 10.1016/j.cjche.2021.03.028.

44. G. D. Stefanidis, B. Merci, G. J. Heynderickx, and G. B. Marin, "CFD simulations of steam cracking furnaces using detailed combustion mechanisms," *Comput. Chem. Eng.*, vol. 30, no. 4, pp. 635–649, Feb. 2006, doi: 10.1016/j.compchemeng.2005.11.010.

AIChE Paper Number: 75a

DETAILED CHARACTERIZATION OF PLASTIC PYROLYSIS OILS AND THEIR CONTAMINANTS BY FT-ICR MS AND GCxGC

Yannick Ureel
Ghent University, Ghent, Belgium

Marvin Kusenberg
Dr.
Ghent University, Ghent, Belgium

Martha L. Chacón-Patiño
Dr.
National High Magnetic Field Laboratory, Tallahassee, United States
iC2MC, Harfleur, France

Ryan P. Rodgers
Prof. Dr.
National High Magnetic Field Laboratory, Tallahassee, United States
iC2MC, Harfleur, France
Florida State University, Tallahassee, United States

Maarten K. Sabbe
Prof. Dr.
Ghent University, Ghent, Belgium

Kevin M. Van Geem
Prof. Dr.
Ghent University, Ghent, Belgium

Prepared for Presentation at the 2024 Spring National Meeting
New Orleans, LA, March 24 – March 28, 2024

AIChE Paper Number: 75a

DETAILED CHARACTERIZATION OF PLASTIC PYROLYSIS OILS AND THEIR CONTAMINANTS BY FT-ICR MS AND GCxGC

Yannick Ureel
Ghent University, Ghent, Belgium

Marvin Kusenberg
Dr.
Ghent University, Ghent, Belgium

Martha L. Chacón-Patiño
Dr.
National High Magnetic Field Laboratory, Tallahassee, United States
iC2MC, Harfleur, France

Ryan P. Rodgers
Prof. Dr.
National High Magnetic Field Laboratory, Tallahassee, United States
iC2MC, Harfleur, France
Florida State University, Tallahassee, United States

Maarten K. Sabbe
Prof. Dr.
Ghent University, Ghent, Belgium

Kevin M. Van Geem
Prof. Dr.
Ghent University, Ghent, Belgium

Abstract:
The current linear plastic lifecycle is detrimental to the environment as it leads to plastic waste pollution, increased greenhouse gas emissions and is unsustainable due to its dependence on fossil resources. Therefore, it is essential to recycle plastic waste to alleviate these issues and move to a more sustainable and climate-friendly industry. The thermal pyrolysis of polyolefins is one of the most promising technologies to chemically recycle mixed polyolefin waste which cannot be recycled mechanically. However, the obtained pyrolysis oils are unsuitable for further processing in steam cracking without thorough upgrading due to impurities present in the post-consumer waste. In this context, upgrading technique require in-depth knowledge of the complex hydrocarbon matrix. Therefore, in this work the molecular hydrocarbon composition and oxygenates present in four different plastic pyrolysis oils (originating from virgin polyethylene (PE), as well as post-consumer PE, polypropylene (PP), and mixed polyolefins (MPO)) has been investigated by combining two separate high-performance analytical techniques, namely, two-dimensional gas chromatography and Fourier-

transform ion cyclotron resonance mass spectroscopy. This allowed to clearly see the effect of the type of polymer feedstock on the molecular composition of the pyrolysis oils. A higher share of aromatics was found in the waste PE and waste MPO oils compared to virgin PE or waste PP oils and complex heptaaromatics could be identified in the former samples. The source of the oxygenates could be elucidated for all pyrolysis samples, originating either from plastic additives, organic residue, or secondary reactions between oxygenates and hydrocarbons. The obtained fundamental knowledge will help in designing improved decontamination strategies and enable to close the loop for polyolefin waste.

Introduction

Plastics are an incredible lightweight material which have replaced other materials in different applications because of plastics' superior properties [1, 2]. Moreover, plastics are used in many technologies required for a climate-neutral economy: such as lightweight packaging, insulation, or composites for windmills to name a few [3]. However, at present end-of-life plastics are often discarded, landfilled, or incinerated. With worldwide less than 9% being recycled [4]. The recycling of plastic waste is therefore a necessity to move to a sustainable and climate-neutral industry. In this way, a sustainable circular plastics economy can be achieved which is less or even not dependent on fossil sources anymore. Plastic waste streams can be reduced, avoiding spills in the oceans or leaching of chemicals in landfills and lastly it will reduce the emission of greenhouse gas emissions during the plastics lifecycle. It is estimated that 1.87 tons of CO_2 per ton of plastic can be avoided by chemical recycling as this prevents the emissions from the incineration of end-of-life plastics [5].

At present, mechanical recycling is the dominant pathway to recycle plastics with a global share of 9% [4]. Nevertheless, mechanical recycling is not feasible for all types of plastic waste streams. For example, mechanical recycling cannot be employed for mixed polyolefin waste [6]. It is challenging to completely separate polyethylene (PE) and polypropylene (PP) waste, which results in a recycled polymer containing both PE and PP. Due to polymer immiscibility this recycled polymer is of inferior quality than virgin polymer and hence cannot be used for its original purpose. making this an example of downcycling instead of recycling [7-9]. Furthermore, thermal stress and mechanical shear are inevitable during mechanical recycling resulting in thermomechanical degradation of the recycled polymer, also leading to inferior quality compared to the virgin polymer [10]. Chemical recycling of plastic waste can alleviate these two issues by first breaking down the plastic waste to its monomers and again polymerizing to virgin-grade polymers.

At present, one of the most mature and promising technologies is the thermal pyrolysis of plastic waste to pyrolysis oils [6, 11-13]. Thermal pyrolysis is especially valuable for mixed polyolefin waste (mixtures of PE and PP), which comprise of over 60% of the plastic waste globally [1]. Ideally, plastic pyrolysis oils can be converted with the existing petrochemical infrastructure by feeding it to a steam cracker, where it is then converted to polymer monomers i.e., ethylene and propylene [14-16]. However, plastic pyrolysis oils differ markedly from conventional crude oil or naphtha fractions, for which petrochemical processes were originally designed. Firstly, these recycled oils have a different hydrocarbon composition with a much higher share of unsaturated hydrocarbons such as olefins. The presence of these olefins will result in a higher share of cyclization or Diels-Alder reactions which form monoaromatics during thermal pyrolysis and steam cracking [17, 18]. Subsequently, these monoaromatics react further to polyaromatic hydrocarbons which in the end coke the cracking tubes and hinder the continuous operation of the cracking furnace [19, 20]. Next to the elevated presence of unsaturated hydrocarbons, plastic pyrolysis oils contain a plethora of contaminants stemming from polymer additives, ill-sorted polymers, and waste accumulated during the plastic lifetime [21, 22]. Kusenberg et al. investigated the elemental composition of different polyolefin waste pyrolysis oils by inductively-coupled plasma optical emission spectroscopy (ICP-OES) [21]. This analysis showed that nearly all elements of the periodic system are present within the recycled pyrolysis oils, hampering their use as steam cracking feed [21, 23]. For example, high concentrations of oxygen, nitrogen, chlorine, and various metals were found which exceeded the maximum threshold for use in steam crackers by orders or magnitude [16, 24]. These elements should be avoided as they pose several risks such as increased coking rates, fouling of downstream equipment and the poisoning of downstream catalysts [25, 26]. To facilitate the further processing of this valuable carbon source, it is essential to acquire a proper understanding of their molecular composition regarding both hydrocarbons and contaminants.

Therefore, in this work the results from both two-dimensional comprehensive gas chromatography with flame ionization detector (GC × GC-FID) and Fourier transform ion cyclotron resonance mass spectroscopy (FT-ICR MS) [27-29] were combined to characterize different post-consumer plastic waste pyrolysis oils. GC × GC-FID is one of the workhorses of the petrochemical industry as it allows for a detailed quantitative hydrocarbon composition [30, 31]. FT-ICR MS has an improved resolution and limit of detection compared to other MS techniques as it traps the ions to analyze in a cyclotron achieving much stronger signals [28, 29, 32]. Here, an ultra-high-resolution 21 tesla FT-ICR MS has been used for the first time on plastic pyrolysis oils [33]. Four different waste and virgin polyolefin pyrolysis oils with varying polymer composition have been examined. These oils stem from virgin low-density polyethylene (LDPE), waste PE, waste PP, and mixed polyolefin waste (MPO). The detailed characterization of these pyrolysis oils allows to shed light on the influence of plastic feedstock on their hydrocarbon composition and provide insights into the source of contaminants.

Materials and Methodology

Materials

Four polymer samples have been pyrolyzed, of which three post-consumer waste blends and one virgin sample. The post-consumer waste consisted of a waste PE, waste PP, and waste MPO sample, while the virgin LDPE was used as well. The post-consumer waste pyrolysis oils [21], and virgin pyrolysis oil [34] were produced in earlier work where more details can be found. The post-consumer plastic waste was received from the recycling company Ecoo (Belgium). The post-consumer waste is representative of real plastic waste, as it went through the authentic pre-processing undergoing shredding, washing, drying, and sorting. As mentioned, these sorting procedures are challenging for polyolefin waste. Therefore, the plastic blends contain traces of other ill-sorted polymers.

All plastic samples were pyrolyzed in a pilot-scale pyrolysis unit as reported by Kusenberg et al. [21]. All four pyrolysis oils were obtained at the same process conditions of 450 °C and 1 bar. Further details can be found in literature [21, 34]. The samples have been characterized by pyrolysis elemental analysis, ICP-OES, and GC × GC-FID for which the methodology can be found in previous work [21, 34].

21-tesla FT-ICR MS

The four plastic pyrolysis oils were analyzed with two different ionization modes. Positive-atmospheric pressure photoionization [(+)APPI] is an excellent technique to analyze hydrocarbon samples, as it offers a high reproducibility and a rather uniform ionization efficiency across different species [35-37]. On the other hand, negative-electrospray ionization [(-)ESI] is valuable for the identification of acidic contaminants and oxygenated compounds [38]. In this way, the combination of both ionization modes offers a complete image of the pyrolysis oil composition with a focus on the hydrocarbon matrix with (+)APPI and oxygenated contaminants (-)ESI.

The analysis by FT-ICR MS allows to determine the molecular formulas of all compounds present in the plastic pyrolysis oils with high certainty due to its ultra-high-resolution. Data visualization was done by displaying the double bond equivalent (DBE) vs. the carbon number as this has a one-on-one relation to the molecular formula. The DBE of a compound is defined by Equation (1), with C the number of carbon atoms, H the number of hydrogen atoms and the number of oxygen atoms detected not influencing this value.

$$DBE = C - \frac{H}{2} + 1 \tag{1}$$

Results and Discussion

Hydrocarbons

The hydrocarbon composition of the pyrolysis oils has been investigated via GC × GC-FID and (+)APPI FT-ICR MS. As mentioned, GC × GC-FID is excellently suited to quantify hydrocarbons in complex hydrocarbon mixtures. Even though GC × GC-FID can have challenges eluding heavy hydrocarbons, compounds with up to 76 carbon atoms have been identified in the PP waste pyrolysis oil. For the virgin LDPE, the smallest carbon range was found ranging from C_5-C_{52}. The detailed PIONA-results obtained by GC × GC-FID are presented in Figure 1, combining results from previous work [21, 34]. The results presented in Figure 1, confirm literature findings with PE pyrolysis oils mainly comprising of linear hydrocarbons i.e., n-paraffins and α-olefins. In addition, waste PE pyrolysis oil consists of more aromatic hydrocarbons compared to the waste PP pyrolysis oil. Furthermore, the PP waste pyrolysis oil consists of more branched compounds like isoolefins and isoparaffins. For the MPO waste, a mix between branched compounds and linear species is found. However, a strong increase in aromatic constituents is observed. This can be explained by a higher share of contaminants such as ill-sorted polymers. Within the MPO pyrolysis oil a larger fraction of polystyrene (PS) was found compared to the waste PE and PP. This PS will dissociate in styrene monomer or dimers resulting in a larger share of aromatics.

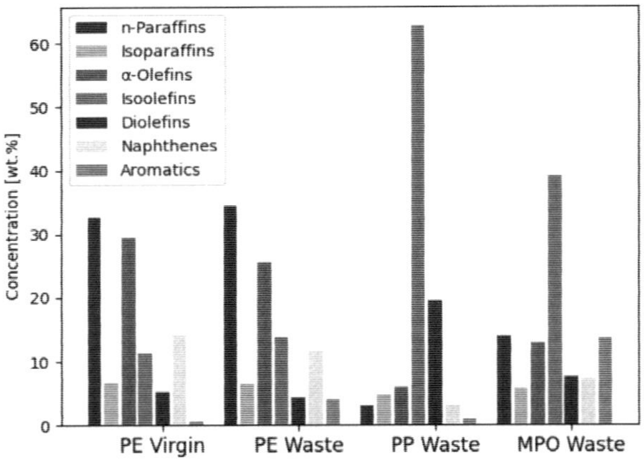

Figure 1. Detailed PIONA-results obtained by analysis of the four pyrolysis oils by GC × GC-FID. Results adapted from [21, 34]

GC × GC-FID is a powerful method but has challenges discerning heavy hydrocarbons and especially complex aromatics due to the maximum allowable temperature of a GC-column. Therefore, (+)APPI FT-ICR MS has been employed to investigate the heavy fraction of all pyrolysis oils. Figure 2 displays the DBE vs. carbon number plots of the four pyrolysis oils obtained by (+)APPI FT-ICR MS. The size of the dots scales with the determined relative abundance of the molecular formulas. It should be stressed that MS is a semi-quantitative methodology, meaning that the observed relative abundances do not scale linearly with the actual concentration due to a difference in ionization efficiency between compounds. At a DBE of zero, linear and branched paraffins are found while a DBE of one corresponds with olefins or cycloalkanes, and monoaromatics correspond to a DBE of four.

Compared to GC × GC-FID a much wider carbon range is detected, up to C_{85}. Furthermore, a wide range of DBEs is found up to a DBE of 22 corresponding to heptaaromatics. This clearly demonstrates how (+)APPI FT-ICR MS can complement GC × GC-FID for the analysis of plastic pyrolysis oils both in carbon range and DBE. One advantage FT-ICR MS allows is an excellent differentiation between different polycyclic aromatics. In the PE and MPO waste $C_{28}H_{16}$ was found corresponding to a heptaaromatic such as for example dibenzoperylene. It is however not possible with FT-ICR MS to determine the exact molecular structure aside from the molecular formula. When comparing the difference in composition with (+)APPI FT-ICR MS between the waste and virgin PE (Figure 2), a remarkable difference is the share of polycyclic aromatic hydrocarbons. Within the virgin PE, the maximum DBE was 16 while for the waste PE pyrolysis oil this was 22 corresponding to pentaaromatics and heptaaromatics respectively. This difference in aromatics is also found in Figure 1, with the virgin pyrolysis oil containing a smaller fraction of aromatics. The reason for this difference is threefold. First, the waste PE is contaminated with aromatic polymers like

PS and polyethylene terephthalate (PET). In addition, the presence of metal particles in the waste PE can catalyze the formation of aromatics and coke. Lastly, the waste PE will have more defects and oxygenates which also promote the formation of aromatics.

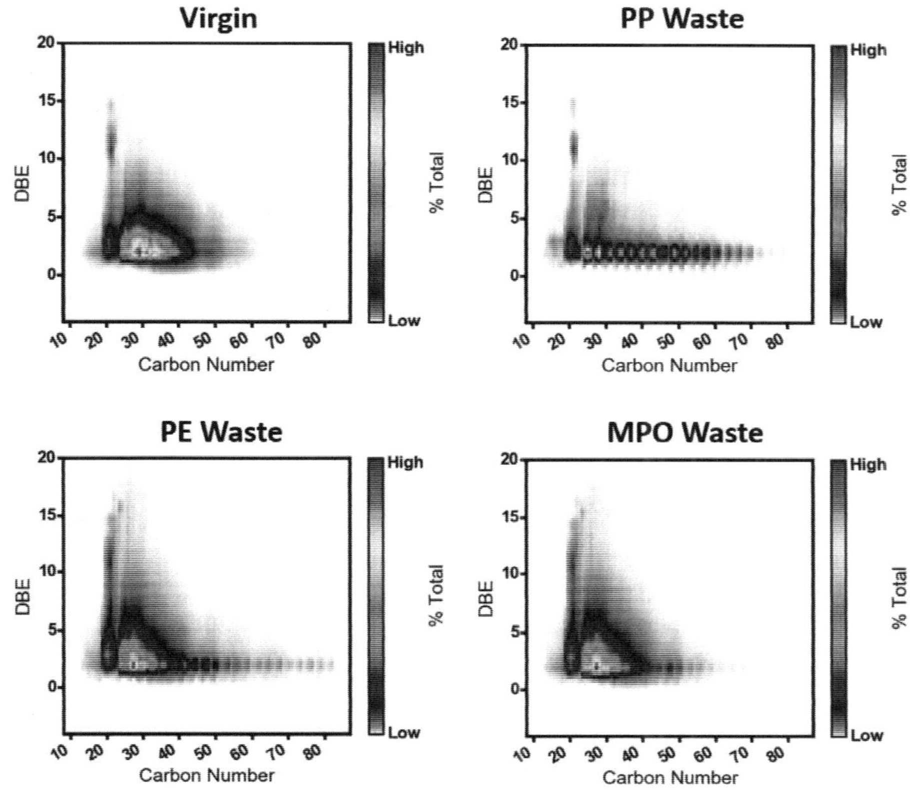

Figure 2. DBE vs. carbon number plots of hydrocarbons for virgin PE, waste PE, PP and MPO obtained by (+)APPI FT-ICR MS with the color indicating the measured relative abundance.

Similar to the virgin PE, the waste PP oil consists of less complex aromatics than the waste PE oil. However, the reason for this distinction is different for the waste PP oils. As PP is a branched polymer and produces many branched products it is less favorable for these products to undergo cyclization reactions due to the sterical hindrance of the branches. Aromatics are typically formed via a Diels-Alder mechanism or internal cyclization reactions of diolefins but because of the side-chains of isoolefins or diolefins these reactions are suppressed. As a consequence, the share of diolefins (DBE=2) in PP oils is much higher than for PE oils.

The DBE vs. carbon number plot (Figure 2) of the waste MPO pyrolysis oil resembles a combination of the waste PP and PE. Both complex heptaaromatics (DBE=22) and a pronounced share of diolefins (DBE=2) are found within the MPO oil. However, as these results are not quantitative, a wrong perception can be created that the MPO composition is just an average of PE and PP. From the GC × GC-FID analysis it is known that the waste MPO contains by far the highest fraction of aromatics. Additionally, when pyrolyzing mixtures of virgin PE and PP an interaction between the PE and PP can be observed resulting in a slightly higher share of aromatics [39].

Oxygenates

The most prevalent contaminant element in plastic pyrolysis oils is oxygen. This oxygen originates from ill-sorted polymers such as PET, polyurethanes (PUR), or ethylene vinyl alcohol (EVOH). In addition, polymer additives such as antioxidants, stabilizers, or colorants will contribute to the presence of oxygen within the polymers. Lastly, during the usage of these plastics they often get in contact with food, cosmetics or other products resulting in an accumulation of organic and inorganic oxygen in the post-consumer plastic. The presence of oxygen is detrimental for their processability in steam crackers or other petrochemical infrastructure as the oxygen will be converted to carbon monoxide or carbon dioxide resulting in a decrease in carbon efficiency.

Figure 3 displays the DBE vs. carbon number plots determined by (+)APPI for O1, O2, and O3 oxygenates with the oxygen concentration in every pyrolysis oil depicted on top. The limit of detection (LOD) here is 1000 mass ppm. The classes O1, O2, and O3 resemble compounds with respectively exactly one, two, or three oxygen atoms in their molecular structure. The relative abundance of every class is shown in the top right corner of every plot. This relative abundance resembles the share of O1, O2, and O3 detected in the respective oil compared to all species detected in that pyrolysis oil. The relative abundance shows that O1 and O2 are present in comparable quantities in all oils while O3 is less present. Overall the DBE vs. carbon number of the oxygenates resembles a similar pattern as the hydrocarbons detected with (+)APPI FT-ICR MS. However, the higher DBEs (DBE $\geq$ 4) are more pronounced for the oxygenates than the hydrocarbons. This indicates that most oxygen is present in combination with aromatic groups corresponding to phenol, benzaldehyde, or benzoic acid groups. As during pyrolysis, most oxygen is converted to carbon dioxide, carbon monoxide, and water this indicates the presence of secondary reactions between oxygen and the hydrocarbon matrix. Especially because of the wide variety of oxygenates detected, secondary reactions between oxygen containing intermediates and hydrocarbons are likely to occur. These secondary reactions can, for instance, be carbon monoxide insertions on radical hydrocarbons. With FT-ICR MS, a higher selectivity of reactions with aromatic hydrocarbons instead of aliphatic structures is found. This effects stems from the higher stability of the carbonyl group on aromatic hydrocarbons compared to aliphatic hydrocarbons. There is a difference in bond dissociation energy of around 60 $kJ.mol^{-1}$ in C-C bond between carbonyl and aliphatic or aromatic structures, favoring aromatics [40]. A similar effect is found for alcohol groups, making hydroxyl groups favor aromatic structures as well [41]. Due to this enhanced stability more aromatic oxygenates will be formed during thermal pyrolysis.

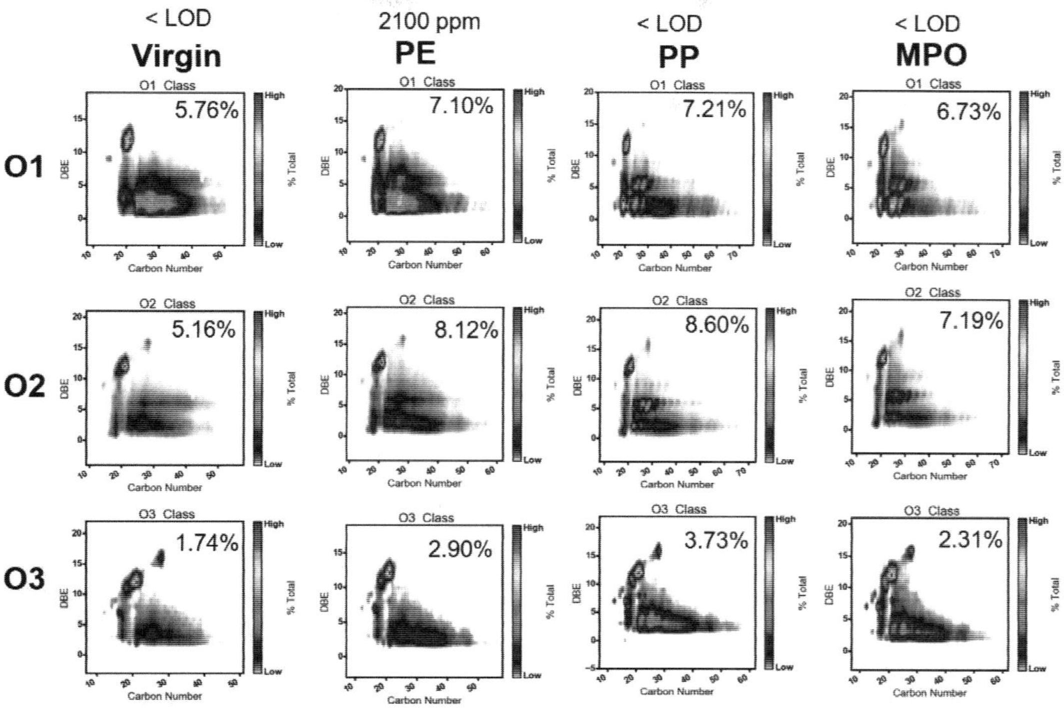

Figure 3. DBE vs. carbon number plots of oxygenates for virgin PE, waste PE, PP and MPO obtained by (+)APPI FT-ICR MS with color indicating the measured relative abundance. The oxygen concentration in mass ppm determined by pyrolysis elemental analysis is shown on top, and the relative abundance (%) of every type of compound is depicted in the top right of the plot.

While (+)APPI has a more uniform ionization selectivity among all molecules, (-)ESI specifically targets oxygenates. Figure 4 illustrates the DBE vs. carbon number plots of oxygenates detected by (-)ESI, complemented with the relative abundance in the top right of every graph. While the type of observed compounds is similar, clear peaks of oxygenates can be discerned. Especially for O1, three clear peaks are found in all samples. These stem from contaminants which are not fully dissociated upon thermal pyrolysis. As the three peaks are also present in the virgin LDPE pyrolysis oil, this indicates that the source of these peaks are plastic additives and not organic waste or other polymers. The lowest peak originates from compounds with a DBE of four and carbon number of 14 and 15, corresponding to $C_{14}H_{22}O$ and $C_{15}H_{24}O$. The most likely compounds with these compositions are 2,6-di-tert-butylphenol ($C_{14}H_{22}O$) and butylated hydroxytoluene ($C_{15}H_{24}O$). These are products of the thermal dissociation of Irganox antioxidants such as the most commonly employed Irganox 1010 [42, 43]. These antioxidants are needed to avoid polymer degradation by terminating free radicals. This finding, for the first time, shows that contaminants in pyrolysis oils are not only originating from the polymer lifetime but are also intentionally added during the manufacturing process. As such components are hampering chemical recycling and are cumbersome to remove, knowledge generated by FT-ICR-MS can be used to develop guidelines for the plastics manufacturing industry to facilitate chemical recycling in the future.

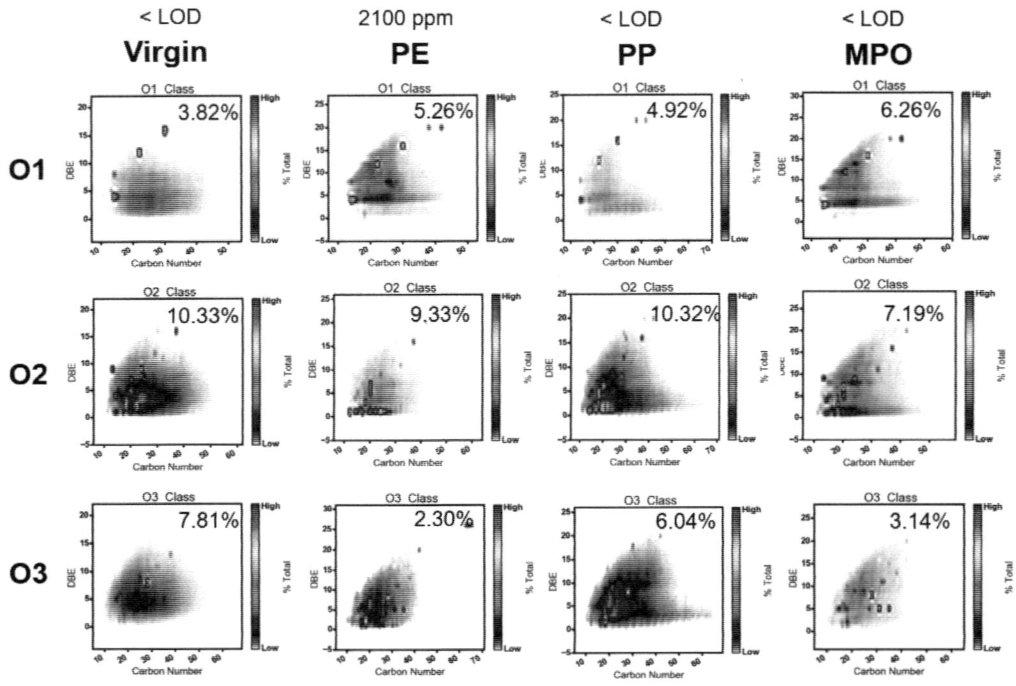

Figure 4. DBE vs. carbon number plots of oxygenates for virgin PE, waste PE, PP and MPO obtained by (-)ESI FT-ICR MS with the color indicating the measured relative abundance. The oxygen concentration in mass ppm determined by combustion elemental is shown on top, and the relative abundance (%) of every type of compound is depicted in the top right of the plot.

The O2 compounds detected are mostly prevalent in the PE and PP waste. These are mostly molecules at a DBE=1, corresponding to fatty acids. These fatty acids are accumulated during the use of the plastic as these are prevalent in food and cosmetics.

Conclusion

The combination of GC × GC-FID and FT-ICR MS is shown to be a powerful tool for the detailed characterization of plastic pyrolysis oils. The quantitative nature of the GC × GC-FID, the extended carbon range of FT-ICR MS and the differentiation between complex aromatics in FT-ICR MS allows for a complete image of the pyrolysis oil composition. This study clearly shows the effect of polymer feedstock on the pyrolysis oil composition and the differences and similarities between virgin and post-consumer plastic pyrolysis oils. The presence of ill-sorted aromatic polymers and metals in post-consumer plastic waste has been shown to increase the formation of aromatics in their corresponding pyrolysis oils. The effect of plastic feedstock is also demonstrated by the difference in pyrolysis oil composition of waste PE, PP, and MPO. Within PE and especially MPO, a higher fraction of aromatic hydrocarbons was found with also a higher complexity, identifying polycyclic aromatics up to heptaaromatics. Next to a broader carbon number range, FT-ICR MS also allows to investigate the heteroatom contaminants present in plastic waste pyrolysis oils. These results indicate that during pyrolysis, carbon monoxide and hydroxyl radicals have a tendency to react with aromatic hydrocarbons over aliphatic ones. The formed oxygenated aromatic hydrocarbons are more difficult to break down or remove in later decontamination strategies due to their more stable nature. Therefore, it is beneficial to remove as much oxygen as possible prior to the thermal pyrolysis by improved waste sorting and separation and improved washing processes to remove residues. The source of oxygen varied from organic and inorganic residue in the post-consumer pyrolysis oils to plastic additives detected in all four samples including the virgin-grade one. These should be a priority to avoid in the design of polymers or in the processing of plastic waste. The detailed characterization of the different pyrolysis oils will allow us to optimize existing decontamination strategies and enable us to further close the loop for polyolefin waste.

Acknowledgements

Yannick Ureel acknowledges financial support from the Fund for Scientific Research Flanders (FWO Flanders) through the doctoral fellowship grant 1185822N. A portion of this work was performed at the National High Magnetic Field Laboratory, which is supported by National Science Foundation Division of Materials Research and Division of Chemistry through DMR-2128556 and the State of Florida. The authors acknowledge financial support by the Catalisti-ICON project (HBC.2018.0262) MATTER (Mechanical and Thermochemical Recycling of mixed plastic waste) funded by Flanders Innovation & Entrepreneurship (VLAIO). The authors acknowledge funding from the European Research Council under the European Union's Horizon 2020 research and innovation program / ERC grant agreement n° 818607.

References

[1] K. Ragaert, L. Delva, and K. Van Geem, "Mechanical and chemical recycling of solid plastic waste," in *Waste Management* vol. 69, ed: Elsevier Ltd, 2017, pp. 24-58.

[2] R. Francis, *Recycling of polymers: methods, characterization and applications*. John Wiley & Sons, 2016.

[3] A. Andrady, M. Bomgardner, D. Southerton, C. Fossi, and A. Holmström, "Plastics in a sustainable society," *Background paper, Stockholm,* 2015.

[4] OECD. Plastic waste by end-of-life fate - projections [Online] Available: https://www.oecd-ilibrary.org/content/data/3f85b1c2-en

[5] "Documentation for greenhouse gas emission and energy factors used in the waste reduction model (WARM)," U.S. Environmental Protection Agency - Office of Resource Conservation and Recovery, 2016.

[6] O. Dogu *et al.*, "The chemistry of chemical recycling of solid plastic waste via pyrolysis and gasification: State-of-the-art, challenges, and future directions," in *Progress in Energy and Combustion Science* vol. 84, ed: Elsevier Ltd, 2021, pp. 100901-100901.

[7] J. P. Lange, "Towards circular carbo-chemicals – the metamorphosis of petrochemicals," *Energy & Environmental Science,* vol. 14, no. 8, pp. 4358-4376, 2021/8// 2021, doi: 10.1039/D1EE00532D.

[8] J.-P. Lange, "Sustainable development: efficiency and recycling in chemicals manufacturing," *Green Chemistry,* vol. 4, no. 6, pp. 546-550, 2002/12// 2002, doi: 10.1039/B207546F.

[9] I. A. Ignatyev, W. Thielemans, and B. Vander Beke, "Recycling of polymers: a review," *ChemSusChem,* vol. 7, no. 6, pp. 1579-1593, 2014.

[10] A. Merrington, "Recycling of plastics," in *Applied plastics engineering handbook*: Elsevier, 2017, pp. 167-189.

[11] M. S. Abbas-Abadi *et al.*, "Challenges and opportunities of light olefin production via thermal and catalytic pyrolysis of end-of-life polyolefins: Towards full recyclability," *Progress in Energy and Combustion Science,* vol. 96, p. 101046, 2023.

[12] S.-H. Jung, M.-H. Cho, B.-S. Kang, and J.-S. Kim, "Pyrolysis of a fraction of waste polypropylene and polyethylene for the recovery of BTX aromatics using a fluidized bed reactor," *Fuel Processing Technology,* vol. 91, no. 3, pp. 277-284, 2010/03/01/ 2010, doi: https://doi.org/10.1016/j.fuproc.2009.10.009.

[13] J. M. Encinar and J. F. González, "Pyrolysis of synthetic polymers and plastic wastes. Kinetic study," *Fuel Processing Technology,* vol. 89, no. 7, pp. 678-686, 2008/07/01/ 2008, doi: https://doi.org/10.1016/j.fuproc.2007.12.011.

[14] M. Kusenberg, A. Eschenbacher, M. R. Djokic, A. Zayoud, K. Ragaert, S. De Meester, and K. M. Van Geem, "Opportunities and challenges for the application of post-consumer plastic waste pyrolysis oils as steam cracker feedstocks: To decontaminate or not to decontaminate?," *Waste Management,* vol. 138, pp. 83-115, 2022/02/01/ 2022, doi: https://doi.org/10.1016/j.wasman.2021.11.009.

[15] M. Kusenberg *et al.*, "Contaminant removal from plastic waste pyrolysis oil via depth filtration and the impact on chemical recycling: A simple solution with significant impact," *Chemical Engineering Journal,* p. 145259, 2023.

[16] M. Kusenberg *et al.*, "Assessing the feasibility of chemical recycling via steam cracking of untreated plastic waste pyrolysis oils: Feedstock impurities, product yields and coke formation," *Waste Management,* vol. 141, pp. 104-114, 2022.

[17] F. D. Kopinke, G. Zimmermann, G. C. Reyniers, and G. F. Froment, "Relative rates of coke formation from hydrocarbons in steam cracking of naphtha. 2. Paraffins, naphthenes, mono-, di-, and cycloolefins, and acetylenes," *Industrial & Engineering Chemistry Research,* vol. 32, no. 1, pp. 56-61, 1993/01/01 1993, doi: 10.1021/ie00013a009.

[18] G. C. Reyniers, G. F. Froment, F.-D. Kopinke, and G. Zimmermann, "Coke Formation in the Thermal Cracking of Hydrocarbons. 4. Modeling of Coke Formation in Naphtha Cracking," *Industrial & Engineering Chemistry Research,* vol. 33, no. 11, pp. 2584-2590, 1994/11/01 1994, doi: 10.1021/ie00035a009.

[19] F. D. Kopinke, G. Zimmermann, and S. Nowak, "On the mechanism of coke formation in steam cracking—conclusions from results obtained by tracer experiments," *Carbon,* vol. 26, no. 2, pp. 117-124, 1988.

[20] F. D. Kopinke, G. Zimmermann, G. C. Reyniers, and G. F. Froment, "Relative rates of coke formation from hydrocarbons in steam cracking of naphtha. 3. Aromatic hydrocarbons," *Industrial & Engineering Chemistry Research,* vol. 32, no. 11, pp. 2620-2625, 1993/11/01 1993, doi: 10.1021/ie00023a027.

[21] M. Kusenberg *et al.*, "A comprehensive experimental investigation of plastic waste pyrolysis oil quality and its dependence on the plastic waste composition," *Fuel Processing Technology,* vol. 227, p. 107090, 2022/03/01/ 2022, doi: https://doi.org/10.1016/j.fuproc.2021.107090.

[22] V. Marturano, P. Cerruti, and V. Ambrogi, "Polymer additives," *Physical Sciences Reviews,* vol. 2, no. 6, p. 20160130, 2017.

[23] Y. Zhang, G. Ji, C. Chen, Y. Wang, W. Wang, and A. Li, "Liquid oils produced from pyrolysis of plastic wastes with heat carrier in rotary kiln," *Fuel Processing Technology,* vol. 206, p. 106455, 2020/09/01/ 2020, doi: https://doi.org/10.1016/j.fuproc.2020.106455.

[24] M. Kusenberg *et al.*, "Towards high-quality petrochemical feedstocks from mixed plastic packaging waste via advanced recycling: The past, present and future," *Fuel Processing Technology,* vol. 238, p. 107474, 2022/12/15/ 2022, doi: https://doi.org/10.1016/j.fuproc.2022.107474.

[25] G. Caeiro, J. M. Lopes, P. Magnoux, P. Ayrault, and F. Ramôa Ribeiro, "A FT-IR study of deactivation phenomena during methylcyclohexane transformation on H-USY zeolites: Nitrogen poisoning, coke formation, and acidity–activity correlations," *Journal of Catalysis,* vol. 249, no. 2, pp. 234-243, 2007/07/25/ 2007, doi: https://doi.org/10.1016/j.jcat.2007.04.005.

[26] L. L. Hegedus and R. W. McCabe, "Catalyst poisoning," in *Studies in Surface Science and Catalysis,* vol. 6: Elsevier, 1980, pp. 471-505.

[27] M. B. Comisarow and A. G. Marshall, "Fourier transform ion cyclotron resonance spectroscopy," *Chemical physics letters,* vol. 25, no. 2, pp. 282-283, 1974.

[28] A. G. Marshall, C. L. Hendrickson, and G. S. Jackson, "Fourier transform ion cyclotron resonance mass spectrometry: A primer," *Mass Spectrometry Reviews,* vol. 17, no. 1, pp. 1-35, 1998/01/01 1998, doi: https://doi.org/10.1002/(SICI)1098-2787(1998)17:1<1::AID-MAS1>3.0.CO;2-K.

[29] A. G. Marshall and R. P. Rodgers, "Petroleomics: The Next Grand Challenge for Chemical Analysis," *Accounts of Chemical Research,* vol. 37, no. 1, pp. 53-59, 2004/01/01 2004, doi: 10.1021/ar020177t.
[30] H. E. Toraman, T. Dijkmans, M. R. Djokic, K. M. Van Geem, and G. B. Marin, "Detailed compositional characterization of plastic waste pyrolysis oil by comprehensive two-dimensional gas-chromatography coupled to multiple detectors," *Journal of Chromatography A,* vol. 1359, pp. 237-246, 2014/9// 2014, doi: 10.1016/j.chroma.2014.07.017.
[31] H. Dao Thi, M. R. Djokic, and K. M. Van Geem, "Detailed Group-Type Characterization of Plastic-Waste Pyrolysis Oils: By Comprehensive Two-Dimensional Gas Chromatography Including Linear, Branched, and Di-Olefins," *Separations,* vol. 8, no. 7, p. 103, 2021. [Online]. Available: https://www.mdpi.com/2297-8739/8/7/103.
[32] R. P. Rodgers and A. G. Marshall, "Petroleomics: Advanced characterization of petroleum-derived materials by Fourier transform ion cyclotron resonance mass spectrometry (FT-ICR MS)," in *Asphaltenes, heavy oils, and petroleomics*: Springer, 2007, pp. 63-93.
[33] D. F. Smith, D. C. Podgorski, R. P. Rodgers, G. T. Blakney, and C. L. Hendrickson, "21 Tesla FT-ICR Mass Spectrometer for Ultrahigh-Resolution Analysis of Complex Organic Mixtures," *Analytical Chemistry,* vol. 90, no. 3, pp. 2041-2047, 2018/02/06 2018, doi: 10.1021/acs.analchem.7b04159.
[34] M. S. Abbas-Abadi *et al.*, "Thermal pyrolysis of waste versus virgin polyolefin feedstocks: The role of pressure, temperature and waste composition," *Waste Management,* vol. 165, pp. 108-118, 2023/06/15/ 2023, doi: https://doi.org/10.1016/j.wasman.2023.04.029.
[35] R. Olcese, V. Carré, F. Aubriet, and A. Dufour, "Selectivity of Bio-oils Catalytic Hydrotreatment Assessed by Petroleomic and GC*GC/MS-FID Analysis," *Energy & Fuels,* vol. 27, no. 4, pp. 2135-2145, 2013/04/18 2013, doi: 10.1021/ef302145g.
[36] Y. Cho, A. Ahmed, A. Islam, and S. Kim, "Developments in FT-ICR MS instrumentation, ionization techniques, and data interpretation methods for petroleomics," *Mass Spectrometry Reviews,* vol. 34, no. 2, pp. 248-263, 2015/04/01 2015, doi: https://doi.org/10.1002/mas.21438.
[37] M. P. Barrow, K. M. Peru, and J. V. Headley, "An Added Dimension: GC Atmospheric Pressure Chemical Ionization FTICR MS and the Athabasca Oil Sands," *Analytical Chemistry,* vol. 86, no. 16, pp. 8281-8288, 2014/08/19 2014, doi: 10.1021/ac501710y.
[38] R. P. Rodgers *et al.*, "Combating selective ionization in the high resolution mass spectral characterization of complex mixtures," *Faraday discussions,* vol. 218, pp. 29-51, 2019.
[39] S. Klaimy, J. F. Lamonier, M. Casetta, S. Heymans, and S. Duquesne, "Recycling of plastic waste using flash pyrolysis – Effect of mixture composition," *Polymer Degradation and Stability,* vol. 187, p. 109540, 2021/05/01/ 2021, doi: https://doi.org/10.1016/j.polymdegradstab.2021.109540.
[40] A. K. Vasiliou *et al.*, "Biomass pyrolysis: thermal decomposition mechanisms of furfural and benzaldehyde," *The Journal of chemical physics,* vol. 139, no. 10, 2013.
[41] E. Furimsky, "Catalytic hydrodeoxygenation," *Applied Catalysis A: General,* vol. 199, no. 2, pp. 147-190, 2000/06/12/ 2000, doi: https://doi.org/10.1016/S0926-860X(99)00555-4.
[42] J. N. Hahladakis, C. A. Velis, R. Weber, E. Iacovidou, and P. Purnell, "An overview of chemical additives present in plastics: Migration, release, fate and environmental impact during their use, disposal and recycling," *Journal of Hazardous Materials,* vol. 344, pp. 179-199, 2018/02/15/ 2018, doi: https://doi.org/10.1016/j.jhazmat.2017.10.014.
[43] M. Bolgar, J. Hubball, J. Groeger, and S. Meronek, *Handbook for the chemical analysis of plastic and polymer additives.* CRC Press, 2015.

AIChE Paper 80b

IMPACT OF α-OLEFINS AND DIOLEFINS FROM POLYOLEFIN PYROLYSIS OIL TO THE CATALYTIC CRACKING MECHANISTIC PATHWAYS

Bahman Goshayeshi
Ir.
Ghent University, Aristotle University of Thessaloniki

Stavros Alexandros Theofanidis
Dr. ir.
Aristotle University of Thessaloniki, AristEng S.à r.l.

Mehrdad Seifali Abbas-Abadi
Dr. ir.
Ghent University

Ehsan Mahmoudi
ir.
Aristotle University of Thessaloniki, KU Leuven

Kevin M. Van Geem
Prof. Dr. ir.
Ghent University

Angeliki Lemonidou
Prof. Dr. ir.
Aristotle University of Thessaloniki, CPERI/CERTH

Statements:

Prepared for Presentation at the 2024 Spring Meeting and 20th Global Congress on Process Safety
New Orleans, Louisiana, Date (Meeting March 24 – 28, 2024)

AIChE Paper 80b

IMPACT OF α-OLEFINS AND DIOLEFINS FROM POLYOLEFIN PYROLYSIS OIL TO THE CATALYTIC CRACKING MECHANISTIC PATHWAYS

Bahman Goshayeshi
ir.
Ghent University, Aristotle University of Thessaloniki

Stavros Alexandros Theofanidis
Dr. ir.
Aristotle University of Thessaloniki, AristEng S.à r.l.

Mehrdad Seifali Abbas-Abadi
Dr. ir.
Ghent University

Ehsan Mahmoudi
ir.
Aristotle University of Thessaloniki, KU Leuven

Kevin M. Van Geem
Prof. Dr. ir.
Ghent University

Angeliki Lemonidou
Prof. Dr. ir.
Aristotle University of Thessaloniki, CPERI/CERTH

ABSTRACT

The challenges in ex-situ catalytic pyrolysis of plastic waste, focusing on the selective conversion of higher olefins and diolefins, were addressed through an investigation of the catalytic cracking mechanism using 1-decene (α-olefin) and 1,9-decadiene (diolefin) as model compounds. Zeolite-based catalyst HZSM-5 was employed in a fixed-bed reactor to elucidate reaction pathways leading to light olefin production, aromatization, and coke formation during the pyrolysis of polyolefins. The HZSM-5 catalyst underwent steaming at 800 °C for 4 hours to increase stability and enhance selectivity towards light olefins. The impact of contact time (~65.2-250.7 ms) and temperature (250 °C, 300 °C, 350 °C, and 450 °C) variations on product yields and reaction pathways during catalytic pyrolysis of 1-decene and 1,9-decadiene were investigated. For 1-decene, the sequence initiated with isomerization through double-bond rearrangements at low conversion, followed by skeletal isomerization at higher conversion. Subsequently, intensified cracking reactions led to the generation of light C_{3-4} linear olefins and C_{5-7} nonlinear olefins. These compounds underwent further cracking, resulting in the production of C_{3-4} linear olefins with notably low cyclization rates at elevated temperatures. Conversely, 1,9-decadiene yielded a substantial quantity of naphthenes and lighter diolefins, alongside C_{3-5} linear olefins. Cracking of 1,9-decadiene enhanced aromatics production through the cyclization and aromatization of the produced light conjugated diolefins, acting as precursors for Diels-Alder reactions. Consequently, the presence of diene compounds in plastic pyrolysis oil served as a precursor for significant coke formation. Overall, this study provides valuable insights for the catalytic upgrading of pyrolysis oil derived from polyolefin pyrolysis.

Introduction

The employment of chemical recycling for polyolefin waste management, a promising eco-friendly approach, has gained substantial attention across scientific and industrial sectors. The current emphasis in polyolefin pyrolysis lies in the generation of light olefins through both direct and indirect approaches. The direct method mostly utilizes zeolite catalysts for effective light olefin production, with the ex-situ configuration showing potential for industrial applications [1-6]. In contrast, the indirect method yields pyrolysis oil through thermal pyrolysis. The plastic pyrolysis oil can be blended with naphtha and other hydrocarbon feeds in steam cracking units to yield light olefins or undergo catalytic upgrading via cracking processes. Research has shown that pyrolysis oil derived from polyolefins typically consists of approximately 50-60% mono and di-olefins, with cyclic hydrocarbons (naphthenes and aromatics) comprising around 10-15% and the remaining fraction mainly consists of paraffins [7-9]. For the production of light olefins from pyrolysis oil, the presence of paraffins does not pose significant challenges in terms of secondary reactions and coke formation. The primary hurdles for steam cracking units arise from the olefins and diolefins. Studies on these units have revealed that when the feed contains more than 2% olefins, secondary reactions intensify [10]. In the ex-situ catalytic pyrolysis of polyolefins using ZSM-5, the challenge of coke formation arises as a result of the substantial production of olefins and diolefins [11]. It is important to note that pyrolysis oils contain various olefins, classifiable into linear-olefins and nonlinear-olefins. Based on earlier research [9,12], pyrolysis oil from polypropylene, containing nonlinear-olefins as byproducts, presents fewer challenges in the production of light olefins due to steric hindrance from methyl groups. The main challenges are associated with linear α-olefins and diolefins, the primary components in the thermal pyrolysis of polyethylene waste [13]. Polyolefin pyrolysis entails the presence of numerous species and reactions within expansive and complex free-radical reaction networks. Investigating degradation mechanisms in such a system proves challenging, if not impossible, due to the diverse factors influencing product distribution and reaction networks [14,15]. To navigate this complexity, model compounds are utilized to represent key facets of the intricate network, providing a foundational understanding of underlying reactions and facilitating manageable analysis. In this context, two pivotal compounds, 1-decene (α-olefin) and 1,9-decadiene (diolefin), serve as model compounds for pyrolysis oil, significantly contributing to the pyrolysis reaction network. Therefore, this study aims to meticulously investigate the cracking mechanisms during the catalytic pyrolysis of 1-decene and 1,9-decadiene using HZSM-5 catalyst under varying conditions. The objective is to discern disparities in the cracking mechanisms of olefins and diolefins, with specific attention to secondary reactions such as aromatization and coke formation. This study provides insights into how mid-range carbon number (C_{10}) linear olefins and diolefins, derived from the chain scission of long-chain hydrocarbons, influence the overall polyolefin cracking process, thereby affecting the production of light olefins, cyclic and aromatic compounds, and coke formation.

Material, Equipment, and Methods

A fixed-bed quartz reactor featuring a coaxial thermocouple for temperature monitoring was utilized for thermal and catalytic pyrolysis of model compounds. The setup included an online gas chromatography (GC) with a thermal conductivity detector (TCD) and an offline GC with a flame ionization detector (FID). The electrically heated reactor, equipped with three thermocouples positioned along the reactor length and one thermocouple placed inside the bed for monitoring the catalyst bed temperature. A preheater line was employed to raise the feed temperature to 250 °C before entering the reactor, facilitating vaporization of liquid hydrocarbons feeds. The model compounds, 1-decene and 1,9-decadiene, were provided from the Thermo Fisher Scientific with purities of 96% and 97%, respectively. Liquid hydrocarbon feeds were delivered to the preheater and subsequently to the fixed-bed reactor using a Shimadzu LC-20AD liquid pump. Nitrogen (N_2) served as the carrier gas, introduced via a mass flow controller to maintain a 10% molar composition of 1-decene or 1,9-decadiene and 90% N_2 for all experiments. After passing through the fixed-bed reactor, the effluent stream was directed to a condenser for heavy hydrocarbon separation, then the gas directed to the GC-TCD (Agilent 7890B GC) for detecting light hydrocarbon concentrations. Simultaneously, liquid hydrocarbons were collected from the condenser for offline GC-FID analysis and identification of the compounds were done through a GC-mass spectrometry (MS) analysis.

The following equations were used for calculation of contact time, W/F and product yields.

$$Contact\ Time\ (ms) = \frac{catalyst\ bed\ volume\ (ml)}{carrier\ gas\ volumetric\ flow\ rate\ (\frac{ml}{ms})} \quad \text{Eq. (1)}$$

$$yield_i\,\% = \frac{outlet\ molar\ flowrate_i \times carbon\ number_i}{inlet\ molar\ flowrate_{feed} \times carbon\ number_{feed}} \times 100\ ,\ \ i = species \qquad \text{Eq. (2)}$$

$$\frac{W}{F} = \frac{catalyst\ loading\ (kg)}{total\ molar\ flow\ rate\ (\frac{mol}{s})} \qquad \text{Eq. (3)}$$

The NH_4-ZSM-5 catalyst (sample: CBV 5524G, SiO2/Al2O3= 50) was provided from Zeolyst International. Prior research has demonstrated that subjecting the same ZSM-5 samples to steam under these conditions can further improve the selectivity toward light olefins and enhance the stability of the catalysts [11,16]. Therefore, prior to experiments, the NH_4-ZSM-5 catalyst was steamed at 800 °C for 4 hours under a flow containing 95 vol% deionized H_2O and 5 vol% N_2 as a carrier gas. After steam-treatment, the catalyst was calcined at 550 °C for 5 hours under ambient air to form HZSM-5. The experimental conditions, including catalyst size, loading, and W/F (based on Eq. (3)) (125-150 µm, 14.5 mg of catalyst, W/F= 15.24×10-2 kg·s/mol), N_2 flowrate (116.3 ml/min), liquid flowrate (0.1 ml/min), feed molar composition (10% liquid hydrocarbon, 90% N_2), and reactor pressure (1.3 bara), remained consistent across all experiments, unless specified otherwise. Changes were made only to N_2 and liquid flowrates (and thus W/F) during experiments to examine the impact of contact time on product yields. In each trial, the steam treated catalysts were mixed with a constant quantity of inert medium particles (125-150 µm quartz, 0.99 g) to ensure a uniform catalyst bed height of 9 mm.

Effect of Temperature

Thermal pyrolysis experiments were conducted without a catalyst to assess feed conversion and extent of thermal reactions at different temperatures, aiming to distinguish the role of catalysts in the degradation process. Across the temperature range of 250 °C to 450 °C, there was minimal thermal conversion of 1,9-decadiene. Similarly, 1-decene exhibited little change, but at 450 °C, approximately 50% isomerization occurred. This isomerization comprised approximately two-thirds double bond rearrangement and one-third skeletal isomerization, leading to the formation of the 2-methyl-2-nonene compound. Hence, it can be inferred that thermal degradation has a negligible effect across all temperature ranges, and every cracking reaction observed during catalytic cracking can be attributed to the catalytic effect. Considering this, catalytic cracking of 1-decene and 1,9-decadiene, employed as model compounds for plastic pyrolysis oil, was carried out using steam-treated HZSM-5 catalyst within a temperature range of 250 °C to 450 °C. In the cracking of 1-decene, linear-olefins, nonlinear-olefins, and a few naphthenes were formed, with negligible amounts of paraffins and aromatics . A 4-hour steam treatment at 800 °C reduced the concentration of Brønsted acid sites which are promoting cracking, isomerization, and cyclization reactions [6,17,18].

According to Figure 1, at 250 °C, only isomerization of 1-decene took place with a conversion of ~28%, wherein the selectivity leaned more towards double bond rearrangement than skeletal isomerization, resulting in the production of linear C_{10} olefins (2-decene, 4-decene, and 5-decene). Almost no conversion of 1,9-decadiene was detected at 250 °C (<4%). Increasing the temperature to 300 °C led to more conversion of 1-decene to ~70%. Isomerization and chain scission of 1-decene were observed, with skeletal isomerization being more favorable than double bond rearrangement. Chain scission of 1-decene also led to the production of C_{5-7} nonlinear olefins and C_{3-5} linear olefins. The conversion of 1,9-decadiene was slightly raised to ~6% at 300 °C, favoring the production of cyclic compounds. At 350 °C and 450 °C, full conversion of 1-decene was achieved and almost no 1-decene isomers were detected. Almost all C_{10} olefins were cracked into lighter olefins via β-scission to produce C_{5-7} nonlinear olefins and C_{3-5} linear olefins. Elevating the temperature to 450 °C notably influenced the cracking of C_{5-7} nonlinear-olefins, leading to increased yields of propylene and butene, while the yields of C_{6-7} olefins gradually decreased. Additionally, a slight increase in pentene yield was observed, suggesting that higher temperatures promote the formation of nonlinear olefins with lower carbon numbers, which are subsequently cracked to produce propylene and butene, with ethylene yield being very low [19]. The amount of naphthene produced was low which suggests the low cyclization reaction rates during 1-decene catalytic cracking. The temperature-dependent behavior of 1-decene involved isomerization, chain β-scission, and minor cyclization reactions. Double-bond rearrangement of 1-decene occurred rapidly compared to skeletal isomerization, with higher conversion favoring skeletal isomerization through alkyl migration or protonated cyclopropane. In the produced skeletal isomers, the yield of mono-branched isomers, particularly 2-methyl-2-nonene, exceeded that of di-branched isomers. As temperature increased, more cracking of 1-decene took place, resulting in increased production of C3-7 olefins, with a preference for butene and propylene.

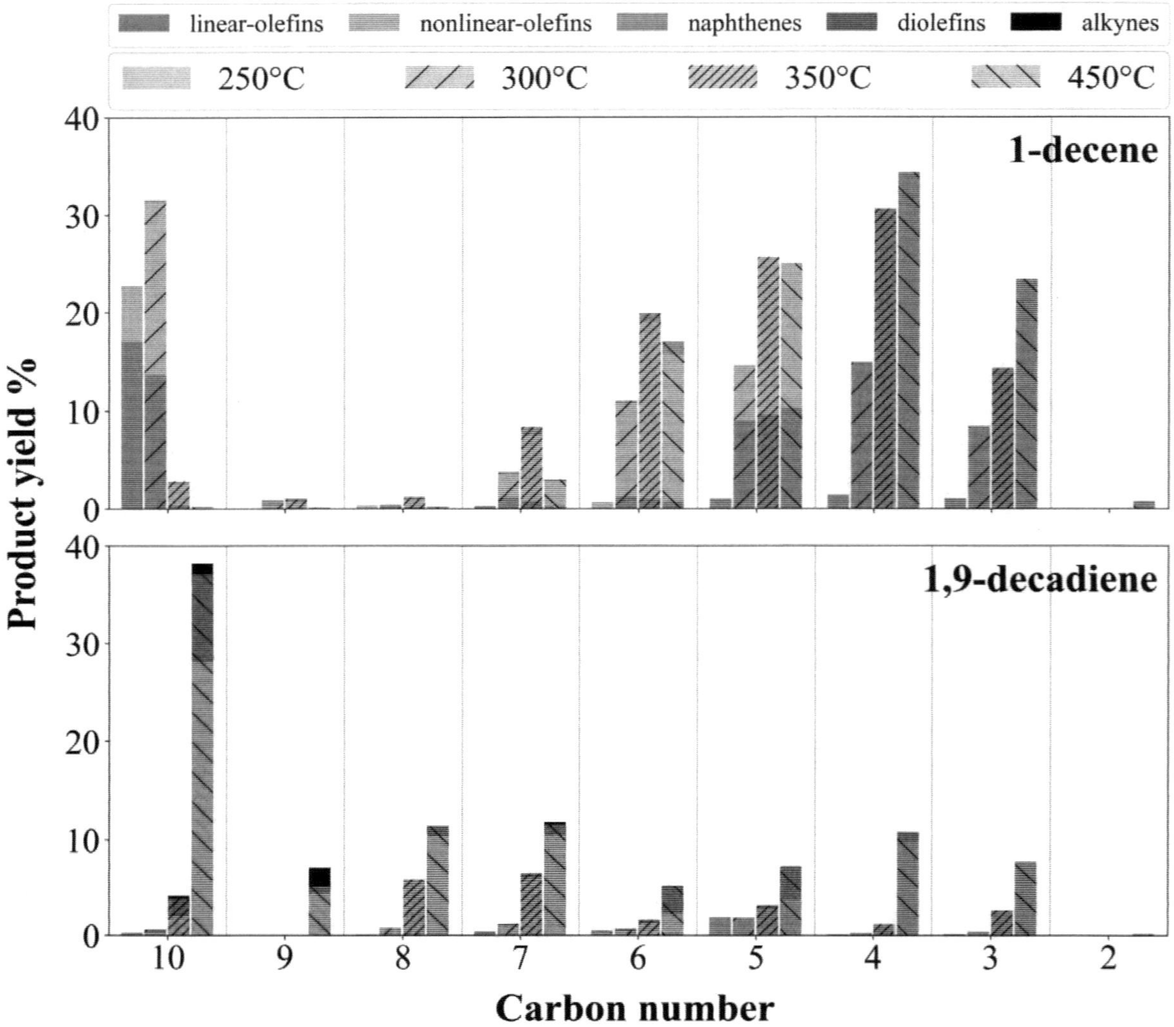

Figure 1: Catalytic pyrolysis of 1-decene and 1,9-decadiene over steam treated HZSM-5 catalysts under different temperatures (P=1.3 bara, W/F=15.24×10-2 kg·s/mol, contact time: 250.7 ms).

At 350 °C, the conversion of 1,9-decadiene exceeded 26%, primarily leading to the production of naphthenes. C_{3-5} linear olefins and C_{10} diolefins were produced in relatively small amounts. Full conversion of 1,9-decadiene was only achieved at 450 °C. Figure 1, shows an increase in 1,9-decadiene conversion with temperature, leading to higher production of naphthenic products, primarily consisting of C_{10} naphthenes, followed by C_7, C_8, and C_9 naphthenes. Predominantly identified as cyclohexene with varying alkyl groups, these naphthenes were formed within the carbon number range of 6 to 10. The progression continued with the generation of diolefins, mainly comprising C_{10}, C_5 (pentadiene), and C_6 (hexadiene and pentadiene with a methyl group). At 300 °C and 350 °C, the predominant naphthenes products were C_8 and C_7, while at 450 °C (full conversion), the C_{10} naphthenes selectivity substantially increased. Catalytic pyrolysis of 1,9-decadiene exhibited high selectivity towards naphthenes and diolefins, with almost no nonlinear-olefins produced at various temperatures. The primary products of 1,9-decadiene pyrolysis were naphthenes, C_{3-5} linear olefins, diolefins, and alkynes (mostly C_9, C_{10}, and C_7). In contrast to 1-decene, 1,9-decadiene pyrolysis showed significantly lower isomerization. Although alkynes can be considered functional isomers of diolefins, the overall yield of C_{10} alkynes at high temperatures (450 °C) was low (3.5%). The average carbon number of products from 1,9-decadiene cracking increased with temperature (~6.5 @250 °C – ~7.5 @450 °C), unlike 1-decene cracking (~9.1 @250 °C - ~4.3 @450 °C), which decreased. Cyclic compounds, mostly C_{10}, were predominantly produced at high conversion levels, and C_{4-6} diolefins, especially conjugated dienes like 1,3-pentadiene and 1,3-butadiene, played a crucial role in facilitating the Diels-Alder reaction, leading to cyclic compound production, which could undergo aromatization through dehydrogenation. 1,3-butadiene and other diene compounds were exclusively identified in the products resulting from the catalytic pyrolysis of 1,9-decadiene, with minimal traces in the products of 1-decene as feed. Even though in low amounts, 1,9-decadiene catalytic pyrolysis resulted in a higher production of aromatic compounds (~0.8% at 450 °C) than 1-decene catalytic pyrolysis (~0.1% at 450 °C).

Effect of Contact Time

For a thorough exploration of catalytic reaction pathways, we systematically investigated sequential reaction steps by varying the contact times between feed hydrocarbon vapors and the steam-treated catalyst. At 350 °C, employing four different contact times ranging from 65.2 to 250.7 ms, the 1-decene vapors were passed over the HZSM-5 steam-treated catalyst. As depicted in Figure 2, prolonging the contact time resulted in a decrease in the yield of C_{10} isomers due to their cracking into lighter compounds, mainly C_{3-5} olefins. There was an ascending trend in 1-decene conversion, increasing from 93.62% at the shortest contact time to reaching full conversion at the longest. Throughout all the tests, ethylene production remained minimal (<0.1%). The primary light olefin products were butenes (~30.6%), followed by propylene (~14.3%), 2-butene, 2-methyl- (~16.1%), and pentene (~9.6%) at the longest contact time (250.7 ms). Extending the contact time from 65.2 ms to 128.7 ms led to an increased production of C_{3-5} olefins and a decreasing trend in C_6 olefins, mostly nonlinear olefins. There was no change in the product yields of C_{7-9} olefins. Almost doubling the contact time from 128.7 ms to 250.7 ms resulted in full conversion and a significant reduction in the yields of C_{10} olefins. Moreover, there was a substantial increase in the yield of C_{4-6} olefins. There were no significant variations observed in the yield of naphthenes concerning contact time. All the evidence suggests that when 1-decene is used as a feedstock, the initial process involves C_{10} isomerization, followed by the cracking of C_{10} olefins to yield C_{3-5} and C_{6-7} olefins, with subsequent cracking of C_{6-7} olefins into C_{3-4} olefins. As the contact time nearly doubled from 128.7 ms to 250.7 ms, almost all the C_{10} compounds underwent cracking into lighter species. Simultaneously, there was an increase in the yield of C_6 nonlinear olefins, potentially attributed to the β-scission of C_{10} nonlinear olefins into butene and C_6 nonlinear olefins [18]. Nevertheless, as demonstrated, with an increase in temperature, C_6 compounds experienced further cracking into butene and propylene.

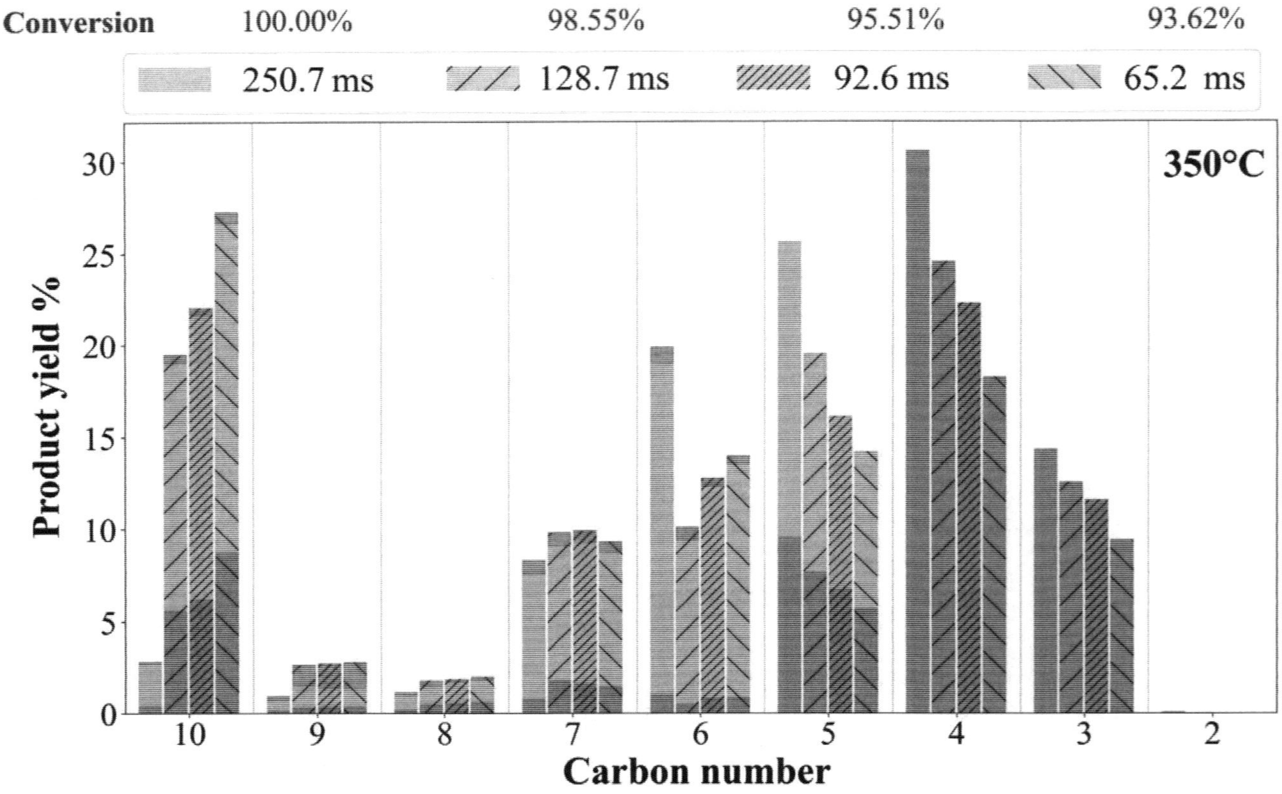

Figure 2: Catalytic pyrolysis of 1-decene over steam treated HZSM-5 catalyst under various contact times (T=350 °C, P=1.3 bara). Yields of each product type based on carbon number and conversion (in red on top of each contact time legend) are provided.

The impact of contact time on the degradation of 1,9-decadiene at 350 °C was investigated, and the results are presented in Figure 3. Elongating the contact time at 350 °C led to increased conversion rates, although full conversion was not attained. Unlike 1-decene catalytic pyrolysis, where the primary products were olefins, the dominant outcomes of 1,9-decadiene catalytic pyrolysis were naphthenes. The extension of contact time resulted in a higher yield of naphthenes, with no notable shift in the average carbon number. Additionally, there was an increase in the yield of C_{10} diolefins and C_{3-5} linear olefins. Notably, propylene and pentene yields were significantly higher compared to other compounds. While the amount of aromatics remained relatively low (<0.05%), a prolonged contact time led to an increased yield of aromatic compounds. The production of paraffins (C_6) was only observed at the longest contact time. In summary, at 350 °C, the extended contact time primarily resulted in the production of naphthenes due to increased conversion, with C_{7-8} compounds (mostly cyclohexene rings with alkyl groups) being the major naphthenic compounds.

At 450 °C, catalytic pyrolysis of 1,9-decadiene was conducted with two distinct contact times: 65.2 ms and 250.7 ms. In both cases, 1,9-decadiene conversion reached 100%. Elevating the contact time led to a minor reduction in the yield of C_{6-8} naphthenes from ~27.3% to ~24.1%, while the yield of C_{9-10} (~30.9%) cyclic compounds remained unaffected. This reduction resulted in a slight increase in the yield of butene (~8.5% to ~9.9%) and propylene

(~5.8% to ~7.8%). No significant changes in the yield of diolefins and alkynes were observed due to variations in contact time. The primary constituents of C_{6-8} naphthenes included cyclohexene with either an ethyl, dimethyl, or methyl group, as well as cyclopentene with a methyl group. In conclusion, at 450 °C, the extension of contact time led to a slight increase in the production of light olefins (C_{3-4}) at the expense of C_{6-8} naphthenes.

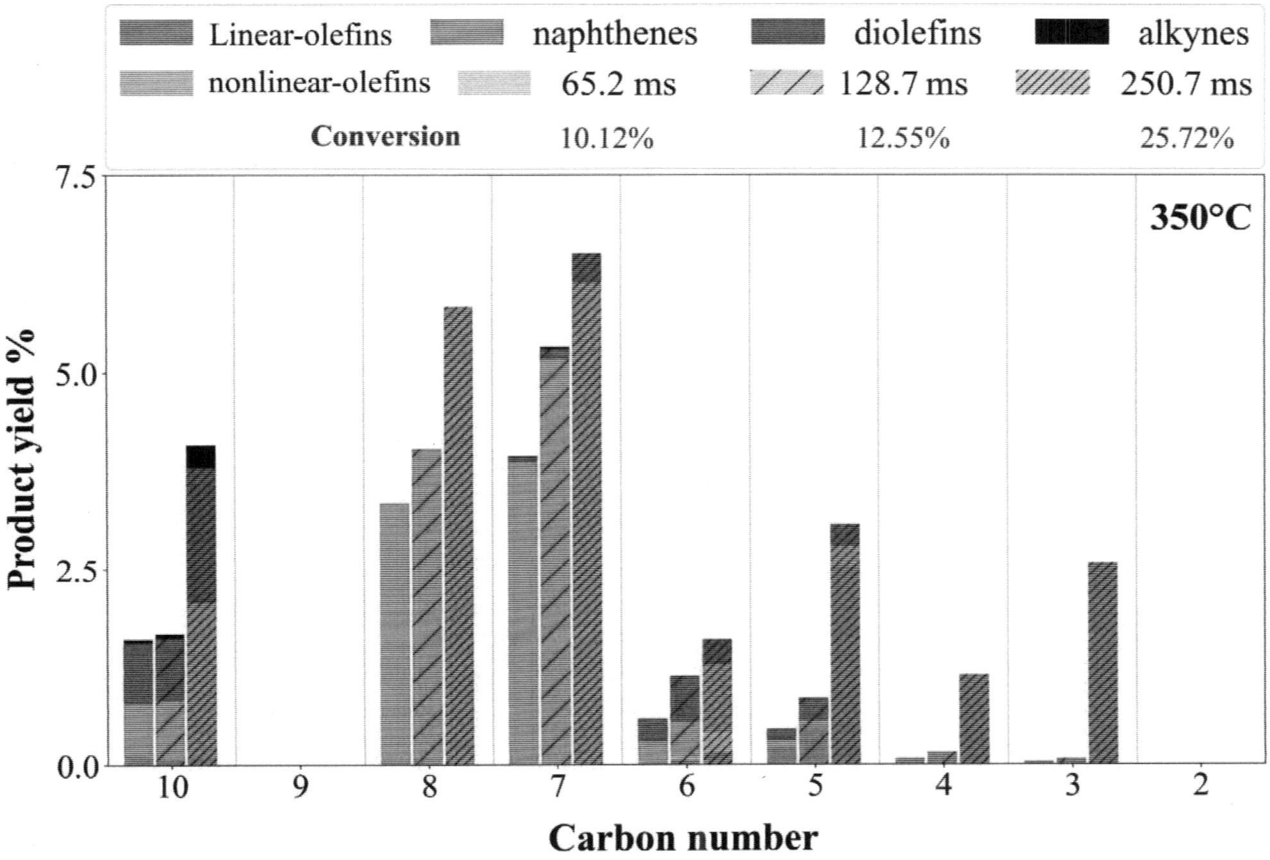

Figure 3: Catalytic pyrolysis of 1,9-decadiene over steam treated HZSM-5 catalyst under various contact times (T=350 °C, P=1.3 bara). Yields of each product type based on carbon number, conversion (in red below each contact time legend) are provided.

During the catalytic pyrolysis of 1-decene, which occurred within the temperature range of 250 °C to 450 °C at varying contact times, the conversion process included isomerization followed by chain scission. This led to the production of lighter olefins. The dominant reaction pathway was skeletal isomerization, generating branched C_{10} isomers (nonene with a methyl group), which was notably more prevalent than double bond rearrangement, particularly at higher conversion levels. With increasing conversion, the chain scission of 1-decene isomers became more pronounced, generating lighter olefins like butene and propylene. In the context of 1-decene catalytic pyrolysis, the initial reactions included isomerization, succeeded by chain scission, ultimately yielding light olefins. Under intensified reaction conditions, cyclization of olefins can occur, although it does not represent the primary reaction step (Figure 4).

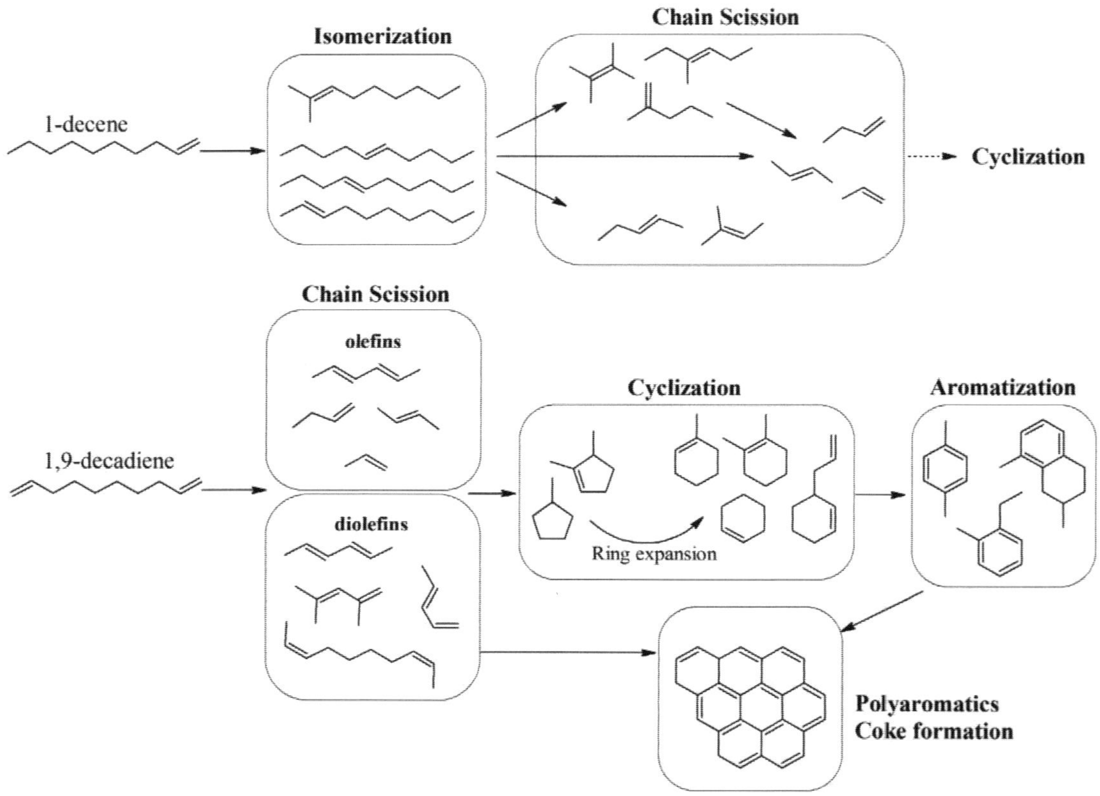

Figure 4: Reaction pathways of 1-decene (top) and 1,9-decadiene (bottom) catalytic pyrolysis (temperature ranges from 250 °C to 450 °C and varied contact times) over steam-treated HZSM-5 catalyst

In contrast, during the catalytic pyrolysis of 1,9-decadiene (diolefin), the major products were cyclic compounds, particularly cyclohexenes with an alkyl group, and light olefins. More severe reaction conditions, such as increased temperature and contact time, resulted in greater conversion of 1,9-decadiene into cyclic compounds and chain scission into lighter diene compounds. Thermal Gravimetric Analysis (TGA) of spent catalysts utilized in the pyrolysis of 1-decene and 1,9-decadiene demonstrated a significant rise in the generation of heavy hydrocarbons or polyaromatic compounds, both on the catalyst's surface and within micropores, during the catalytic pyrolysis of 1,9-decadiene in comparison to 1-decene. The TGA results showed that the amount of produced coke was ~0.003% g_{coke}/g_{feed_inlet} for 1-decene and ~0.02% g_{coke}/g_{feed_inlet} for 1,9-decadiene, indicating a significantly higher coke production during the catalytic pyrolysis of 1,9-decadiene. Additionally, the presence of hard coke, characterized by more structured polyaromatics, was observed exclusively in the case of 1,9-decadiene. The presence of lighter conjugated diolefins generated through chain scission of 1,9-decadiene, facilitated the Diels-Alder reaction, promoting cyclization, aromatic production, and, consequently, coke formation (Figure 4). This phenomenon was less prevalent during the cracking of 1-decene, as it did not produce conjugated diene or any diene compounds, thus limiting the cyclization process. Hence, the reaction pathways of 1,9-decadiene catalytic pyrolysis, as illustrated in Figure 4, included chain scission, cyclization, ring expansion, and the formation of cyclohexene rings, which subsequently converted to aromatic compounds [20].

In conclusion, the examination of the structural characteristics of these olefins and diolefins reveals a notably higher tendency for the formation of butadiene derivatives within the diolefins category compared to α-olefins. As a result, diolefins display an increased susceptibility to Diels-Alder reactions, giving rise to a diverse array of cyclic compounds, ultimately leading to the generation of aromatics and polyaromatics. Therefore, the presence of diene compounds in plastic pyrolysis oil acted as a precursor for significant coke formation. This occurred as they underwent transformation into cyclic compounds and lighter conjugated dienes at a significantly higher rate compared to olefins during catalytic cracking. In this regard, a thorough evaluation that includes the proportion of polypropylene in the pyrolysis feedstock, along with the measurement of linear diolefins and α-olefins, as well as their respective ratio, provides valuable insights into the operational dynamics of the subsequent light olefin production unit.

REFERENCES

1. Eschenbacher A, Goodarzi F, Varghese RJ, et al. Boron-Modified Mesoporous ZSM-5 for the Conversion of Pyrolysis Vapors from LDPE and Mixed Polyolefins: Maximizing the C2–C4 Olefin Yield with Minimal Carbon Footprint. *ACS Sustainable Chemistry & Engineering.* 2021/11/01 2021;9(43):14618-14630.

2. Inayat A, Inayat A, Schwieger W, Sokolova B, Lestinsky P. Enhancing aromatics and olefins yields in thermo-catalytic pyrolysis of LDPE over zeolites: Role of staged catalysis and acid site density of HZSM-5. *Fuel.* 2022/04/15/ 2022;314:123071.

3. Xue Y, Johnston P, Bai X. Effect of catalyst contact mode and gas atmosphere during catalytic pyrolysis of waste plastics. *Energy Conversion and Management.* 2017/06/15/ 2017;142:441-451.

4. Eschenbacher A, Varghese RJ, Abbas-Abadi MS, Van Geem KM. Maximizing light olefins and aromatics as high value base chemicals via single step catalytic conversion of plastic waste. *Chemical Engineering Journal.* 2022/01/15/ 2022;428:132087.

5. Kanduri PK, Seethamraju S. Effect of process mode on product yields and selectivity in the catalytic co-pyrolysis of peanut shells and polypropylene using spent fluid catalytic cracking (FCC) catalyst. *Bioresource Technology Reports.* 2023/06/01/ 2023;22:101478.

6. Serrano DP, Aguado J, Escola JM. Developing Advanced Catalysts for the Conversion of Polyolefinic Waste Plastics into Fuels and Chemicals. *ACS Catalysis.* 2012/09/07 2012;2(9):1924-1941.

7. Belbessai S, Azara A, Abatzoglou N. Recent Advances in the Decontamination and Upgrading of Waste Plastic Pyrolysis Products: An Overview. *Processes.* 2022;10(4):733.

8. Soják L, Kubinec R, Jurdáková H, Hájeková E, Bajus M. High resolution gas chromatographic–mass spectrometric analysis of polyethylene and polypropylene thermal cracking products. *Journal of Analytical and Applied Pyrolysis.* 2007/03/01/ 2007;78(2):387-399.

9. Kusenberg M, Zayoud A, Roosen M, et al. A comprehensive experimental investigation of plastic waste pyrolysis oil quality and its dependence on the plastic waste composition. *Fuel Processing Technology.* 2022/03/01/ 2022;227:107090.

10. Baumgartner AJ, Engineers AIoC. *Feedstock Contaminants in Ethylene Plants--an Update*: American Institute of Chemical Engineers; 2004.

11. Akin O, Varghese RJ, Eschenbacher A, et al. Chemical recycling of plastic waste to monomers: Effect of catalyst contact time, acidity and pore size on olefin recovery in ex-situ catalytic pyrolysis of polyolefin waste. *Journal of Analytical and Applied Pyrolysis.* 2023/06/01/ 2023;172:106036.

12. Abbas-Abadi MS, Kusenberg M, Zayoud A, et al. Thermal pyrolysis of waste versus virgin polyolefin feedstocks: The role of pressure, temperature and waste composition. *Waste Management.* 2023/06/15/ 2023;165:108-118.

13. Abbas-Abadi MS, Zayoud A, Kusenberg M, et al. Thermochemical recycling of end-of-life and virgin HDPE: A pilot-scale study. *Journal of Analytical and Applied Pyrolysis.* 2022/09/01/ 2022;166:105614.

14. Obali Z, Sezgi NA, Doğu T. PERFORMANCE OF ACIDIC MCM-LIKE ALUMINOSILICATE CATALYSTS IN PYROLYSIS OF POLYPROPYLENE. *Chemical Engineering Communications.* 2008/10/08 2008;196(1-2):116-130.

15. Levine SE, Broadbelt LJ. Detailed mechanistic modeling of high-density polyethylene pyrolysis: Low molecular weight product evolution. *Polymer Degradation and Stability.* 2009/05/01/ 2009;94(5):810-822.

16. Eschenbacher A, Varghese RJ, Delikonstantis E, et al. Highly selective conversion of mixed polyolefins to valuable base chemicals using phosphorus-modified and steam-treated mesoporous HZSM-5 zeolite with minimal carbon footprint. *Applied Catalysis B: Environmental.* 2022/07/15/ 2022;309:121251.

17. Dong Z, Chen W, Xu K, Liu Y, Wu J, Zhang F. Understanding the Structure–Activity Relationships in Catalytic Conversion of Polyolefin Plastics by Zeolite-Based Catalysts: A Critical Review. *ACS Catalysis.* 2022/12/16 2022;12(24):14882-14901.

18. Lin LF, Zhao SF, Zhang DW, Fan H, Liu YM, He MY. Acid Strength Controlled Reaction Pathways for the Catalytic Cracking of 1-Pentene to Propene over ZSM-5. *ACS Catalysis.* 2015/07/02 2015;5(7):4048-4059.

19. Li C, Vidal-Moya A, Miguel PJ, Dedecek J, Boronat M, Corma A. Selective Introduction of Acid Sites in Different Confined Positions in ZSM-5 and Its Catalytic Implications. *ACS Catalysis.* 2018/08/03 2018;8(8):7688-7697.

20. Joshi YV, Bhan A, Thomson KT. DFT-Based Reaction Pathway Analysis of Hexadiene Cyclization via Carbenium Ion Intermediates: Mechanistic Study of Light Alkane Aromatization Catalysis. *The Journal of Physical Chemistry B.* 2004/01/01 2004;108(3):971-980.

80c Advances in Dechlorination of Plastic Pyrolysis Oil

Daniel Miskin

The world is quickly shifting from dependance on fossil fuels and feedstocks to use of bio-based and circular feedstocks. This will help combat growing landfills and environmental pollution as well as minimize anthropogenic emissions. One key technology in this area is chemical recycling. In particular, pyrolysis has the potential to convert used plastics which are not suitable for mechanical recycling back into a feedstock for the petrochemical industry. As with any new technology, there are many hurdles to overcome in bringing this to scale. One of the biggest hurdles today is organic chlorides. These are primarily brought in with the pyrolysis reactor feed in the form of PVC. Several technologies exist to reduce the chloride concentration within the pyrolysis reactor, but even the cleanest pyrolysis oil products will still have tens of ppms of organic chlorides. This is still high enough to cause significant corrosion risks within petrochemical facilities. While organic chloride sorbents have existed for decades, they have proven to be minimally effective on this class of higher molecular weight organic chlorides. To combat this, a new generation of pyrolysis oil specific adsorbents is being developed. With these new adsorbents, it is now possible to meet the chloride specifications for even the most stringent petrochemical complexes.

Sulfur As a Contaminant in Ethylene Feedstocks

Vivek Srinivasan, Mahesh Subramaniyam

New information has been developed on the impact of sulfur contamination on furnace efficiency of Ethylene crackers. Conventional thinking is that the primary impact of excess sulfur contamination is mostly in the rate of coking. However, data generated in pilot studies and field experiences are now available to show that excess sulfur can impact furnace efficiency, impacting in best practices for feed and control of sulfiding agents.

This paper presents a review on the current literature for the impact of sulfur on the furnace performance. Pilot plant and field data will also be shared, relating the sulfur addition on furnace with the efficiency of the furnaces, and how to minimize negative effects of excess sulfur.

Management of Silicon Contamination on Pygas Hydrogenation Catalyst for a Steam Cracker

Michael Gaas, Esma Rokia, Marielle Gagniere, Jacob Hilbrich

An increasing number of petrochemical producers are actively exploring the transition to a circular economy as a means to enhance sustainability in plastics production. Within this shift, the incorporation of new feedstocks into steam crackers has raised challenges due to high contamination levels. Notably, pyrolysis gasoline plays a pivotal role in managing the effluent from liquid furnaces, concentrating many of the contaminants found in the steam cracker feedstock. Silicon contamination is one that has emerged as a significant concern due to its detrimental impact on the performance of Pyrolysis Gasoline (PyGas) hydrogenation catalysts. This article investigates the mechanisms of Silicon contamination, and explores its effects on the catalytic activity of PyGas hydrogenation catalyst, first and second stage. The sources of Silicon contamination, including circular feedstock's, are reviewed to understand the pathways through which Silicon impurities enter the Steam Cracker process. The purpose of the PyGas first stage hydrogenation catalyst is to eliminate the highly reactive species such as diolefins, styrenics and indenics. The purpose of the PyGas second stage hydrogenation catalyst is to saturate the olefins and remove the sulfur present in the PyGas. With this knowledge, this paper will explore strategies for mitigating Silicon contamination, including feedstock purification, catalyst modification, and process optimization. The insights provided in this paper will aid in developing effective solutions to address Silicon-related contamination, leading to improved performance and operational reliability.

80f Tar in Quench Systems: Causes, Consequences, and Care Solutions

Bryan Crom, Eric Le, Nimeshkumar Patel, Joop Dees, David Hood, Dwight Hines

During the thermal cracking of hydrocarbon feedstocks (i.e. naphtha, LPG, ethane, shale gas, etc.) to produce ethylene and the subsequent quenching of the cracked feedstock, several secondary substances may also be generated, including a relatively heavy material. This heavy material may be referred to by various names, such as tar or py-tar, and can manifest itself as solids or viscous liquids. This heavy material is condensable in the quench section of the ethylene plant, and can lead to considerable fouling throughout the quench system and dilution steam system. This heavy material may also migrate to the hydrocarbon-water interface in the quench water separator and cause emulsification issues. Fouled quench system components and phase separation issues in the quench water separator can lead to reduced production throughput, operational challenges, reduced run-lengths, and expensive cleaning operations.

When the cracking feedstock is light, such as shale gas, ethane, or LPG, less of this heavy material is generated than when the cracking feedstock is heavy (i.e naphtha), but somewhat counterintuitively, the problems caused by this heavy material can be worse when cracking light feedstock because there is also a reduction in other condensable hydrocarbons that can help solubilize and naturally disperse the heavy material. Management strategies are needed to help mitigate the consequences resulting from this heavy material.

Recently, a novel multifunctional dispersant chemistry has been identified which can assist with the management of this heavy material. By dispersing the heavy material into either the bulk hydrocarbon phase or the bulk caustic phase, the heavy material is not allowed to deposit or foul the quench system. Also, by keeping the heavy material in the bulk phase, it cannot migrate to the oil-hydrocarbon interface of the quench water separator, reducing the heavy material's contribution to emulsion issues.

This paper will review the likely formation mechanisms for this heavy material, the potential consequences resulting from this heavy material, and possible management strategies for reducing or alleviating these consequences, including the use of the aforementioned novel dispersant chemistry.

AIChE Paper 80e

EFFECT OF FEEDSTOCK COMPOSITION AND OPERATIONAL TEMPERATURE ON COKE FORMATION DURING STEAM CRACKING REACTIONS

Hamed Mohamadzadeh Shirazi
ir.
Ghent University

Lucas Dos Santos Vargette
ir.
Ghent University

Georgios Bellos
Dr. ir.
DOW Company

Marie Francoise Reyniers
Prof. Dr. ir.
Ghent University

Kevin M. Van Geem
Prof. Dr. ir.
Ghent University

Statements:
Prepared for Presentation at the 2024 Spring Meeting and 20th Global Congress on Process Safety
New Orleans, Louisiana, Date (Meeting March 24 – 28, 2024)

AIChE Paper 75e

EFFECT OF FEEDSTOCK COMPOSITION AND OPERATIONAL TEMPERATURE ON COKE FORMATION DURING STEAM CRACKING REACTIONS

Hamed Mohamadzadeh Shirazi
ir.
Ghent University

Lucas Dos Santos Vargette
ir.
Ghent University

Georgios Bellos
Dr. ir.
DOW Company

Marie Francoise Reyniers
Prof. Dr. ir.
Ghent University

Kevin M. Van Geem
Prof. Dr. ir.
Ghent University

ABSTRACT

The production of olefins through steam cracking reactors faces a significant challenge due to coke formation, which has a detrimental effect on their performance and productivity. To tackle this issue, it is crucial to enhance our understanding of fouling phenomena and their relationship with feedstock composition and process conditions. While it is widely believed that aromatics contribute to increased coke formation, our research challenges this assumption by conducting steam cracking reactions using naphtha feedstock spiked with various aromatic compounds. It was found that the addition of single-ring aromatics up to 9% resulted in a 12% reduction in the total coke formation rate in the radiant and transfer line heat exchanger (TLE) sections. Conversely, the spiking of 3 wt.% aromatic-olefin and 2 wt.% double-ring aromatic exhibited a minor increase in coking rate, approximately 4 wt.%. However, the introduction of

1.2 wt.% naphtheno-diaromatic showed a notable 35% increase in the coke formation rate. On the other hand, temperature plays a pivotal role in coke formation as higher temperatures promote feedstock conversion and secondary reactions that contribute to coke deposition. In our study, we utilized the "Fouling Assessment Setup (FAST)" experimental unit, which allows for the assessment of fouling in the radiant and TLE sections within a single experimental run. By employing the refinery gas analyzer (RGA) and comprehensive gas chromatography coupled with a flame ionization detector (GC×GC-FID), we also gathered real-time information about the composition of the reactor effluent. In summary, our study underscores the significance of comprehending the underlying mechanisms of coke formation in steam cracking reactors.

INTRODUCTION

Steam cracking plays a crucial role in the industrial sector by facilitating the production of light olefins. Nevertheless, this process yields byproducts that ultimately lead to the accumulation of coke on the interior surfaces of cracker furnaces [1-4]. Although they are usually present in small amounts, typically less than 0.1% by weight, the buildup of these carbonaceous residues in the pyrolysis reactor and transfer line exchanger (TLE) often leads to the need for industrial unit shutdowns and the removal of carbon deposits from the reactors. This process negatively affects the economic efficiency of the process [5, 6].

Extensive research has been done to gain a comprehensive understanding of the determinants affecting coke formation. These investigations have used both theoretical [7, 8] and empirical [6, 9-11] methodologies. Among the various contributing factors, the composition of the feedstock and the operating temperature emerged as pivotal influencers on the coke formation process.

Elevating the temperature exhibits a pronounced impact on the formation of coke within the inner walls of the reactor. This influence can be attributed to its role in promoting the fouling of coke by-products through secondary reactions, thereby diminishing the operational run-length of the furnace [12]. On the other hand, it is widely held that certain hydrocarbons, such as aromatics, present in steam cracker feedstocks, serve as the coke precursors within steam cracking reactors [8, 13-16].

To challenge this hypothesis, the present research investigates how the presence of aromatic compounds and increasing temperature affect the formation of coke. To accomplish this, a series of experiments involving coke formation measurements in steam cracking reactions were carried out using a bench-scale setup that mimics the radiant and TLE sections of a real industrial cracker. These experiments utilized reference naphtha (RN) as a base, which was then spiked with specific aromatics:

- single-ring aromatics (3 wt.%), referred to as AR1
- single-ring aromatics (9 wt.%), referred to as AR2
- an aromatic-olefin (3 wt.%), referred to as AR3
- a double-ring aromatic (2%), referred to as AR4
- a naphtheno-diaromatic (1.2%), referred to as AR5

Moreover, two separate experiments were conducted using two different feedstocks, namely AR1 and AR5, to assess the relative influences of temperature and feedstock composition on coke formation. These feedstocks underwent a temperature elevation of 20°C in the radiant section.

For the purpose of this investigation, this study made use of the Fouling Assessment SeTup (FAST), which facilitates the assessment of fouling independently in the radiant and transfer line exchanger (TLE) sections in a single experimental run through two distinct methods. One method is conducted offline, involving the combustion of coke in a specific section and the detection of resulting CO/CO_2 emissions using infrared (IR) equipment. The other method is carried out online, where a coupon is suspended inside the reactor, and the gradual increase in weight due to coke deposits is monitored using a magnetic suspension balance (MSB). The latter approach enables the calculation of catalytic and asymptotic coking rates [17]. By employing a refinery gas analyzer (RGA) and a comprehensive gas chromatograph coupled with a flame ionization detector (GC×GC-FID), real-time information on the composition of the reactor effluent are also gathered.

FOULING ASSESSMENT SETUP (FAST)

The Fouling Assessment Setup (FAST) is a specially designed experimental apparatus used to evaluate the propensity for fouling in a wide range of feedstocks within the radiant and TLE sections of steam crackers. This setup is structured into four distinct parts: feed section, convection section, reaction section, and analysis section. The reaction section corresponds to the radiant section and the transfer line exchanger (TLE). Each of these sections is clearly labeled on the simplified schematic Figure 1. The operational parameters for every part of the apparatus are established and closely monitored through the use of a Programmable Logical Controller (PLC).

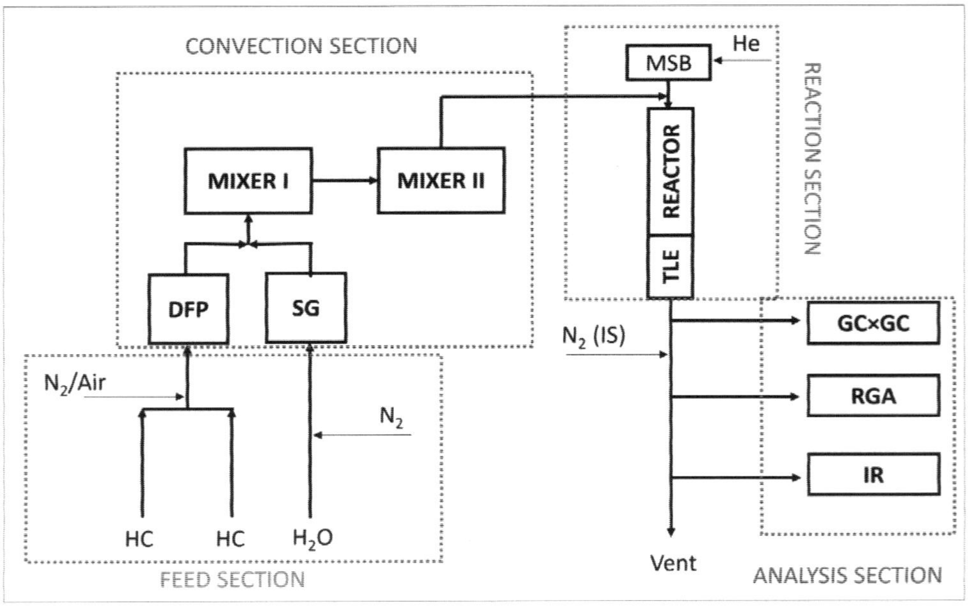

Figure 1. Simplified schematic of the FAST for assessing the coking in steam cracking reactions.

Feeding section

Hydrocarbons (HC), water, nitrogen, and air are supplied to the system. The mass flowrate is managed through Coriolis mass flow controllers in conjunction with control valves for gases and precision rotary pumps for liquids. Nitrogen is used to pressurize the water and hydrocarbon vessels to maintain a consistent suction pressure for the pumps. Nitrogen is introduced separately into the TLE outlet and serves as the primary internal standard for determining product yields in the effluent [18, 19]. Similarly, helium is delivered through a separate line to the Magnetic Suspension Balance (MSB) at the inlet of the radiant section to protect the interior of the MSB from the outflow of the convection section [17].

Convection section

The convection section comprises four distinct zones: Steam Generators (SG), Dry Feed Preheater (DFP), Dilute Feed Preheater I (DFPH I), and Dilute Feed Preheater II (DFPH II). In the DFP, hydrocarbons are introduced without any dilution, while water is supplied to SG. Downstream of the DFP, steam diluent is combined with the hydrocarbon stream, and then the mixture flows through DFPH I and DFPH II. All these heaters are constructed using vertically oriented electrically heated Incoloy 600 tubes, and they incorporate K-type thermocouples on the outer wall. The DFP and SG are filled with quartz beads to facilitate efficient evaporation, while DFPH I and DFPH II are equipped with inner stationary mixers to ensure a consistent blend of feedstock and steam diluent. To prevent cold areas in the convection section's pipelines, all components of the convection section are enclosed within four separate electrically heated insulated enclosures, each under independent control.

Reaction section

The vaporized mixture, consisting of hydrocarbon and steam exits the convection section. This mixture is introduced into the Incoloy 800HT reactor from the top side. The reactor has a length of 0.6 m and an internal diameter of 10 mm. The reactor is divided into four zones, each 0.15 m in length (as shown in Figure 2). The first three zones are considered the radiant sections where steam cracking takes place. The temperature of their outer walls is regulated using electrically heated furnaces. In contrast, the last section functions as the transfer line heat exchanger (TLE), and the temperature in this part is controlled by a combination of an electrically heated furnace and the introduction of air through a coil that is spirally wrapped around it. Thus, the temperature adjustment in this section of the reactor is achieved by simultaneously adjusting the heat input from the furnace and regulating the airflow rate for cooling.

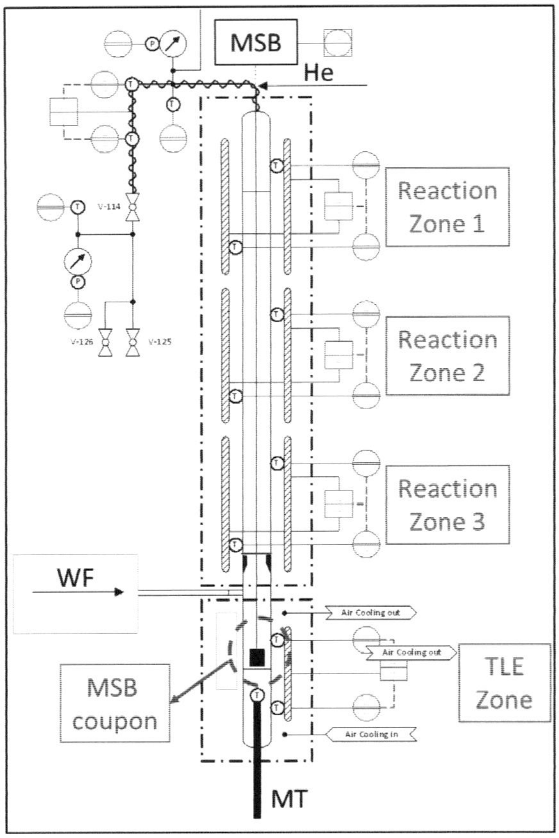

Figure 2. Schematic of the radiant section of the Fouling Assessment SeTup (FAST) [20].

As mentioned, the assessment of coke mass deposition involves the utilization of two distinct methods: online and offline. In the online approach, a coupon made of 35/45 Cr-Ni alloy (10×4×1 mm) is suspended within Reaction Zone 3 of the reactor, specifically located 0.38 meters from the reactor inlet. This is achieved by suspending the coupon using a Platinum wire with a 0.5 mm diameter, which is connected to the MSB. Therefore, the MSB continually tracks the incremental increase in coke mass, which results from fouling on the flat coupon, over time. This monitoring is carried out with a precision of ±5 μg at a frequency of 1 Hz. This enables calculating both the initial (catalytic) and asymptotic coking rates. To ensure the accuracy of these measurements, maintaining precise control over the reactor's pressure is essential due to its significant influence on the performance of the MSB. To address this concern, pressure control is guaranteed in all experimental runs by utilizing a Back Pressure Regulator (BPR) situated downstream of the reaction section.

The total amount of coke on the reactor wall can be assessed offline through a post-cracking decoking process. In this method, the transfer line exchanger (TLE) is initially heated, and air is introduced to support the combustion of deposits. The calculation of the total deposits follows the methodology outlined in the Methods of Fouling Assessment section. Subsequently, the same procedure is duplicated for the radiant section. Although fouling on the suspended coupon is also burnt during this process, the quantity of carbon deposited on the reactor wall far surpasses the negligible amount on the suspended coupon.

Product composition analysis

The analysis section of the experimental setup facilitates the determination of the composition of reactor effluents spanning a broad range of boiling points. Specifically, it allows for the quantification of N_2, H_2, CO, CO_2, and a spectrum of hydrocarbons ranging from methane to polycyclic aromatic hydrocarbons (PAHs). The lower boiling point hydrocarbon products (C_1 to C_4), as well as N_2, H_2, CO, and CO_2, are separated and detected using a Refinery Gas Analyzer (RGA) in conjunction with two Thermal Conductivity Detectors (TCDs) and Flame Ionization Detector (FID). Additionally, all hydrocarbons ranging from methane to PAHs are subjected to online analysis via comprehensive two-dimensional gas chromatography (GC×GC) before being detected with FID. The identification of predominant reaction products in the reactor effluent relies on their chromatographic retention times as established in a database through the injection of standard mixtures or mass spectrometry identification. Conversely, minor products are identified using Kovats retention indices and the roof-tiles principle [18, 21]. The quantification procedure is based on the internal standard methodology [18, 19, 21].

Online and offline coke measurement

In the offline method, to determine the total amount of coke deposited in a specific section over the entire experimental duration, a dedicated carbon oxide Infrared (IR) measurement is employed. This measurement quantifies the volumetric concentrations of carbon monoxide (CO) and carbon dioxide (CO_2), which are the combustion product of decoking and can be correlated to the amount of coke. Light intensity is monitored and compared against a reference sample containing inert nitrogen. As carbon oxides exhibit distinct infrared absorption characteristics, calibration of the measuring cells with a calibration bottle is performed, allowing for the quantification of carbon oxides based on the Lambert-Beer law. Ultimately, the cumulative coke mass is computed over the entire decoking period using the following formula:

$$m_c = MM_c \sum_{i=0}^{N} \frac{\dot{Q}_t \times (y_{CO,i} + y_{CO_2,i}) \times p_i}{R \times T_i} \qquad \text{Eq. (1)}$$

Where MM_c represents the molecular mass of pure coke, $\dot{Q}_t$ stands for the volumetric flowrate, R denotes the gas constant, y represents the volumetric concentration, p signifies the pressure, and T indicates the temperature.

During an experimental run, the quantity of deposits is continuously monitored through online weighing, facilitated by the MSB along with a coupon placed at the designated position. The initial balance signal exhibits high-frequency noise, necessitating the application of filtering for noise reduction. Subsequently, the filtered data is subjected to regression analysis, where parameters A, B, C, and D are optimized to minimize the total sum of squares in Eq. (2).

$$m_t = At + B\left(1 - \frac{1}{2}(e^{-Ct} + e^{-Dt})\right) \qquad \text{Eq. (2)}$$

Equation Eq. (2) is called the fitted mass curve, with m_t the mass of coke on the surface at time t. The corresponding variation in mass over time can be calculated by taking the derivative of Eq. (2).

$$r_{f,c} = \frac{dm_t}{dt} = A + \frac{B}{2}(Ce^{-Ct} + De^{-Dt}) \qquad \text{Eq. (3)}$$

Ultimately, the fouling rate is calculated by dividing Eq. (3) by the coupon's surface area, denoted as S_C, resulting in the final form of the coking rate (Eq. (4)).

$$R_f = \frac{r_{f,c}}{S_c} \qquad \text{Eq. (4)}$$

FEEDSTOCK PREPARATION

The reference naphtha sample was intentionally spiked with varying weight percentages of aromatic compounds. Specifically, the AR1 and AR2 samples were created by incorporating benzene (Sigma-Aldrich, ≥99.7%) and toluene (Sigma-Aldrich, 99.8%) into the reference naphtha, resulting in final aromatic concentrations of 3 wt.% and 9 wt.%, respectively. AR3 sample was prepared by spiking 3 wt.% of styrene into reference naphtha. Furthermore, the AR4 and AR5 samples were formulated by introducing 2 wt.% of naphthalene (Sigma-Aldrich, 99%) and 1.2 wt.% of acenaphthylene (TCI America™, >94%). The chosen concentrations of benzene, toluene, styrene, and naphthalene are attributed to their maximum concentrations found in the feedstock or products of steam cracking of naphtha. Additionally, the selected concentration of acenaphthylene is based on its maximum solubility in naphtha.

OPERATING PROCEDURES AND CONDITIONS

A summary of the operating conditions employed at two different radiant temperatures is provided in Table 1.

Before conducting each cracking experiment, a preoxidation process lasting for 18 hours was carried out to establish an oxide layer on the reactor and coupon surfaces in order to enhance the alloy's resistance to coke formation. Additionally, a presulfidation step was implemented before the cracking, lasting for 30 minutes. It involved using a solution with 500 ppmwt. of dimethyl disulfide (DMDS) at a mass flow rate of 35 g/h. Throughout this step, the temperature in Reaction Zone 3 was maintained at 827°C. The presulfidation process is believed to establish a protective sulfur layer on the reactor surface, reducing direct contact between the hydrocarbon gas phase and the metal surface, especially during the initial stages of the reaction where catalytic mechanisms are prevalent [22-24].

Table 1. Summary of the operating conditions in FAST.

		Feed section [g/h]		
HC	Presulfidation DMDS solution	Continuous sulfidation DMDS solution	N_2-IS	He
140	35	49	20	5
		Convection section		
DFP [°C]	SG [°C]	DFPH-I [°C]	DFPH-II [°C]	HTTL [°C]
150	200	260	400	450
		Radiant section [°C]		
Zone I	Zone II	Zone III	TLE	
600/620	660/680	950/970	400	
		Analysis section [bar a]		
BPR	RGA	IR		
1.8	1.5	1.2		

During the cracking phase, hydrocarbons were supplied to the Dry Feed Preheater (DFP) at a flow rate of 140 g/h, while DMDS solution contained 60 ppmwt. was fed into the Steam Generator (SG) at a flow rate of 49 g/h, resulting in a dilution of 0.35 g of water per gram of hydrocarbon. The steam generated in the SG was mixed with the hydrocarbon downstream of the DFP and upstream of DFPH I. To achieve a temperature close to the steam cracking temperature, the mixture passed through DFPH I and DFPH II in series. The feed was then transferred to the topside of the reactor through a high-temperature transfer line (HTTL), resembling the crossover temperature (XOT) observed in real steam crackers.

The temperature profile for the reaction was controlled by adjusting the temperature of three zones. Within this study, two temperature profiles were designated for two different types of experiments. In the case of high-temperature experiments (HT experiments), a temperature elevation of 20°C was applied across all three radiant zones. Throughout the experiments, the first, second, and third reaction zones were maintained at temperatures of 600°C (620°C for HT), 660°C (680°C for HT), and 950°C (970°C for HT), respectively. In all experiments, the temperature of the TLE was set to 400°C. Each cracking run lasted for 6 hours at a coil outlet pressure (COP) of 1.8 bar, which was controlled by a back-pressure regulator (BPR). Subsequently, the tubular reactor was cooled down and purged to remove any remaining hydrocarbon.

EFFECT OF FEEDSTOCK COMPOSITION ON FOULING

Figure 3 displays the asymptotic coking rates, which have been derived from the coking rate curves of various feedstocks. The interval of 5 to 6 hours was pinpointed as the asymptotic fouling regimes. In contrast to catalytic coking rates, which are influenced by surface conditions such as aging and roughness—factors not addressed in this study—asymptotic coking, deemed the most crucial factor for evaluating fouling tendencies in industrial steam crackers, primarily depends on the composition of the feedstock [25]. Figure 3 uncovers that the introduction of single-ring aromatics, namely benzene and toluene, at levels up to 9 wt.%, resulted in a 6%

decrease in the asymptotic coking rate. Adding styrene to the reference naphtha (RN) led to a negligible 1.6% increase in the asymptotic coking rate. Significantly, an increase in the presence of naphthalene and acenaphthylene notably raised the asymptotic coking rates by 9% and 23%, respectively. This occurrence can be attributed to the heightened reactivity of aromatics, which increases with the number of aromatic rings, thereby contributing to fouling within the cracking environment.

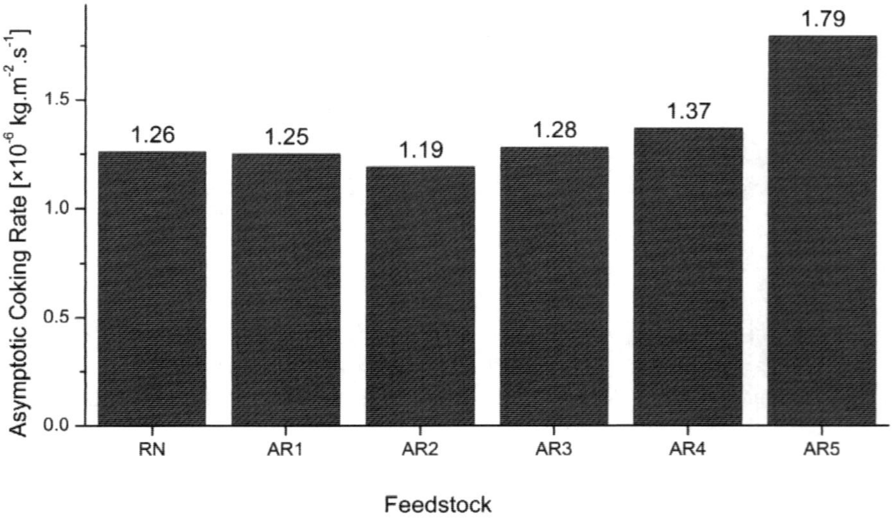

Figure 3. Asymptotic coking rates on the surface of a suspended coupons.

The offline assessment of coke buildup in both the radiant and TLE sections was conducted using a post-cracking decoking procedure. To calculate the average fouling rate in each section, the total quantity of coke deposits was divided by the duration of the six-hour cracking experiment and the area of the relevant section (4869.47 mm^2 for TLE and the third zone of the radiant section), as illustrated in Figure 4. It is assumed that coking only occurs in the third zone of the radiant section, as the temperatures in the other two zones are lower than the threshold of the minimum steam cracking temperature. Based on the findings from offline coking measurements, it can be determined that the introduction of benzene and toluene as spiking components did not result in an elevated fouling level within the TLE section. When compared to the RN sample, the AR5 samples exhibited a 14% higher rate of coke deposition in TLE. This observed increase in fouling can be attributed to the greater production of PAH, signifying that the condensation of heavier hydrocarbons stands as the principal mechanism driving fouling in the TLE section [26, 27].

The inclusion of single-ring aromatics of benzene and toluene in the radiant section resulted in a reduction in coke formation, thereby corroborating the observations made through the online coke measurements. Like the TLE coking, there is a noticeable rise in the rate of coke formation when shifting from single-ring aromatics to double-ring aromatic and naphthene-diaromatic compounds within the radiant section. This pattern is especially significant in the instance of the AR5 sample, which exhibits a coking rate that is 43% higher than that of the RN sample.

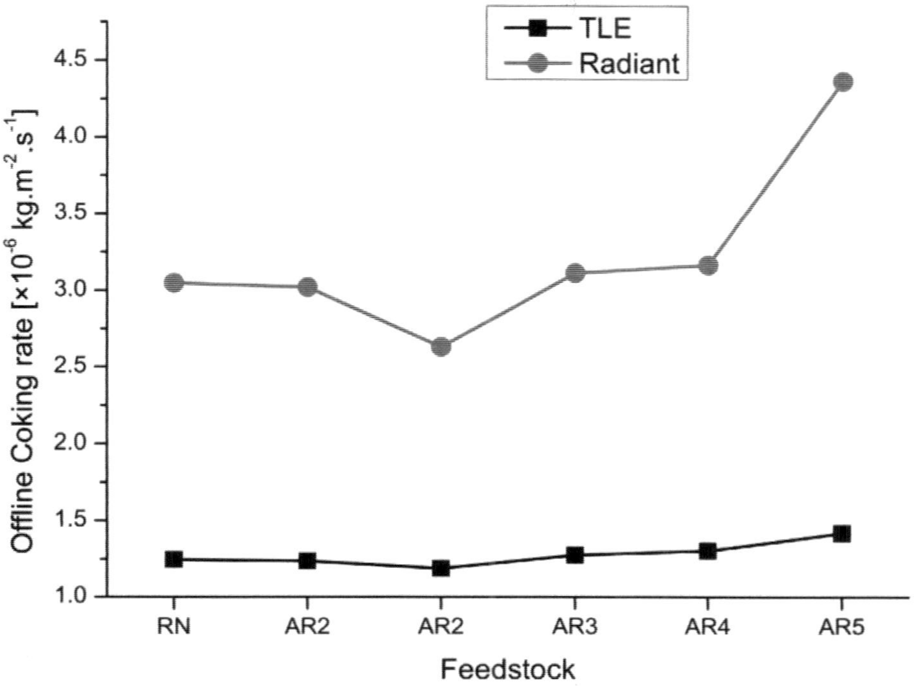

Figure 4. Offline Coking rate in the radiant and TLE sections.

EFFECT OF TEMPERATURE ON COKE FORMATION

When comparing the coking rates derived from offline measurements of coke (Figure 5), it becomes apparent that the increase in temperature had a significant impact on the formation of coke in the TLE section. The deposition of coke showed a substantial increase of 71% for AR1 and 97% for AR5 feeds, respectively, owing to the 20°C temperature elevation. This occurrence is intricately linked to the higher PAH formation in the cracking environment which further condensate and deposit on the TLE inner wall [27-29].

As illustrated in Figure 5, a similar trend is consistently observed in the radiant section. Both the AR1 and AR5 samples exhibited a rise in coke formation of 24% and 22%, respectively, when exposed to higher temperatures. This phenomenon can be ascribed to the increased activity of coke precursors and radical formation at elevated temperatures, thereby facilitating a greater production of coke.

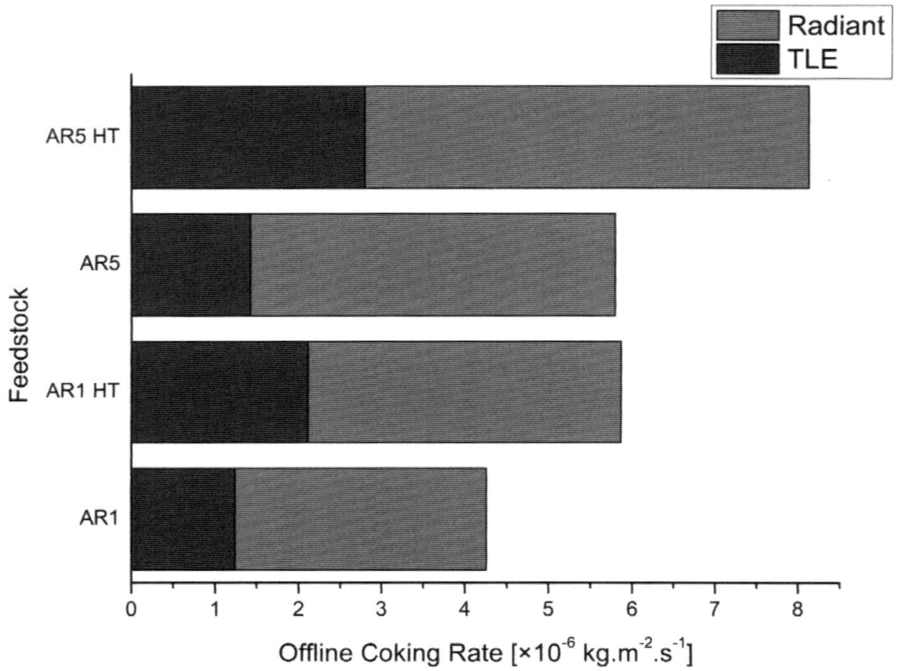

Figure 5. Impact of temperature on coke formation.

CONCLUSION

The assessment of fouling in various sections of a steam cracker, namely the radiant and TLE sections, was conducted using both online and offline methods in the Fouling Assessment SeTup (FAST). The study focused on the impact of four types of aromatics: single-ring aromatic, aromatic-olefin, double-ring aromatic, and naphthene-diaromatic. The results indicated that the introduction of single-ring aromatics had little to no significant effect on coke formation in both the TLE and radiant sections. However, as the number of aromatic rings in the hydrocarbon feedstock increased, fouling became more pronounced, especially in the TLE section. This heightened fouling was primarily attributed to the condensation mechanism of polyaromatic hydrocarbons (PAH) produced in the process. Furthermore, the study revealed that temperature played a pivotal role in the rate of coking. Even a modest temperature increase of 20°C in the radiant section led to a 71% and 97% increase in coking rate for AR1 and AR5, respectively, in the TLE section. In the radiant section, a 24% increase in coking rate was observed for AR1 and a 22% increase for AR5.

NOMENCLATURE

A, B, C, D	Parameters in coking curve fitting
AR1	Naphtha feed containing 3 wt.% single-ring aromatics (benzene and toluene)
AR2	Naphtha feed containing 9 wt.% single-ring aromatics (benzene and toluene)

AR3	Naphtha feed containing 3 wt.% aromatic-olefin (styrene)
AR4	Naphtha feed containing 2 wt.% double-ring aromatic
AR5	Naphtha feed containing 1.2 wt.% naphthene-diaromatic
BPR	Back Pressure Regulator
DFP	Dry Feed Preheater
DFPH	Dilute Feed Preheater
DMDS	Dimethyl disulfide
FAST	Fouling Assessment SeTup
FID	Flame Ionization Detector
GC×GC	Comprehensive Two-Dimensional Gas Chromatography
HTTL	High Temperature Transfer Line
IR	Infrared gas analyzer
m_c	Mass of deposited coke, kg
m_t	Mass of deposited coke at time t, kg
MM	Molecular mass
MSB	Magnetic Suspension Balance
PAH	Polycyclic Aromatic Hydrocarbons
PLC	Programmable logical controller
$\dot{Q}_t$	Volumetric flow rate, m^3/s
R	Gas constant, J/mol/K
r_f	Rate of mass deposition, Kg/s
R_f	Rate of mass deposition per surface area, $Kg.s^{-1}.m^{-2}$
RGA	Refinery Gas Analyzer
RN	Reference Naphtha
SG	Steam generator
S_c	Surface area of coupon, m^2
T	Temperature, K
TCD	Thermal Conductivity Detector
TLE	Transfer Line Exchanger
$y_{x,i}$	Volumetric percentage concentration of component x at time i
XOT	Crossover Temperature

Subscript

c	coke
i	time
x	component

REFERENCES

1. Zimmermann, H. and R. Walzl, *Ethylene*. Ullmann's Encyclopedia of Industrial Chemistry, 2009.
2. Pyl, S.P., et al., *Biomass to olefins: Cracking of renewable naphtha*. Chemical engineering journal, 2011. **176**: p. 178-187.
3. Geerts, M., et al., *Crude to olefins: effect of feedstock composition on coke formation in a bench-scale steam cracking furnace*. Industrial & Engineering Chemistry Research, 2020. **59**(7): p. 2849-2859.
4. Lederer, J., et al., *A study of industrial pyrolysis of the high-bouling products from hydrocracking of petroleum vacuum distillate*. Fuel, 1994. **73**(2): p. 295-299.

5. Mohamadzadeh Shirazi, H., et al., *Carburization of High-Temperature Alloys during Steam Cracking: The Impact of Alloy Composition and Temperature.* Industrial & Engineering Chemistry Research, 2023.
6. Kopinke, F., G. Zimmermann, and S. Nowak, *On the mechanism of coke formation in steam cracking—conclusions from results obtained by tracer experiments.* Carbon, 1988. **26**(2): p. 117-124.
7. Wauters, S. and G. Marin, *Computer generation of a network of elementary steps for coke formation during the thermal cracking of hydrocarbons.* Chemical Engineering Journal, 2001. **82**(1-3): p. 267-279.
8. Wauters, S. and G. Marin, *Kinetic modeling of coke formation during steam cracking.* Industrial & engineering chemistry research, 2002. **41**(10): p. 2379-2391.
9. Muñoz Gandarillas, A.s.E., K.M. Van Geem, M.-F.o. Reyniers, and G.B. Marin, *Influence of the reactor material composition on coke formation during ethane steam cracking.* Industrial & Engineering Chemistry Research, 2014. **53**(15): p. 6358-6371.
10. Sarris, S.A., et al., *Optimization of the in situ pretreatment of high temperature Ni–Cr alloys for ethane steam cracking.* Industrial & Engineering Chemistry Research, 2017. **56**(6): p. 1424-1438.
11. Froment, G., *Coke formation in the thermal cracking of hydrocarbons.* Rev. Chem. Eng., 1990. **6**: p. 293-328.
12. Ranzi, E., et al., *Coking simulation aids on-stream time.* Oil & gas journal, 1985. **83**(35): p. 49-52.
13. Cai, H., A. Krzywicki, and M.C. Oballa, *Coke formation in steam crackers for ethylene production.* Chemical Engineering and Processing: Process Intensification, 2002. **41**(3): p. 199-214.
14. Geem, K.M.V., et al., *Coke formation in the transfer line exchanger during steam cracking of hydrocarbons.* Industrial & engineering chemistry research, 2009. **48**(23): p. 10343-10358.
15. Zhang, J., et al., *Combined Catalytic and Pyrolytic Coking Model for Steam Cracking of Hydrocarbons.* Industrial & Engineering Chemistry Research, 2022. **61**(11): p. 3917-3927.
16. Towfighi, J., M. Sadrameli, and A. Niaei, *Coke formation mechanisms and coke inhibiting methods in pyrolysis furnaces.* Journal of chemical engineering of Japan, 2002. **35**(10): p. 923-937.
17. Muñoz Gandarillas, A.E., K.M. Van Geem, M.-F. Reyniers, and G.B. Marin, *Influence of the Reactor Material Composition on Coke Formation during Ethane Steam Cracking.* Industrial & Engineering Chemistry Research, 2014. **53**(15): p. 6358-6371.
18. Van Geem, K.M., et al., *On-line analysis of complex hydrocarbon mixtures using comprehensive two-dimensional gas chromatography.* Journal of Chromatography A, 2010. **1217**(43): p. 6623-6633.
19. Au - Ristic, N.D., M.R. Au - Djokic, K.M. Au - Van Geem, and G.B. Au - Marin, *On-line Analysis of Nitrogen Containing Compounds in Complex Hydrocarbon Matrixes.* JoVE, 2016(114): p. e54236.
20. dos S. Vargette, L., et al., *Comprehensive Fouling Assessment in Steam Crackers: Fossil versus Renewable NEXBTL Naphtha.* Industrial & Engineering Chemistry Research, 2023.
21. Schoenmakers, P.J., et al., *Comparison of comprehensive two-dimensional gas chromatography and gas chromatography – mass spectrometry for the characterization of complex hydrocarbon mixtures.* Journal of Chromatography A, 2000. **892**(1): p. 29-46.
22. Singh, A., et al., *Role of Presulfidation and H2S Cofeeding on Carbon Formation on SS304 Alloy during the Ethane–Steam Cracking Process at 700° C.* Industrial & Engineering Chemistry Research, 2018. **57**(4): p. 1146-1158.
23. Symoens, S.H., et al., *State-of-the-art of coke formation during steam cracking: Anti-coking surface technologies.* Industrial & Engineering Chemistry Research, 2018. **57**(48): p. 16117-16136.
24. Jakobi, D. and P. Karduck. *Behavior of High-Temperature Tube Materials in Sulfur-Containing Steam-Cracking Conditions.* in *NACE International Corrosion Conference Proceedings*. 2018. NACE International.
25. Reyniers, M.-F.S.G. and G.F. Froment, *Influence of Metal Surface and Sulfur Addition on Coke Deposition in the Thermal Cracking of Hydrocarbons.* Industrial & Engineering Chemistry Research, 1995. **34**(3): p. 773-785.
26. Bach, G., et al., *Transfer-line heat exchanger fouling during pyrolysis of hydrocarbons. 1. Deposits from dry cracked gases.* Industrial & engineering chemistry research, 1995. **34**(4): p. 1132-1139.
27. Kopinke, F.D., G. Bach, and G. Zimmermann, *New results about the mechanism of TLE fouling in steam crackers.* Journal of Analytical and Applied Pyrolysis, 1993. **27**(1): p. 45-55.
28. Kopinke, F.D., G. Zimmermann, G.C. Reyniers, and G.F. Froment, *Relative rates of coke formation from hydrocarbons in steam cracking of naphtha. 3. Aromatic hydrocarbons.* Industrial & Engineering Chemistry Research, 1993. **32**(11): p. 2620-2625.
29. Kopinke, F.D., G. Zimmermann, G.C. Reyniers, and G.F. Froment, *Relative rates of coke formation from hydrocarbons in steam cracking of naphtha. 2. Paraffins, naphthenes, mono-, di-, and cycloolefins, and acetylenes.* Industrial & Engineering Chemistry Research, 1993. **32**(1): p. 56-61.

AIChE Paper Number: 80d

AMMONIA IN ETHYLENE PLANTS: A REVISIT AND A WAR STORY

Dwight Hines
Principal Chemical Process Industry Advisor
Veolia Water Technologies & Solutions

David Hood
Senior Engineer
Veolia Water Technologies & Solutions

Prepared for Presentation at the 2024 Spring National Meeting
New Orleans, LA, March 24 – March 28, 2024

AIChE Paper Number: 80d

AMMONIA IN ETHYLENE PLANTS: A REVISIT AND A WAR STORY

Dwight Hines
Principal Chemical Process Industry Advisor
Veolia Water Technologies & Solutions

David Hood
Senior Engineer
Veolia Water Technologies & Solutions

Abstract:

Ammonia has long been a contaminant of concern in ethylene units. It can enter as an impurity in cracked and uncracked feeds, as well as be generated inside the battery limits via the cracking/decomposition of nitrogen compounds. Ammonia's impact on the unit is multifaceted: it can elevate the pH within the quench and process water system, act as a temporary poison on catalyst, participate in the formation of dangerous solids in the cold box, and ultimately result in off-spec ethylene and propylene. Fortunately, over the years, the industry has developed methods to monitor ammonia concentrations and techniques to eliminate or at least mitigate the associated negative consequences.

This paper will review the ethylene industry's understanding of ammonia, including where it comes from, how it impacts unit operations, how it can be monitored, and what solutions have been successfully deployed.

Introduction

This paper revisits several prior papers on the subject, including the three major papers below. These papers contain information about what happens to ammonia in the ethylene plant, analytics, and mitigation while most other papers regarding ammonia centered on sources of ammonia.

- "Ammonia in the BASF FINA Petrochemicals Limited Partnership Steam Cracking Plant at Port Arthur Texas," Dwight Hines (2006, 218c)[1]
- "Minor Components, Major Issues: An Overview of Contaminants in Ethylene Plants," David Hood (2007, 51c)[2]
- "Feedstock Contaminant War Stories – Ammonia," Dwight Hines (2007, 51d)[3]

A "War Story" is also included concerning a recycle stream from a Butadiene Unit. As mentioned above, other papers have reported sources of ammonia or many nitrogen compounds that make ammonia in furnaces. These include additives such as DEHA (diethyl hydroxylamine), SFR's (Stable Free Radicals), extraction solvents in recycle streams to the furnaces, and a variety of amines. Some of these papers are found in the Reference section of this paper.[4]

Review of Papers

The 2006 paper, "Ammonia in the BASF FINA Petrochemicals Limited Partnership Steam Cracking Plant at Port Arthur Texas," Dwight Hines (2006, 218c)[1] chronicles the ammonia problem that resulted in the serious contamination of ethylene in the product pipeline. The paper shows nitrogen and ammonia (NH_3) data on the feeds, recycle streams from inside and outside the plant, and additives that were added at that time to all sections of the plant. The Root Cause Analysis showed the most likely culprit was Light Cycle Oil (LCO) used for level control in the Gasoline Fractionator, quench oil viscosity control and coke settler clean-out. At times high nitrogen levels were found (from 100 wppm to 1000 wppm) in LCO coming into the ethylene plant from the refinery. Further investigation showed that some of the nitrogen compounds distilled overhead in the Gas Fractionator to the Quench Water Tower and into the gasoline stream. The nitrogen compounds in the gasoline then proceed to a Pygas Unit, where in the second stage hydrotreater, they are converted to ammonia. This ammonia returns to the cold section in hydrogen vent streams and then to the quench water.
In the quench water, ammonia exist as ammonium hydroxide (NH_4OH and other ammonium salts) if the pH is neutral or acidic. However, as the pH increases these ammonium salts release NH_3 into the vapor, especially at elevated temperatures of the Process Water Stripper and Dilution Steam Generators. NH_3 in the vapor of quench water has several paths it can take:

1. NH_3 can strip out in the Process Water Stripper and go back to the quench tower.

2. NH_3 can return to the furnaces in dilution steam.
3. NH_3 can go with process gas to the Charge Gas Compressor and ultimately distill with ethylene.

Options 1 and 2 are both recycles that allow ammonia as NH_3 to simply build up in the system. Option 3 is the only real purge point for NH_3, and products are not good contaminant purge points. The normal specification for ammonia in ethylene is 1 ppm.
Since we do not want NH_3 to go to the pipeline in ethylene product, we would like to take ammonia as NH_3 out of the quench and process water system. To take ammonia out, we must keep ammonia in the quench water as a salt, remove it from the charge gas vapor, or preferably both.

Ammonia (NH_3) Mitigation in the Vapor

In David Hood's 2007 paper[2], he comments on several adsorbents that could be used for NH_3 removal. He mentions molecular sieves or adsorbents such as 5A, 13X or modified molecular sieves as a layer in charge gas dryers. One of the most popular is UOP's 3A-EPG-N. This sieve gives you a roughly ten-fold capacity boost in ammonia adsorption and does not decrease water adsorption. Per UOP, this sieve is currently only available in 1/16 inch and can be added as a layer in that section of the Charge Gas Drier. These are very large driers and can remove large amounts of NH_3.

We would be remiss if we did not look at where this ammonia goes during a regeneration cycle. Most dryers are regenerated with regen gas that then returns to the fuel system. Ammonia in the fuel system would therefore end up in burners of fired equipment such as furnaces and boilers. Some of this NH_3 could be converted to NOx in the burners and returned to the environment. This would be at very low concentrations and most people have not had a problem. However, if your fired equipment contains SCR (Selective Catalytic Reduction) catalysts in the flue gas for NOx removal, NH_3 is actually injected for the catalyst to work ($NH_3 + NOx \rightarrow H_20 + N_2$). Therefore, NH_3 converted to NOx in fired equipment would be removed by the SCR and any NH_3 that is not converted in burners could be beneficial. Though no site that we know of does this, the next obvious step would be to use this 3A-EPG-N in the small drier upstream of the ethylene fractionator as a last level of removal of NH_3. This small drier is installed in most plants with a back-end acetylene converter to remove the small amount of water formed in the reactor. However, at present UOP only makes the 3A-EPG-N in the 1/16-inch size. Most ethylene driers use 1/8-inch 3A mole sieve to minimize pressure drop in this system. In conversations with UOP, they indicated that they may make this sieve in 1/8-inch on request and for a small surcharge. This would be a great "last line of defense."

The other way to limit the amount of NH_3 in the vapor is to force it into the quench water by keeping the pH low or neutral. This can be done by injecting a weak acid such as acetic or citric acid if the pH is high. However, as we will see later, this may only be a temporary solution or other steps must be taken to prevent even worse problems.

Mitigation of Ammonia in Quench Water

As NH_3 is absorbed into the quench water, much of it will form ammonium hydroxide (NH_4OH). Ammonium hydroxide is a fairly strong base and will raise the pH of the water. As the pH goes up, the reaction will reverse evolving NH_3 and leaving behind water (H_2O). Therefore, if a weak acid is used to keep the pH low or neutral, the water has much more capacity for ammonia. This ammonia can then be "blown down" in the dilution steam system to waste water. However, in the wastewater system, most plants have a tower with a variety of names such as Wastewater Stripper, NESHAP Tower, or Benzene Stripper. The tower is there to remove benzene from the wastewater before the BIOX system. In this tower, much of the ammonia will disassociate and go overhead. In most plants, the overhead containing benzene goes to a Wet Flare Drum. This drum, which contains some water (and now ammonia) but mostly benzene-containing hydrocarbon, is periodically pumped back to the Quench Tower to recover the benzene. However, in sending back this stream, ammonia now absorbed in the water phase also goes back to the Quench Tower. This stream can have very high ammonia content, even into the thousands of ppm. This can result in off-spec ethylene for two reasons. First, this material from the Wet Flare Drum can have a very high pH that can affect the pH of the quench water and liberate NH_3 to the vapor. Second, it will increase the ammonia load of the quench water which again can increase the NH_3 in the vapor. A specification can be set from the Wet Flare Drum. If the ammonia is over the spec, it is loaded into tank trucks and sent out for incineration, thus protecting the Quench Water System (Figure 1).

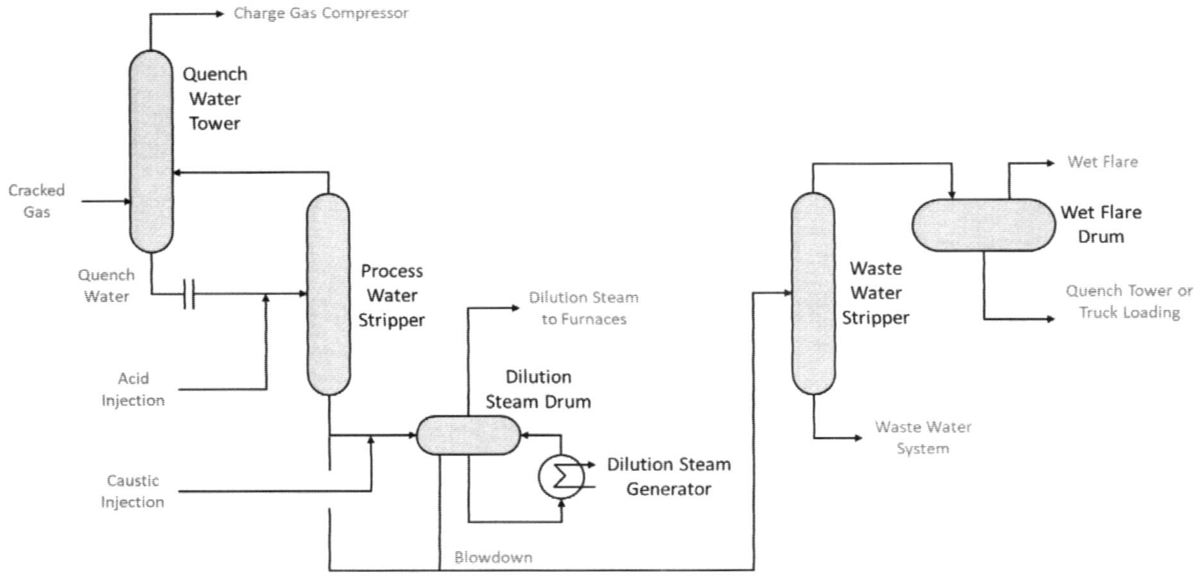

Figure 1: Typical Dilution Steam System Overview

The other stream that can affect ammonia in the quench water is the stream from the caustic system that usually comes back to the Quench Tower. In the caustic system, red oil is formed and is removed in the condensed gasoline through skim lines. This gasoline/red oil/caustic is usually sent to a drum where more gasoline is added to "wash" the red oil out of the caustic. The hydrocarbon phase of this drum is sent to a second drum where it is water washed to remove the small amount of remaining caustic. The water washed hydrocarbon phase is then sent to the Quench Tower. In Figure 2 below, a typical caustic drum set up is shown from a 2009 paper by Dave Payne at Westlake[8]. If there is an upset in this system (and there quite often is) caustic can end up in the hydrocarbon phase to the quench tower. Caustic (sodium hydroxide) is a very strong base and even a small amount of caustic can raise the pH of the Quench Water and thereby release ammonia (as NH_3) to the vapor. This vapor goes with the overhead of the Quench Tower to the Charge Gas Compressor and again ultimately to the ethylene and even propylene products. See Figure 1 for the effect of pumping back a Wet Flare Drum containing ammonia and a caustic upset as stated above.

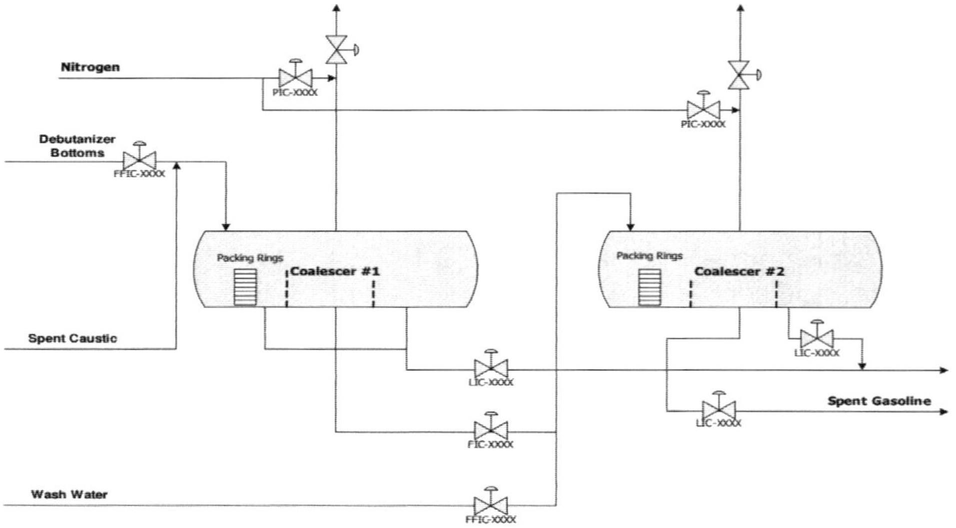

Figure 2: Typical Caustic Drum Overview (Reference #8, Dave Payne, Westlake)

Impact of Ammonia in Ethylene Plants

Ammonia impacts the ethylene plant in many ways; some we have already covered, such as contamination of ethylene and even propylene products. Unfortunately, there are other effects as well. First, as we mentioned above, when NH_3 is absorbed into the water it forms ammonium hydroxide (NH_4OH) and other salts. The ammonium hydroxide is a relatively strong base and raises the pH of the Quench Water. This increase in pH stabilizes emulsions and causes phase separation in the Quench Water Tower and the Quench Settler to be very difficult. Anyone operating Quench Systems can attest to the problems the emulsions cause as they foul the system.

Ammonia can also act as a temporary poison on many catalysts. The ones most often affected are the palladium catalysts in acetylene (C_2) and MAPD (Methyl Acetylene and Propadiene, C_3) converters. This can cause the plant to be off spec on acetylene or MAPD in addition to NH_3.

Another important reaction of ammonia occurs in the Cold Box. If you have heard Dane Grenoble give updates from the NOx Subcommittee[5], you have not only heard about NOx gums but also ammonium nitrate and ammonium nitrite salts in the Cold Box. As Dane pointed out numerous times, these salts that have been found in Cold Boxes are unstable and can lead to detonations.

Measurement and Sampling of Ammonia

There are several ways to measure ammonia and they differ depending on the matrix and where the sample is collected. In ethylene product, companies have used several measurement techniques. At one time, most plants used the Ion Mobility technology. It was very accurate and could measure down to the low ppb level. However, this technology is not available currently to our knowledge and other techniques have replaced it. A variety of Gas Chromatographic methods are available such as GC/PID, GC/DID and GC/MSD[7]. Tape methods such as the model from C.I. Analytics[7] are still used. The newest and probably most used is the TDALS (Tunable Diode Array Laser Spectroscopy) technology. A paper was given on the system by Gary Langford from SpetraSensors Inc. in 2020[6]. Figure 3 shows several methods used in 2005 for ammonia evaluation.

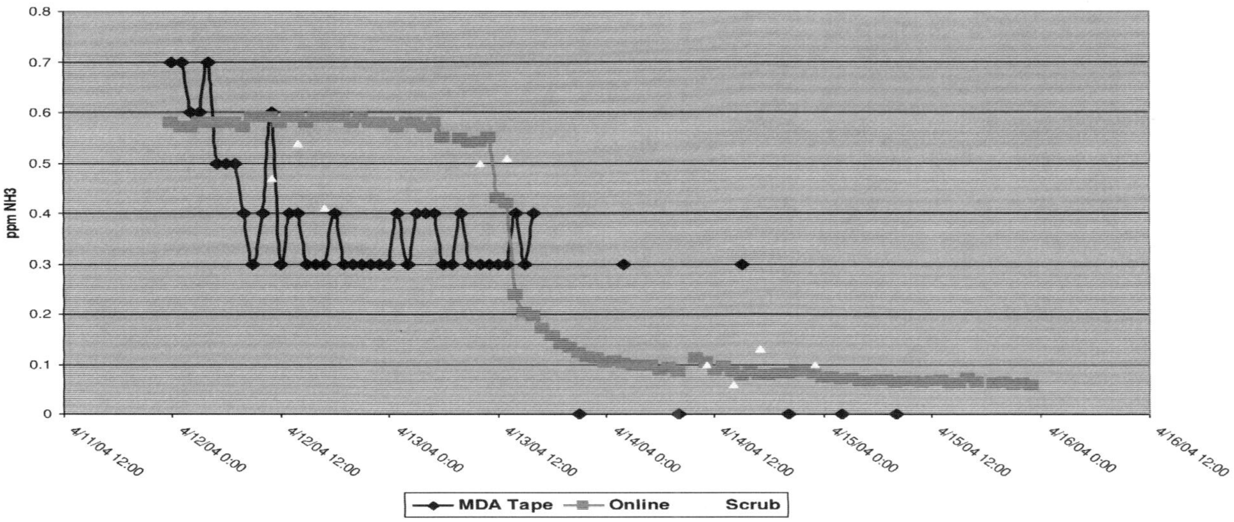

Figure 3: Ammonia Measurements by Multiple Methods

Inside the ethylene plant there are several places that ammonia should be sampled. The first place to measure ammonia is in the vapor. This is necessary because you must know how much NH_3 is in the vapor to know if NH_3 saturation of the Charge Gas Drier will be reached. To decide where to take your sample, several items must be considered. First, you want to take the sample downstream of the Quench System because that is where NH_3 is inventoried. It is the ammonia "sink" that contains the ammonia that must be dealt with from the vapor and the water systems. Unless you have an on-line analyzer, you will probably use Dräger Tubes. Most Dräger Tubes are sensitive to large amounts of H_2S, and the safety of the person sampling dictates a sample downstream of the caustic tower where there H_2S is low. You would like the sample upstream of the drier because even basic 3A sieve removes some NH_3. Therefore, a good spot would be just upstream of the Charge Gas Driers. Then as incidents happen in the plant this information can be used to correlate with NH_3 in the vapor (see the example below in Figure 4). Other than Dräger Tube sampling, samples of vapor can be collected, and a Lab Tape analyzer such as the MDA Analyzer can be used. However, caution must be used to collect samples only in special inert cylinders such as SilicoSteel Cylinders. Typical stainless-steel cylinders will adsorb ammonia and other contaminants on the surface of the cylinder and low results will be found.

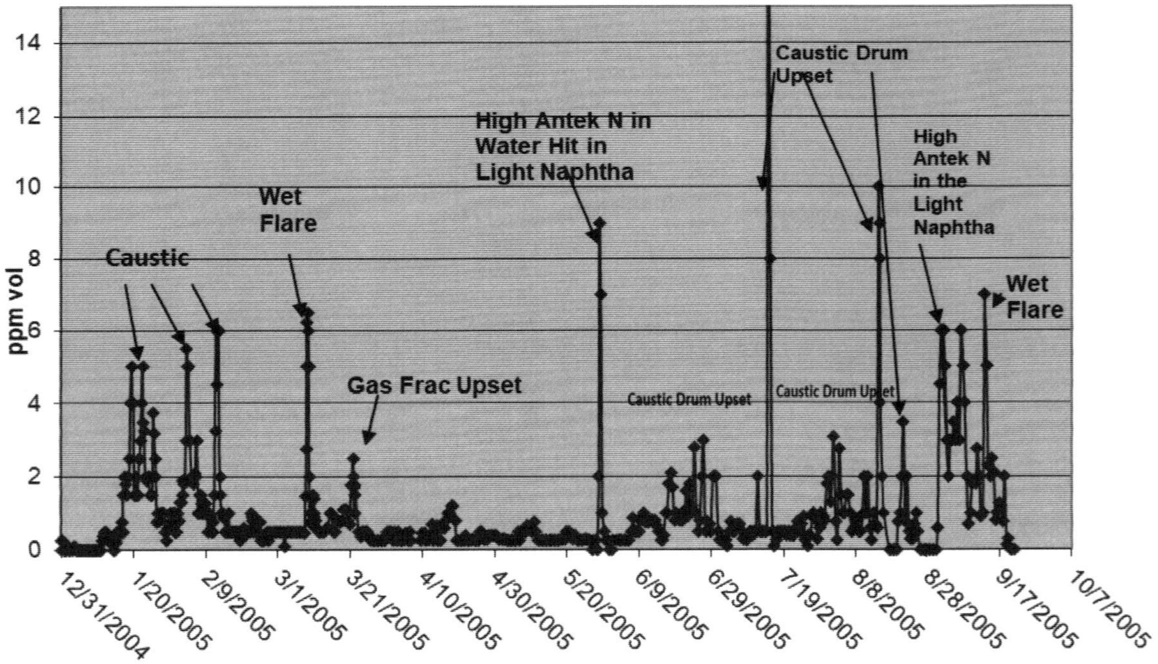

Figure 4: Dräger Results at Drier Inlet

For ammonia in the water samples, a variety of methods that work well are available. The colorimetric Nessler Method and Ion Specific Electrode Method are examples. In the dilution steam circuit, several samples are useful in understanding where ammonia is going. The quench water (1) upstream of the Process Water Stripper gives us a good general understanding of the total ammonia entering the system. The difference between the Process Water Stripper inlet (2) and outlet (3) tells us how much NH_3 is stripped out in that tower, only to return to the Quench Tower. The Dilution Steam Drum Blowdown (4) gives us an idea of how much ammonia is going toward the wastewater system and may come back when the Wet Flare Drum is pumped back to the Quench Tower. The Dilution Steam Condensate (5) will tell us how much ammonia is going to the furnace and will return to the Quench Tower in the cracked gas. Probably the most important sample is the Wet Flare Drum water (6) before it is pumped back to the Quench Tower. This sample will tell us the pH and ammonia in the water that will influence the Quench water when pumped back and we can ultimately decide if sending this stream back to the Quench Tower is prudent (see Figure 5 below for stream locations).

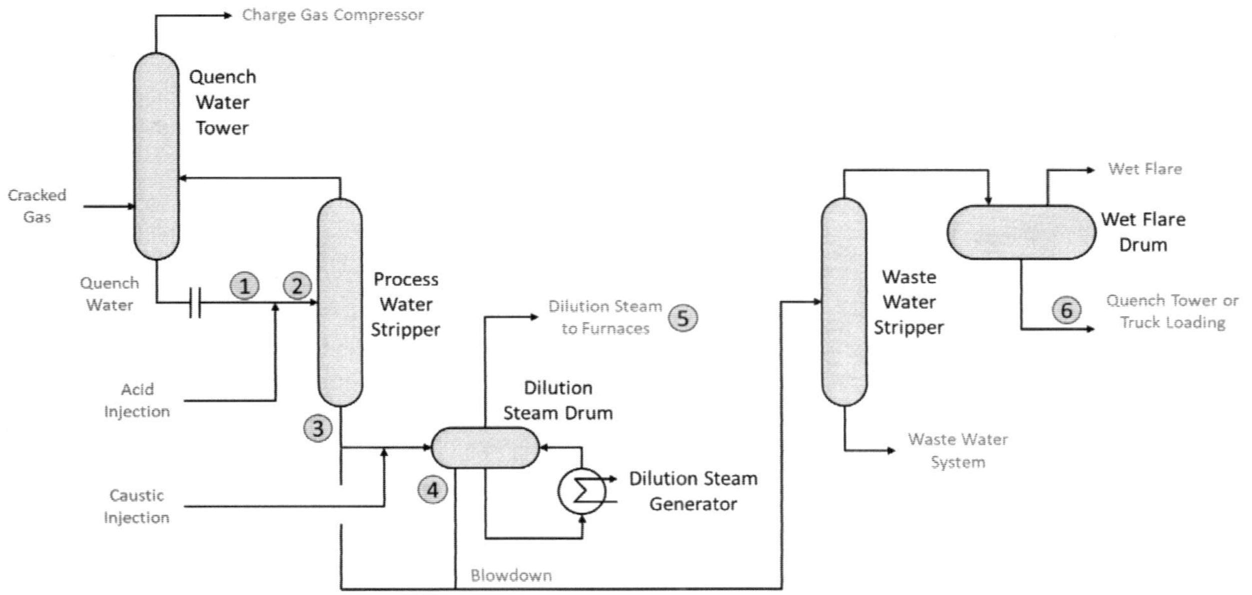

Figure 5: Typical Dilution Steam System Overview with Sample Locations

In hydrocarbon liquid feeds and streams, the analytical method of choice is usually Antek Nitrogen. In these liquid streams we are usually not looking for ammonia but for nitrogen compounds that might form ammonia in furnaces or other high temperature units such as the second stage hydrotreaters. Antek N measures nitrogen compounds other than N_2. It breaks them down in a high temperature furnace and in an O_2-rich atmosphere to NO_2 and this NO_2 is measured by chemiluminescence.

Summary of Ammonia Mitigation

Mitigation techniques for ammonia are interesting because we are not actually taking ammonia out of the plant but ushering it to places where it is destroyed or in high enough concentration that we can dispose of a large amount of ammonia in a small amount of water. Mitigation techniques can be broken down into those in the vapor and those that are in the water.

The ammonia mitigation techniques for the vapor are those we can add to a drier or other vessel to remove ammonia and put it in regen gas. In the regen gas it goes to fuel and is burned in fired equipment such as furnaces and boilers. The best places are those that have SCR catalyst because the NH_3 is converted to N_2 over the catalyst. To greatly enhance the removal of NH_3 in driers a media such are 3A-EPG-N can be added as a layer in charge gas driers and as a final line of defense in ethylene driers.

In quench water, we want to keep enough ammonia in the water so as not to overwhelm the vapor mitigation techniques. If we only have a basic 3A molecular sieve in our driers we must keep most of the ammonia in the water phase because the capacity of the driers will be only a few ppm. The capacity of ammonia on a basic 3A sieve depends on the size of the driers, their orientation, and their regeneration frequency. Under dire circumstances, the driers can be regenerated upon NH_3 breakthrough instead of water breakthrough. However, if this is done very often, it can severely affect the drier lifetime. Ammonia mitigation in the quench water and dilution steam circuit depends on pH and temperature. If we can keep pH low or neutral, ammonia can be directed into the blowdowns. These blowdowns then go through the wastewater circuit where it can be sent to the wet flare drum. Testing can direct us whether this stream can be pumped back to the quench tower or sent out for disposal. Another interesting point for concentrating ammonia might be the wastewater stripper tower reflux, though no work has been done on this that we are aware of. No work using molecular sieves or other traps in water has been tried to our knowledge.

War Story – Ammonia and COS from NMP from the Butadiene Unit Recycle Stream

Like any really good story, this one starts " Once upon a time, in an ethylene plant not so far away, the was an old not too bright chemist. He came home from the ethylene plant one day very tired. He wife, who is always right, told to eat something, take a nice long shower and go to bed early. So, he did. However, about 2 am, the chemist got a call from the Operations Manager at the plant where he worked, who said that they had an upset in the BD Unit and a "lot" of NMP(n-methyl-2-pyrrolidone) was sent back to, mostly, one furnace. They were seeing ammonia, as expected, and CO as expected but also the COS analyzers were high and the MAPD Converters were dying. Even the acetylene converters were beginning to have problems. The question was why COS? There is no sulfur in NMP, so he didn't have an answer. Could we be seeing two problems that didn't have anything to do with each other?

What was happening? "Sample the Refinery Gas for COS just in case that is where the COS was coming from" he said.

Well, at 2 am, he sat straight up in bed out of a dead sleep. His exact words were, "Oh crap"…. except he did not say crap. He thought he knew what happened. He got up, put his clothes on and went to the plant. He called up the DCS on his computer to see if his theory might be correct and it looked like it was. What did he look at? First let's look at the NMP molecule.

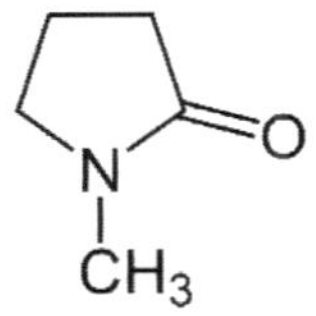

Figure 6: NMP

There is a nitrogen molecule that would crack easily in the furnace to ammonia. There is a methyl group on the nitrogen that most likely would crack to methane. There is a C=O that would easily make CO and we are left with a C_3 that could form propane, propylene or crack to smaller molecules including C_2's or methane. However, as mentioned earlier, there is no sulfur, so where does the COS come from? Let's look at a classic equation (Equation 1):

CO + H2S	<=>	COS + H2
Steam Cracking		Catalytic Cracking

Equation 1: COS Production

The "oh crap moment" was when the old not too bright chemist was a young, not too bright chemist and went down the hallway to ask his friend and mentor, Dr Ed Mohundro why Cat Crackers made COS and steam crackers H2S. Ed new the young chemist was not too bright but he was after all,,, a chemist. So he showed him this equation that shows why catalytic cracking yields COS and steam cracking yields H_2S. It all has to do with the hydrogen concentration in the reactors (in steam cracking, the reactor is the radiant section of the steam cracking furnace). In a low H_2 environment such as catalytic cracking, the equilibrium of this equation lies to the right and therefore COS is formed. However, in a high H_2 environment where H_2 is in the 10-12 mole percent level (as in steam cracking), the equilibrium shifts to the left and H_2S is formed.

Considering Equation 1, let's look at this equation for the cracking of NMP in a steam cracking furnace if propane is formed from the remaining hydrocarbon ring (Equation 2):

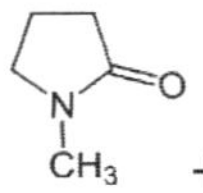

$$+ \ 3H_2 \rightarrow NH_3 + CO + CH_4 + C_3H_8$$

Equation 2: Cracking of NMP

So, this means that if the 3 carbons on the ring crack to propane, it takes 3 moles of H_2 per mole of NMP. What if the hydrocarbons on the ring all crack to methane? Then it takes 5 moles of H_2 per mole of NMP cracked. The bottom line is that cracking a lot of NMP lowers the H_2 content of the radiant section and raises the CO level. Both changes are enough to change the equilibrium of COS and H_2S in Eq 1. Also, the equilibrium towards COS does not have to be complete or, another words, all the H_2S does not have to be replaced by COS. Only enough COS must be formed to affect the MAPD Converter and/or Acetylene Converter. Poisoning of these palladium catalysts only requires 1 to 2 ppm in the C_2 or C_3 streams. Calculations and plant data showed that the amount of NMP recycled to the furnace was sufficient to bring the H concentration low enough to affect the equilibrium of Equation 1.

Now he had a story and data to support it. So, he ran (as much as possible anyway) upstairs to his bosses office. His boss had a great saying: "In God we trust, all others bring data". The old, not too bright chemist thought he had met that demand and they sat down and went through the chemistry. His boss agreed and now they had a story and they were going to stick to it. His boss went to let management know they understood what had happened and the old not too bright chemist went to explain the event to operations.

The old, not too bright chemist lived through another day fighting windmills and the ethylene plant and they all live happily ever after. OR at least until noon when something else happened.

References:

1. "Ammonia in the BASF FINA Petrochemicals Limited Partnership Steam Cracking Plant at Port Arthur Texas," Dwight Hines (2006, 218c)
2. "Minor Components, Major Issues: An Overview of Contaminants in Ethylene Plants," David Hood (2007, 51c)
3. "Feedstock Contaminant War Stories – Ammonia," Dwight Hines (2007, 51d)
4. "Kinetic Study of Thermal Decomposition of DEHA in Steam Crackers", Hangyao Wang, Dow Chemical (2017, 124f)
5. "Safety Tutorial-Management of NOx in Cryogenic Section of Ethylene Plants", Dane Grenoble (2007, Session 71)
6. "TDLAS Analysis of Ammonia in High-Purity Ethylene". Gary Landford, SpetraSensors Inc. (2020, 81D)
7. "Current Methods of Measurement of NO, NO2 and NH3 in Olefins Plants" Dwight Hines (2010, 135e)
8. "Caustic Operation at Westlake", David Payne, Westlake (2009,85)

Reducing Olefin Contaminant Removal Costs Via a Hybrid Adsorbent Approach, AZ-500

Rodrigo Lobo, Stephen Caskey, Allan Hatami

As the petrochemical industry tries to minimize production costs, the use of cost advantaged feeds becomes more and more attractive. However, such feeds may contain a larger amount of contaminants, many of which can poison current generation catalysts. Removal of the contaminants is effected by adsorbents, which need to combine high capacity for contaminants, with low reactivity with the highly reactive feeds (e.g. ethylene) found in petrochemical plants.

A breakthrough hybrid adsorbent, AZ-500, was developed to remove acid gas and polar components in a synergistic manner leading to an adsorbent with 20-45% higher capacity and lower polymerization potential. These characteristics provide solutions at lower Capital costs associated with smaller equipment size, plus reduced regeneration cost and CO2 footprint due to less frequent regeneration for the purification process.

Overview of Alternative Ethylene Production Feedstock and Their Contaminants - Part I

Omar Hamid

The steam cracking process, a significant contributor to global greenhouse gas emissions, is undergoing a substantial transformation in the sourcing of feedstocks for ethylene production. The exploration of alternative feedstocks offers a promising avenue to diversify raw material inputs and reduce environmental impacts in the near future. For Part I of this paper, we consider the following alternative feedstocks:

- *Renewable naphtha from HEFA (Hydrotreated Esters and Fatty Acids).*
- *Plastic pyrolysis oil*
- *Fischer-Tropsch products as a feed (derived from biomass or other sources).*

This transition raises concerns regarding the impact of contaminants and impurities in both the hot and cold sections of the ethylene process from these alternative feedstocks.

This consortium paper provides an overview of biomass and circular economy-based pathways for ethylene production. It includes block flow diagrams and descriptions of each process. The potential for integration into brownfield petrochemical plants is also discussed. The paper offers commentary on the composition, characteristics, and potential contaminants associated with these alternative feedstocks, emphasizing their implications for product quality, process efficiency, and environmental considerations. Key aspects discussed include the presence of impurities such as tar, sulfur, halogens, oxygenates, nitrogen compounds, filler materials, and heavy metals, which can be derived from various alternative feedstock sources and may impact catalyst performance and downstream processing

Contaminants Present in Ethanol and the Impact on Green Ethylene Production

Eder Luis Felipetto

Braskem is commercially producing Ethylene from alcohol since 2010, with a 260 KTA Plant in Brazil. Hydrous ethanol, which is commercially available as fuel-grade alcohol, is the raw material for the Green Ethylene industrial plant.

Hydrous ethanol contains water at approximately 7,5 % by weight. Since steam is used as a thermal fluid in the reaction medium, water is not considered an impurity for the Green Ethylene Process. Ethanol contaminants are typically compounds present in sugar cane or formed during its fermentation. These contaminants can harm the ethanol dehydration reaction. An example is the acetaldehyde, which can favor the condensation of molecules increasing the coke formation. Another contaminant example is methanol that generates ethers as by-products.

Other route of contamination for the ethanol can be related to the transportation matrix (pipelines, road, railway, and waterway). In Brazil, one of the possible problems in ethanol logistics is the potential for contamination with diesel or biodiesel during railway transport. Sharing of tanks and cars to transport these fuels without appropriate cleaning between loadings, increases the chance of contamination. These components cause operational problems by accumulating and clogging the equipment if they are not removed and product contamination should be minimized. However, when unavoidable, as in the case of diesel or biodiesel, the plant must include safeguards to operate even with the presence of these contaminants. Ethanol Vaporization and Effluent Treatment are the most affected stages by contamination with diesel and biodiesel.

The paper will discuss in more details the key contaminants in the Ethanol for the Green Ethylene Plant, and some of the impacts that can be expected from them.

Renewable Feedstocks: Challenges and Solutions

Sai Reddy Pinappu, Sankaran Murugesan

Renewable feedstocks are becoming increasingly important for the development of modern green chemistry. It's interesting to note that non-renewable fossil resources provide 86% of our energy and 96% of organic chemicals. Although much research has been conducted on renewable feedstock conversion, there have been few tangible successes and renewable feedstocks are not yet fulfilling their potential. One of the main issues hindering the success of renewable feedstocks is their lack of purity and homogeneity, which makes the purification processes of these streams particularly expensive.

In this paper, a unique solution will be presented that addresses the problems caused by contaminants that can negatively impact the quality of renewable petrochemical feedstocks. Some of the contaminants present in these feedstocks can lead to issues such as corrosion, fouling, and concerns related to health, safety, and the environment. The approach presented in this paper involves using a combination of specialized chemical additives and mechanical equipment to improve the quality of the feedstock and increase its market value. This holistic approach ensures a synergistic effect, resulting in a premium quality feedstock for the petrochemical plant.

16a Ethylene at Net Zero

Gunther Kracker

At a critical juncture in the petrochemical industry's evolution, the majority of olefin producers have made unequivocal commitments to ambitious net-zero targets by 2050. Simultaneously, technology companies have recently pioneered essential technological tools for the impending net-zero transition, while governments worldwide are enacting regulations to propel the industry toward a sustainable future. However, despite these significant strides, a transformative shift of monumental proportions is yet to happen.

The keynote speech at the 36th Ethylene Producers' Conference provides an elaborate outlook on the petrochemical industry's journey to achieve net-zero carbon emissions by 2050. Presented from the perspective of a combined technology licensor, leading industrial gases producer and plant operator, the conference contribution delves into the "Carbon Management Toolbox for Sustainable Olefin Technologies", showcasing a diverse set of solutions pivotal for realizing net-zero emissions. This toolbox includes among others hydrogen, electrification and molecular management, emphasizing their relevance to the industry's sustainable transformation.

From assumptions based on today's perspective, the speech outlines different pathways and major trends shaping the petrochemical landscape toward 2050 and beyond. It examines the transition from conventional to decarbonized installations, emphasizing challenges and opportunities inherent in this profound shift.

To conclude the outlook on ethylene at net zero examined through the lens of a prominent steam cracking technology licensor, a pivotal segment of the keynote will also deep dive into hydrogen. This section offers the profound vision of the hydrogen economy from a global leader in hydrogen production, who covers every link in the hydrogen value chain from source to service. As a key element in petrochemical decarbonization strategies, alongside the burgeoning global trend of electrification, the strategic adoption of hydrogen will play an indispensable role in driving the petrochemical industry toward a sustainable future.

157a *Policy Perspective: The Resounding Impact of the Inflation Reduction Act on Deployment of Climate Technology and the Global Response to Its Economic Influence*

Adam Green, Shawn Daray

As concerns surrounding climate change increase and widespread embracement of technology as a solution to the issue grows, greenhouse gas emissions-intensive industry is facing significant pressure to commit to ambitious sustainability targets and decarbonize their operations. Industries that fall under the umbrellas of sectors such as energy, chemicals, and resources must navigate the balance between sustainably responsible operations and profitable business. The nascency of some of the technologies that make up the suite of options for emissions reductions ensure that many of these pathways will come at a substantial cost.

It isn't difficult to see that favorable government policy and regulatory frameworks are driving climate technology investment in certain geographic regions. Control strategies, often referred to as "stick" approaches, are legal obligations require certain emitters to comply by paying a tax on their emissions or covering their emissions with allowances allocated from the government or procured from emitters with a surplus. Incentive strategies, which have been dubbed "carrot" approaches, provide financial motivation for voluntarily adopting decarbonization technologies through grants, advantageous financing, or tax benefits.

In August of 2022, The United States Government passed the Inflation Reduction Act (IRA) and immediately changed the climate technology investment landscape. Embedded within this landmark legislation is a portfolio of technology-neutral climate action programs fueled by fiscal appropriations of nearly $400 billion. Though multiple funding instruments are utilized by the IRA, the focus of this paper will be the government's doubling down on the tax credit mechanism and other "carrot" approaches, and how they have positioned the United States to benefit from the economic activity surrounding climate investment.

Tax credits function by reducing the amount a taxpayer owes at the end of a taxable year, and this is not a novel strategy used by the US government to motivate climate action. However, the IRA significantly bolstered existing tax credit programs and introduced several new programs for technologies that weren't already prioritized. Appropriations for grant and loan programs to be administered by the Department of Energy (DOE) are competitive application processes that provide criteria for technology, geography, and other topic areas.

Coupled with the anticipated reduction in country-wide greenhouse gas emissions is an already-realized economic boom that has other jurisdictions taking notice. According to the White House on the one-year anniversary of the enactment of the law, over $110 billion has been announced in clean energy manufacturing investment, greater than 110,000 clean energy manufacturing jobs have been created, and it is estimated that the IRA could create more than 1.5 million additional jobs.

Garnering a keen interest from industry in the US and globally is the tendency for the requirements in the lucrative IRA programs to mobilize a clean energy and clean manufacturing workforce by paying prevailing wages, providing apprenticeship opportunities, and ensuring domestic content assurances are satisfied. There are also stipulations for communities that have been historically marginalized by energy and manufacturing activities. These aspects of the funding programs have been mimicked in major policies of other countries, such as Canada, the European Union, and the United Kingdom. The domestic provisions have also received criticism on the global stage, with the international community claiming them to be protectionist and in violation of the country's obligations to the World Trade Organization.

Other nations with major climate ambitions are, however, responding to the IRA with major policy announcements that seek to draw climate technology and clean energy manufacturing investment back to their respective communities. In addition to the countries previously mentioned, Australia has strengthened its flagship emissions reduction program and individual member states of the European Union are mobilizing funding to encourage deployment within their borders.

This paper will describe the major policy instruments that are currently in use across the globe and identify instances of where they are, and are not, working efficiently and effectively. The various tax credits programs and major funding allocations of the IRA will be highlighted to demonstrate the importance of the bill, as well as the preceding Bipartisan Infrastructure Law, in catalyzing climate technology and manufacturing deployment in the US. Finally, the global response to the IRA will be reviewed and a discussion of the need for cross-border collaboration will be provided to determine how the net zero targets and other sustainability commitments will be achieved most efficiently achieved on time.

157b Combining Electrification, Carbon Capture and Renewable Feedstocks: Large-Scale CO2 Reduction for Olefin Production

Andres Munoz, Torben Hoefel

Effective CO_2 reduction in the course of olefin production is probably the biggest challenge petrochemicals are facing these days. In this contribution, the authors will introduce a simplified and tangible visualization of the different mechanisms currently available for reduction of CO_2 emissions that are most relevant to olefins production through steam cracking.

Examples of emission reduction measures and their impact on the different emission categories will be highlighted. Among the multiple considered approaches, electrification of processes and Carbon Capture and Storage/Utilization (CCS/CCU) are two most promising measures when it comes to direct emissions. Renewable feeds, on the other hand, are among the most effective ways to reduce Scope 3 emissions, in particular the so-called end-of-life emissions.

The paper will then showcase how these technologies can complement each other, what's industrially viable and what level of CO_2 emission reductions is possible. A scenario of simultaneously utilizing renewable feedstocks with either CO_2 capture, or furnace electrification, will be compared to a conventionally operated steam cracker. The conference contribution will conclude how technologies already available today can contribute to olefins and downstream products with minimized direct and indirect CO_2 emissions.

157c Evaluating the Use of Hydrogen As a Cleaner Fuel in a Steam Superheater Application

Jennifer Cunningham

Jennifer Cunningham

Site Production Engineer – Channelview, TX Manufacturing Facility

LyondellBasell

Vivek Sundaram

Fired Equipment Engineer – Global Asset Integrity and Turnaround Department

LyondellBasell

Mitch Gibb

Site Process Specialist – Channelview, TX Manufacturing Facility

LyondellBasell

Rob Fox

Senior Director - Global Process Engineering

LyondellBasell

In September 2021, LyondellBasell announced an ambition to achieve net zero Scope 1 and 2 emissions from global operations by 2050. In an effort to meet decarbonization goals for the company, potential applications for increased hydrogen consumption and other alternate strategies were identified. A recent plant trial was conducted at the LyondellBasell facility in Channelview, Texas, to evaluate the effects of increasing the hydrogen content in the fuel supply to a steam superheater to nearly 90 mol%. This plant trial evaluated the impact of high hydrogen firing on both process and environmental performance and provides insight on the impact of key operating variables. This paper includes the effects of NOx emissions with changes to flame temperature, excess O2, and relative humidity and is a summary of the findings and lessons learned that will help drive future projects for increased hydrogen fueling throughout the company.

157d Strategies to Comply with the Earliest New Hon Requirements at Ethylene Production Facilities

Gary Daves, Joe Ibanez, Larry Poche

In the draft Hazardous Organic NESHAPS (HON) revisions, subject facilities must commence fenceline monitoring (FLM) within one year of the March 2024 publication of the final HON rule revisions in the Federal Register, with the publication deadline set in a consent decree between EPA and several litigants. Thus, HON FLM will begin in a little more than a year from now for HON-affected sites that use, produce, store, or emit benzene, 1,3-butadiene, chloroprene, ethylene dichloride, vinyl chloride, and/or ethylene oxide. The HON revisions, including FLM requirements, will apply to process units subject to the HON at chemical manufacturing complexes including numerous Ethylene Manufacturing facilities. For many of these facilities, these new FLM requirements will come in addition to those already imposed by recent EPA Consent Decrees (CD) and pose new complex challenges with overlapping programs.

As more public and agency scrutiny is applied to FLM data, facilities will find it beneficial to minimize public exposure that comes with elevated fenceline concentrations. As many refineries found in implementing FLM programs and those Ethylene production facilities currently implementing programs required by CDs, emissions affecting fenceline concentrations can come from unexpected sources. Thus, one year from rule promulgation to collecting data to report to the public is a very short timeframe to set up a pilot study and make any changes the site wishes to implement before the FLM data is reported. Therefore, setting up pilot studies and potentially utilizing mobile monitoring technology is something for facilities to consider now, especially considering how laboratory resources were serious constraints during the 2022 HON Section 114 Information Collection Request data collection effort.

Additionally, the draft HON FLM provisions allow for site-specific monitoring plans that can be crucial for HON sites that are adjacent to other facilities that emit one or more of the appliable HON FLM compounds. Collecting initial FLM data and taking appropriate steps to address any elevated concentrations, including determining the contributors to elevated fenceline concentrations and potentially developing an FLM site-specific monitoring plan, can take several months. Therefore, being ready to start pilot studies on or before the March 2024 HON publication date will be important for many HON applicable facilities.

This presentation will provide insights from developing multiple FLM site-specific monitoring plans, setting up more than a dozen refinery and chemical plant FLM pilot studies, including both Method 325A/B and TO-15 summa canisters, and mobile monitoring.

Attendee Takeaways

1. The deadline for implementing the new HON fenceline monitoring requirements is a little more than a year away.
2. Ethylene manufacturing plants with a HON applicable unit(s) will require fenceline monitoring around the entire facility and are likely to include 1,3-butadiene, chloroprene, ethylene dichloride, vinyl chloride, and/or ethylene oxide in addition to benzene which may already be required by an EPA CD.
3. Facilities subject to the new HON fenceline monitoring need concrete plans to conduct pilot studies soon after rule promulgation so they can take steps discussed during the presentation to

identify contributors to elevated concentrations and potentially initiate mitigation strategies.

4. Site-specific monitoring plans may be needed for facilities in close proximity to other refinery and petrochemical facilities.

157e Practical Guidance on How to Prepare and Manage GHG Inventories

Everard Ashworth

Over the past thirteen years, ALG engineers have performed over 1,000 third-party Greenhouse Gas (GHG) report audits or attestations. These GHG inventories have been prepared for a broad spectrum of manufacturing, refining, fuel and electricity retailers, and alternative fuel producers, and provide Scope 1 and Scope 2 emissions consistent with corporate sustainability reporting, applicable state cap-and-trade regulations, and third party carbon trading programs. Based on this practical experience, a technical presentation and associated technical paper will provide a focused review of the how to design an accurate and defensible GHG emissions inventory for your operations. These essential design elements are important given the increased focus on GHG inventories to support corporate sustainability and financial statements, project funding requirements, and participation in monetized carbon trading schemes. Key elements of this presentation will include data collection, data adequacy, data analysis, GHG inventory methods and quality assurance reviews. The presentation will track the Scope 1/Scope 2 inventory analyses over time for specific source categories and provide practical guidance on how best management practices can be implemented to improve the accuracy and reproducibility of corporate GHG inventories. The presentation will also provide practical recommendations on how best to manage an external GHG audit.

157f *Issues during Startup of Steam Cracker Unit*

Holt Lacy, Michael Lopez, Terry Gordy, Paul T. Scott

Indorama Ventures Olefins LLC (IVOL), an indirect subsidiary of Thailand's Indorama Ventures PCL (IVP), Bangkok, owns and operates what was previously a dormant butane cracker in Westlake, LA. The refurbished cracker is designed to process attractively priced ethane and propane feedstock from US shale to produce ethylene and propylene.

Understanding issues – The reduction of flaring during the startup of a steam cracker is not primarily a cost reduction activity. The primary objective is reduction in emissions. In addition, sometimes there are other important process issues and difficulties which need to be overcome. This paper explores in some detail one such issue; high CO2 concentration in the startup gas, and how that is now routinely circumvented during the startup to prevent contamination of final ethylene product.

Authors:

Michael Lopez – IVOL, Operations Manager

Terry Gordy – IVOL, Gate Keeper/OMC

Holt Lacy – IVOL, Production Specialist

Paul T. Scott – KBR, Chief Technical Advisor

33a PREDICTING THE REMAINING LIFE OF CRACKING COILS: AN APPROACH BASED ON MATERIAL DEGRADATION

Rafael Andrade, MSc. João G. Dessi, Wellington A. Silva, Daniel M. Guarnieri

This paper presents a case study of a predicting the remaining life of cracking coils of an ethylene cracking furnace based upon material degradation. ENGEMASA has developed a new approach for predicting the remaining life of a cracking coil furnace, based on the perspective of the tube material degradation due to the damage accumulated, as illustrated in Equation 1.

Equation (1)

where,

REL: Remaining Equipment Life

L_{new}: *Design Life of New Equipment*

D_{acc}: *Accumulated Damage*

In this approach, the remaining coil tube life is determined by the material accumulated damage taking into account three main thermoactivated degradation mechanisms acting synergistically, which are, Carburization, Creep, and Oxidation, as indicated in Equation 2.

Equation (2)

where,

$D_{carburization}$: *Accumulated Damage Due to Carburization*

D_{creep}: *Accumulated Damage Due to Creep*

$D_{oxidation}$: *Accumulated Damage Due to Oxidation*

A sound wall reduction due to damage accumulation approach is introduced to assess the equipment life reduction. Due to the operational temperature and coke deposition during the cracking process, on the inside surface of the tube, the carburization takes place leading to the degradation of the material from the inside diameter inwards the bulk of the material. On the other hand, from the outside surface inwards the volume of the material the oxidation is taken into account, leading to material degradation due to its microstructural depletion at high temperatures and oxidative environment. On the bulk of the material, the creep mechanism is considered through the API 530 formula, where the remaining wall thickness (after disconsidering the wall thickness lost due carburization and oxidation) is evaluated according to operational conditions as pressure and temperature, as well the Larson-Miller curve for the material.

All three mechanisms are time and temperature-dependent, meaning that their influence is determined by the operational conditions.

Real operational condition records were used to evaluate the methodology showing a good correlation. Samples removed from equipment at the end of life were used to validate the results. The obtained results allowed to establish a relationship among the operating conditions, the theoretical remaining life, and the degradation state of the samples.

The methodology presents as a useful supporting tool for cracking furnace monitoring, reliability, and decision-making. The uncertainties in the approach are directly related to the quality and frequency of the information acquired from the field operational controls and could be reduced by combining with NDT methods and material sampling removal guidelines.

33b CARBOWALL SYSTEM CASE STUDY AT PETROBRAS REFINING REPLAN IN 2010 AND RPBC IN 2023

José Carlos Belotti, Luis Henrique Rodriguez de Mattos Gobbo, Mateus Guilherme Camparotto, Leandro Colombo, Misi Gorby, James Smith

This study presents the application and resulting performance data of an innovative insulation lining system, the Alkegen Carbowall System™. This system is an engineered solution utilizing vacuum formed ceramic fiber shapes installed in the convection zones of chemical and petrochemical furnaces. The solution is applicable to furnaces such as oil heater reformers and pyrolysis furnaces, whether new units or for repairs. The Alkegen Carbowall System™ can be installed at the end user location or at external sites.

Traditional project linings consist of refractory castable, Insulating Firebricks (IFB), or Refractory Ceramic Fiber Insulation anchored to the walls of the convection zone. The maintenance of these walls is extremely problematic. In many cases, it is necessary to remove the piping to allow free access to the lining. As well as potentially causing damage to the surrounding lining, this becomes a critical path in the downtime of the unit, costing excess money and time.

Designed and created to replace the conventional insulation lining of the sidewalls of the convection zones, the Alkegen Carbowall System™ consists of customized shapes made of high purity lightweight ceramic fiber and binders that are installed from the external side of the furnace. This eliminates the need to remove piping, dramatically reducing downtime, while improving energy savings.

Petrobras, one of the largest oil and gas producers in the world, operating primarily in exploration, production, refining, generation, and commercialization of energy in Brazil, have installed the Alkegen Carbowall System™ in a number of their facilities since 1998 without need for any maintenance. This study identifies specific results from the installations into the convection section of the Atmospheric Heater at the REPLAN facility in April 2010 as well as into the convection section of the Reformer Furnace at the RPBC facility in April of 2023.

The motivations for utilizing the Carbowall system for these Petrobras refinery projects will be explored, outlining the maintenance benefits of the lining system in addition to the construction options this system provided to Petrobras.

33c ADVANCES IN PSA VESSEL LIFE MANAGEMENT

Steven Altstadt, Annette Karstensen, Kathy Picioccio

Pressure Swing Adsorbers (PSA) are pressure equipment in cyclic service that are used at hydrogen plants for the purification of hydrogen. Due to the inherent nature of cyclic service the PSA vessels have a limiting fatigue life and are typically designed for 20 years or less. In many cases, towards the end of the original design life or during mid-life (50-60% of design life), operators will elect to conduct a life extension study. Usually, the focus is on evaluating the areas of high cyclic stress where fatigue cracks could plausibly grow to the point of fracturing the pressure boundary. The quantitative portion of this life extension study includes a Level 3 Fitness for Service assessment and Finite Element Analysis (FEA). The key outputs from the assessment, minimum reliably detected planar flaw sizes and estimated time intervals for cyclic crack growth to failure, are used to help develop or refine an inspection strategy. Often online inspections are conducted, however inspection during a turnaround or planned outage is an option.

It is not uncommon for PSA vessels to develop cracks during their lifetime. Several locations have been identified as critical, including internal weld attachment points and longitudinal seams welds. This paper is concerned with the analysis of the longitudinal seam welds. There are several factors associated with the fatigue life of PSA vessels; some of these are straightforward such as pressure cycle range, cycles per year, and design pressure - others are not. One of the more nuanced considerations is local stress concentrations associated with shell geometric anomalies, especially at the longitudinal seam welds. Non-ideal geometries such as shell ovality, hi-lo misalignment, and weld peaking are common considerations. To address this, field measurements are typically conducted to measure actual dimensions, which can then be used as a basis for creating the PSA vessel geometry in finite element models. Geometric anomalies and their deleterious effects tend to be more pronounced in vessels fabricated before modern best practices were adopted. A second consideration is fatigue crack growth associated with pressure cycles in a hydrogen environment, which can seriously affect the fatigue life. Laboratory material testing research focused on cyclic crack growth and fracture continues for this specialized topic.

The work described in this paper highlights how it is possible to operate PSAs beyond the original design life by implementing an inspection program that is based, in part, on a combination of laser scanning, other field measurement techniques, finite element analysis, and fracture mechanics. This paper will cover initial considerations before starting a life extension study through final implementation.

AIChE PAPER NUMBER: 36d

SELECTING LOCATION OF IR CAMERA FOR THERMAL IMAGING OF ETHYLENE CRACKER FURNACE IN A BROWNFIELD PROJECT

Rajeev Limaye
Head of Advanced Operations Services US
Linde Engineering Americas

James Cross
Head of Industry Management
AMETEK Land, UK

Prepared for Presentation at the 2024 Spring National Meeting
New Orleans, LA, March 24-28, 2024

SELECTING LOCATION OF IR CAMERA FOR THERMAL IMAGING OF ETHYLENE CRACKER FURNACE IN A BROWNFIELD PROJECT

Rajeev Limaye
Head of Advanced Operations Services US
Linde Engineering Americas

James Cross
Head of Industry Management
AMETEK Land, UK

Keywords: TMT, Tube Metal Temperature, Maintenance, Reliability, IR Camera, Thermal Image Monitoring, Ethylene, Furnace, Edge Computing, Cloud Computing, Image Recognition

Abstract

Real-time measurement of Tube Metal Temperature (TMT) of the coils in the radiant section of an ethylene cracker furnace proves to be extremely beneficial in extending the coil life, ensuring complete decoking, and improving overall maintenance and reliability of the furnace. Thermal image monitoring for ethylene furnaces has been implemented using Ametek near infrared cameras and Linde's image recognition technology called Linde Virtual Furnace Coil Monitor. The cameras are fixed and permanently mounted on the furnace wall. To get the value and projected benefits, it is important to decide the number of cameras needed and the location of those cameras on the furnace. Based on the size of the furnace, number of zones, and coil geometry, the selection of camera locations that give the best field of view of furnace internals is an important engineering step. The number of cameras and the resolution of the images also play a vital role in determining the area of coverage provided by all cameras. This paper explains the process of determining camera location and quantity, special challenges on a brownfield project, and lessons learned from such implementations.

INTRODUCTION

Ethylene cracking furnaces are a key process unit in the ethylene plant. Ethylene is produced by thermal cracking at high temperatures (1300 – 1800 °F) of hydrocarbons such as ethane, methane, and naphtha in the presence of steam in the tubes of an ethylene cracking furnace. This is an endothermic reaction. The heat of reaction is provided by burning fuel gas in the radiant section of the furnace. The mixture of steam and hydrocarbon flows inside the tubes. During the process of thermal cracking, hard deposits of carbon (coke) build up on the inner walls of the tubes. After reaching the threshold amount of coke buildup, the furnace must be taken out of production and into decoking mode. The duration of the furnace run before a decoke is required is known as the "run length", usually reported in number of days. As coke builds up, the Tube Metal Temperature (TMT) increases to maintain the heat transfer to continue the thermal cracking. If the TMT is excessive, it accelerates coke formation, thus reducing furnace run length. Excessive TMT is the single most prominent cause of tube failures leading to unwanted shutdowns and production losses. Accurate measurement of TMT is therefore critical to extending tube life, avoiding the possibility of unplanned shutdowns, and reducing maintenance costs.

BACKGROUND

Linde has analyzed operating data for several years of operations at ethylene cracking furnaces at various sites and concluded that the benefits of real-time automated TMT measurement of the cracking furnaces far outweighed the cost of implementation, in addition to several other benefits.

The solution is built around the installation of fixed near-infrared (NIR) cameras with borescopes, that allow a lens to be situated in the furnace itself. These cameras operate at the same wavelength as handheld infrared pyrometers that have been used by ethylene producers for over thirty years to collect TMTs. The AMETEK Land NIR-B cameras are used to capture temperature data in the form of thermal images (Figure 1). The cameras are connected to an edge device running Linde's computing software to convert the images to JSON files. The JSON files are sent to cloud storage where Linde's proprietary image recognition software, Linde Virtual Furnace (LVF) Coil Monitor translates the image files into temperature profiles for each tube. The user logs into the cloud application to view the temperature profiles, playback history, review alerts, messages, etc.

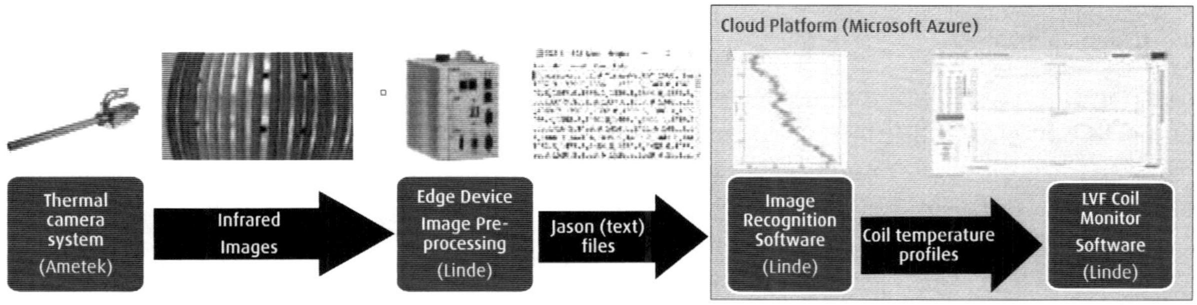

Figure 1: Mechanism of TMT measurement using IR Camera and Linde Coil Monitor software.

Installation

The cameras are permanently mounted on the furnace at selected locations for continuous TMT measurement. Mounting holes are drilled on the wall of the furnace and a nozzle with a flange is welded to the furnace wall. The camera's borescope is inserted through the nozzle with the tip of the nozzle flush with the inside refractory wall of the furnace. Please see Figure 2 for the details of the camera installation assembly.

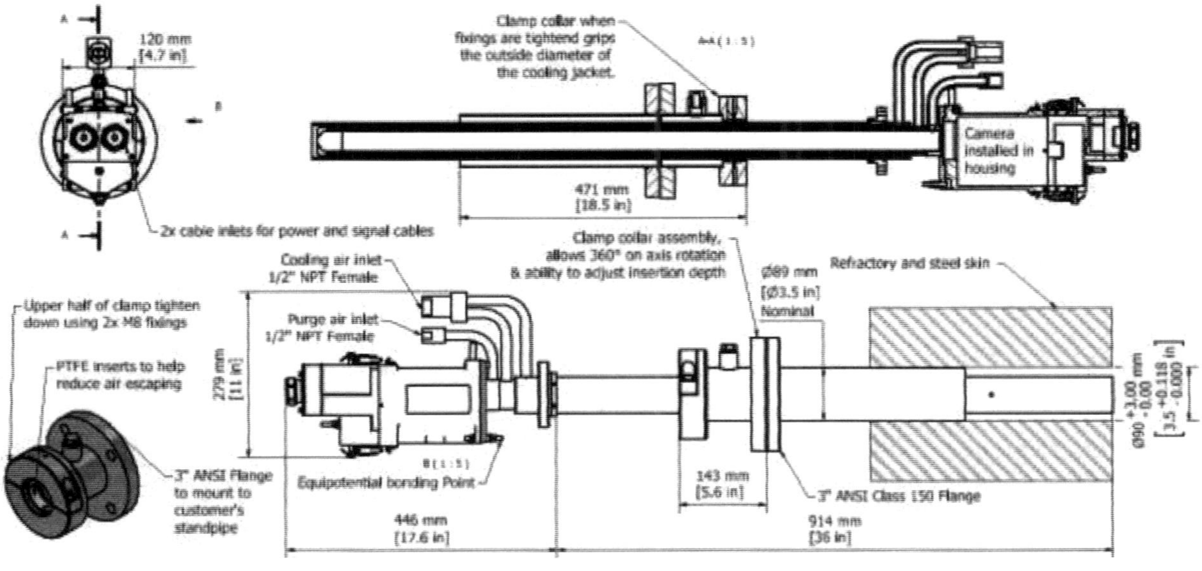

Figure 2: Camera Installation Assembly

Selecting Location of Cameras on the Furnace

While automated TMT measurements are naturally attractive compared to current methods, the installation of cameras, field construction, and IT readiness can be daunting tasks if not carefully planned. Particularly in a brownfield project, where all ethylene furnaces are older, various factors play a role in deciding the location of cameras. These factors are discussed below:

Accurate 3D model of the furnace

A 3D model of the furnace and its internal coil arrangements with accurate dimensions is critical to determine the location of the cameras. For a greenfield project, such models will most likely exist. But for brownfield projects, if an accurate 3D model of the furnace is not available or not updated with changes in coil geometry, the first step will be to build the 3D model. This can be achieved in the most basic form by using existing 2D furnace design drawings, but the more accurate the model, the less likely changes will need to be made later in the project. 3D scans are another great way to capture furnace geometry and to ensure any structural steel, pipe racks, or other ancillary equipment are captured and accounted for.

Accessibility for maintenance

Cameras should be located for easy maintenance accessibility without the need for any ladders and/or scaffolding. Preferred locations will be on the walking platforms about 7 feet above the platform base so that they do not create an obstruction while walking. It is recommended to check the temperature accuracy of the cameras once a year using a portable blackbody furnace that can be brought to the site by a trained engineer. Easy access to cameras will avoid the cost of building scaffolds and other temporary structures to access the cameras. Once accessed, the camera can be easily removed from the cooling jacket without removing the jacket from the furnace.

3D Model of furnace externals (Obstructions due to other structures)

AMETEK LAND cameras come with standard borescope lengths of 1ft, 2 ft, or 3 ft. A 3 ft long borescope offers more flexibility for installation and welding of nozzles on the furnace wall. The camera is removed by retracting the borescope from the nozzle. The total length of the 3 ft borescope camera is about 4 ft long when including camera electronics and the cooled housing. The nozzle welded on the furnace wall can be 6 to 8 inches long. While planning the camera location it is important to make sure that there is a space of about 5 ft from the furnace wall to easily retract and insert the camera from the nozzle. The cameras could be installed at an angle, which is determined during the field of view study. It is important to consider the availability of space with the chosen angle of installation of the camera.

In a brownfield project, there will likely be other structures such as instrument cable trays, conduits, and pipe racks, that may interfere with the cameras. The presence of these structures can limit the locations where the camera can be installed. If possible, some structures may have to be moved to clear space for installation and ensure the cameras are located optimally to both maximize coverage and minimize the required number of cameras.

Having only a 3D model of the furnace and internals is therefore not sufficient. It is important to include in the model all external structures such as walkways, ladders, I-beams, junction boxes, cable trays, pipes, etc. that are within 10 ft of the furnace walls. Taking manual measurements of these structures can be laborious. It is therefore recommended to do a 3D scan of the area for accurate modelling of all external structures. A good model can avoid frequent trips to the field to verify the location. Such an external plant model around the furnace can be useful not just for the cameras but for other maintenance and planning activity.

Initial camera locations

The selection of initial camera locations and the number of cameras require some engineering judgment and experience. Due to the nature of these projects, there is always a compromise somewhere. Initial quantity and locations of the cameras are determined by analyzing the furnace drawings, 3D model of furnace internals as well as externals, areas of interest inside the furnace that are susceptible to overheating, records of past tube failures, etc. These locations must be validated by doing the Field of View (FOV) analysis.

Field of View Analysis

Field of view (FOV) study is done for each camera location by an expert designer. As shown in Figure 3 a, b, c below.

Figure 3a

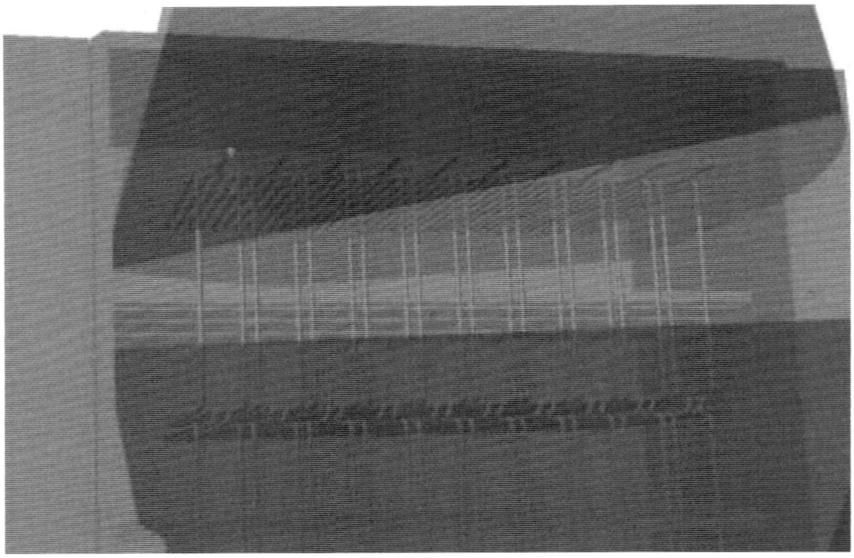

Figure 3b

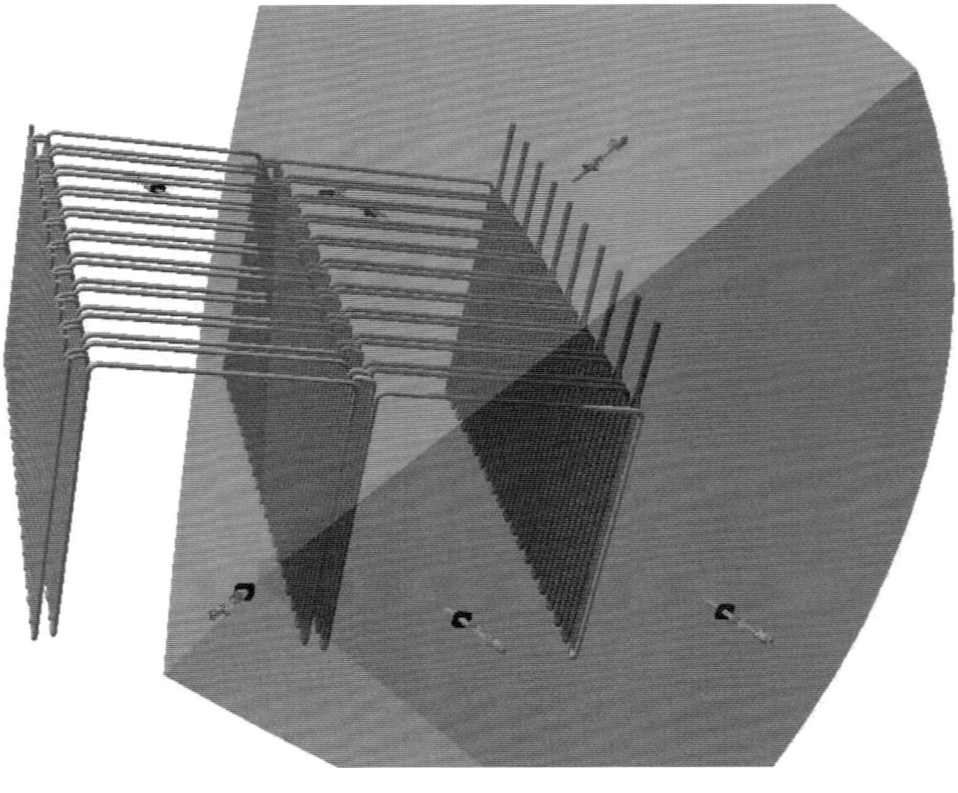

Figure 3c

The FOV analysis provides the area of coverage for the selected location and viewing angle of the camera. The angle and location can be changed to get the best coverage.

The FOV study is usually performed as a group working session with the 3D model expert, furnace expert and the plant mechanical engineer at a minimum.

As a result of the FOV study, it is quite possible to move the location of cameras and viewing angles. The purpose is to optimize the locations to get the best field of view and coverage of the interested areas.

Percentage coverage and number of cameras

The number of cameras required can be a contentious subject. Increasing the number of cameras will increase the coverage area, but will also increase cost. It is not practical or necessary to cover 100% of the coil area with cameras. Engineering judgement, historical data from tube failures and experience with furnace maintenance are some of the key requirements for optimization of camera quantities. After a certain number, incremental benefits of the increase in coverage area do not justify the cost of additional cameras.

To get the best value from TMT monitoring, it is necessary to compare live TMT values with some reference. The high-fidelity furnace model based on the first principles is used to predict the TMT using process measurements and run length prediction. The predicted TMT is compared with the measured TMT, and this comparison is a valuable diagnostic tool. When a reliable model exists, the reliance on higher coverage area by IR cameras is reduced.

Constructability and safety review

After an FOV study, the locations and viewing angles must be field verified to ensure there are no obstructions and that all OSHA and safety requirements are met. If there are any findings in the field verification, it may be necessary to go back to design phase and tweak the FOV study again for different locations. This is an iterative process until all locations have been revised by furnace expert for the coverage area and by the constructability team.

Once the locations are finalized, it is highly recommended to spray paint the location on the furnace wall and produce final installation drawings for construction along with the pictures of the locations.

Nozzle fabrication and installation

Fabrication and installation of the nozzle is a crucial step in ensuring the cameras will provide the intended FOV. The camera viewing angle is determined by how accurately the nozzle is fabricated and installed. As shown in Figure 4a, 4b, the nozzle is fabricated with a baseplate that is welded to the furnace wall. A jig file of each nozzle, which can be 3D printed, is provided so that fabricators can check for correct depth and angle. This minimizes errors. Furthermore, welding a bracing plate to the standpipe ensures that any welding of dis-similar metals is done in a fabrication shop environment rather than at the plant site.

Figure 4a: Camera installation example

Figure 4b: Camera installation example

Wall thickness and additional supports

During the constructability review, it is also important to assess the structural integrity of the installation. The camera, its electronics, and dry borescope weigh about 30 pounds. The nozzle welded to the furnace wall should be capable of supporting this weight with no sagging or deformation. The thickness of the nozzle baseplate should be designed according to a structural analysis of the installation. In rare cases, it may be necessary to provide additional support to the camera. If additional support is required, it is important to know that upfront, as the support should not impede walkway passage and should pass the safety and constructability review.

Hazardous Area classification and Camera resolution

Since the cameras are installed in an area where a flame is always present, the area may be a general-purpose area for hazardous area classification. However, it is important to get this verification at the very beginning of the project. In an area with general classification, higher resolution cameras (1968 X 1476 pixels) can be used whereas lower resolution cameras, with a lower operating temperature, have to be used in a hazardous area. Higher-resolution cameras provide improved spatial resolution on tubes furthest from the installation location, and thus fewer cameras are required.

The hazardous area-certified IR cameras that are certified for ATEX/IECEx and Class 1 Div 2 area classification have lower resolution (640 X 480 pixels). More cameras are required to get the same level of coverage found with the higher resolution units. In such cases, it might be worthwhile to review the hazardous area classification and reclassify the area near camera location as general purpose if possible.

Cooling medium

The cameras require a reliable cooling medium. Instrument air, nitrogen or cooling water are options. If instrument air is used, it should meet the specification of an instrument air quality standard such as ISA-S7.0.01 which establishes four elements of instrument air quality:

1. Pressure dew point: At least 15-20 deg C below the minimum ambient temperature
2. Particle size: Maximum 40 micron
3. Lubricant content: Less than 1 ppm w/w or v/v.
4. Contaminants: Free from corrosive contaminants and hazardous gases

It should be possible to route the instrument air tubing from the nearest air headers to the selected location of cameras. It is also possible to use other inert gases such as plant nitrogen which meets the specifications above for cooling medium.

It typically takes 250 liters/minute (15 Nm3/h) cooling gas per camera for applications up to 1,200°C, but depends on use of air cooling only or combination of air and water. Depending on the number of cameras, one should confirm the availability of cooling media at the site. If the required flowrate is not available, arrangements for reliable cooling gas supply must be made.

If cooling water is used as a lens cooling medium, the water should follow recognized and generally accepted good quality standards, especially around particulate, contaminants, corrosivity, etc. The maximum inlet cooling water temperature is 40°C (104°F). Typically, a camera will require 11 liters/min for applications up to 1,200 °C flue gas temperature. If cooling water is used, there are additional considerations such as winterization of the water tubes, which can add to the installation cost. Small amount of cooling gas will still be required to purge the lens of the borescope camera. The decision on whether water or air cooling or combination of air and water will be utilized for a project should be made early on to ensure the correct parts are quoted as the water- and air-cooled housings have similar footprints, but functionally operate quite differently.

Expert opinion

An invaluable element is engineering judgment from experts in multiple fields in all aspects of decision-making while selecting camera location and quantity. It is important to consider the objective of TMT measurement, areas of interest based on the history of coil failures, accuracy and maturity of the furnace model, and several other factors where the operation and maintenance experience of ethylene furnaces plays a vital role.

Other considerations for Field construction

Once the location and number of cameras are decided several other field installation /construction items can be planned:

Location of PSU

The Power Supply Unit (PSU) provides power and fiber optic connections to the camera. The connection between the camera and PSU (Up to 25 meters) is a special cable and comes with the camera. While the camera is located on the furnace wall, the PSU can be mounted at grade level for convenience of wiring. The distance between the PSU to camera can be a maximum of 25 meters. It is important to estimate the length needed and order accordingly. Otherwise, the extra cable must be coiled up and kept near the PSU.

Power source

The cameras should be powered from a clean power source from an Uninterruptible Power Supply (UPS). UPS power distribution will most likely be in a nearby equipment building such as Remote Instrument Enclosure (RIE), Electrical Building or Control Room. The cameras can accept multiple voltages. It is most practical to route the 110V AC cables from UPS power distribution and specify the voltage requirements for the PSU when ordering the camera hardware.

Fiber Optic

Gigabit fiber optic (FO) is the preferred wiring for carrying the image data from camera to the edge device. FO is immune to electromagnetic interference in the plant and offers noise free data transmission. The routing distance between the camera to edge device could be long; hence single mode fiber should be used. The type of communication and the end connectors for FO cables must be specified while ordering the camera hardware. The small form-factor pluggable (SFP) transceiver that converts copper signal to fiber in the PSU is supplied accordingly to match the specifications of connection type and FO mode. Single mode FO cable and "LC" connectors are most commonly used.

Location of Edge Cabinet

Edge computing makes it possible to capture data at the site, perform minimum pre-processing and transfer the data to the cloud. The applications running in the cloud such as thermal image recognition are processor-intensive. Cloud computing coupled with edge devices allows for simple and minimum hardware and software at the site.

Up to six cameras can be connected to one edge device through a switch. The number of edge devices needed is dependent on the number of cameras. Edge devices are designed for installation in a climate-controlled atmosphere. Multiple edge devices are mounted on DIN rails and integrated into one single free-standing cabinet with redundant 24 VDC bulk power supplies, switches, and fiber optic patch panel. The edge device cabinet should be located in an area where an IT network connection is available. Availability of UPS power, connection to DCS I/Os for hard wiring of alarms, and climate control are other factors to be considered while deciding the location of the edge cabinet.

Planning for spares and future growth

It is highly recommended to plan for 50 to 100% spares for power and fiber cables. Laying power and fiber cables is expensive. If spare camera cables are pulled during initial field construction, later addition of more cameras will be much easier and less expensive.

It is also recommended to plan for at least 25% spare ports while deciding the number of edge devices to allow for future growth. The cost of hardware is much lower compared to labor costs of later additions or modifications to the edge cabinet, fiber/power routing, etc.

CONCLUSION

The continuous measurement of Tube Metal Temperature (TMT) of ethylene furnace tubes shows a lot of promise in reduction of maintenance costs and unplanned shutdowns, and increasing the profitability by increasing the run length. These benefits can be realized when the right engineering decisions ae taken to decide location and number of cameras. Field of view study with 3D model of furnace internal and externals performed by a team consisting of a 3D model designer, furnace expert, plant mechanical/reliability engineer is essential to determine initial location, viewing angle and number of cameras. Various other factors such as TMT measurement objective, cooling medium, hazardous area classification, constructability and safety review must be considered to finalize camera locations.

33e Enhancing Ethylene Cracker Radiant Tube Integrity through Innovative Robotic Inspection

Minghao Zhang

The integrity of ethylene cracker radiant tubes is a critical concern, as they are vulnerable to unexpected failures causing from various factors, including carburization, overheating, thermal shock, and more. These failures are often complex and can occur at diverse locations within the tubes. Traditional inspection methods, such as visual inspections, radiographic testing (RT), and portable carburization measurement devices, have demonstrated limited effectiveness. Consequently, tube failures trigger unscheduled shutdowns, causing substantial economic losses for petrochemical companies

In response to this challenge, we have developed a robotic inspection system, integrating cutting-edge technologies. This system makes proactive inspections not just a possibility but a reality. Field demonstrations have confirmed its proficiency in detecting defects along the entire length of radiant tubes, effectively identifying issues like bulging, carburization, and cracks.

A recent inspection of a two-year-serviced cracker unit has yielded unprecedented results. We acquired comprehensive data on circumferential strain in the entire length of the tubes and established a clear correlation between bulging and carburization. Notably, our findings indicate that carburization is not solely influenced by the temperature field generated by burners but also by feedstock turbulence within the tube, in this case, associated with welding joints. The carburization map obtained for each tube serves as a valuable guideline for the recoil process, particularly given the material's reduced ductility when carburization occurs.

This novel inspection system not only enhances equipment management and maintenance efficiency but also bolsters the sustainability and safety of industrial production. Our application marks a pivotal shift toward a more intelligent, precise, and effective approach to ethylene plant management. This includes risk reduction, optimization of process parameters, accurate tube replacement, and significant economic benefits, ultimately boosting overall competitiveness in the industry.

33f FAILURE OBSERVED AT CROSSOVER PIPING OF PYROLYSIS FURNACE AND PATH FORWARD

Bhadresh Prajapati, Jimmy Cleavinger

304H Stainless Steel has long been known as the material of choice for furnace convection sections. However, exposure to elevated temperatures for long durations can lead to higher temperature degradation. In this case study, we represent the unique failure observed at the crossover piping, 304H metallurgy, at small bore connections after four and half years of service. This systematic failure was observed on 100+ small-bore connections at the same process location on three furnaces. The location of cracks on individual small-bore connections was between the 5 o'clock position and the 7 o'clock position. The crossover piping was operating in the range of 1300°F to 1350°F. The failure analysis has indicated presence of thermal as well as mechanical stresses, however, without sensitized microstructure. Cracks were observed in all three directions: longitudinal, circumferential, and radial. At the toe of the weld, cracks were stepwise, however were not interconnected. Thermal fatigue is identified as the principal root cause. In addition, unexplained chromium dilution in the weld was observed. The probable cause of the chromium dilution is "Self-healing". This presentation is intended for inspectors, engineers and managers explaining the complex failure mode, root cause assessment and effective resolution.

158a Current Trends – Global Olefins

Jeffery Nichols

The past two decades have witnessed considerable change within the global olefins market. That change has not been confined to any single characteristic, either. Facility size, energy efficiency, turnaround duration and cost, and feedstock composition are just some of the areas which have marked notable dynamics, both regionally and globally. In addition, regulatory pressures related to greenhouse gas emissions, carbon footprint, and overall sustainability have encouraged significant modification to both design and operation of facilities.

As global leaders in benchmarking and performance improvement, HSB Solomon Associates, LLC (Solomon Associates) have compiled and summarized data to exhibit global and regional trends related to the following:

- Turnaround key indicators, including durations, intervals, and relative costs
- Energy efficiency, including Solomon's proprietary energy intensity and energy sustainability indices (EII and ESI)
- Olefins plant reliability
- Maintenance costs

Solomon Associates will also present commentary and insight relative to these areas.

158b Ethane Feed Vaporizers Startup Difficulties

Samraat Paul, Andre Bernard, Manuel Dreise

Changes in configuration and operating conditions of parallel ethane feed vaporizers integrated with the ethylene refrigeration system resulted in startup difficulties. While one of the exchangers was vaporizing and superheating the ethane feed stream, the other one seemed stalled. Fouling was initially speculated as the problem cause. However, this predicament was initiated by the exchangers outlet vapor to liquid transition that resulted in a hydraulic imbalance between the 2 exchangers. The condensing side of the exchanger's outlet is characterized by vertical pipe risers, which ties into a long horizontal header. Different liquid accumulation rates between the 2 risers and/ or preferential liquid back flow from the horizontal header to one of the risers caused the vapor supplying the exchangers to follow the path of least resistance, hence stalling one of the exchangers. This paper discusses how operating data complemented with process simulations was used for identifying this problem along with the proposed solution. A review of typical project workflow is also discussed to highlight potential design pitfalls that could result in similar situations.

158c TLE Water Quality: Leveraging Cross-Organizational Collaboration to Improve Chemistry Control and Increase Steam System Reliability

Anna Akker, Jacob Tilley, Carl Peterson

A Transfer Line Exchanger (TLE) is used to rapidly cool steam cracking furnace effluent while generating high pressure steam. Due to extreme conditions present in the TLE's and potential for challenging water circulation, the TLE's often experience steam drum chemistry reliability issues. Poor steam drum chemistry can lead to furnace equipment damage and downstream turbomachinery issues, ranging from inefficient operation to potential ethylene cracker shutdown. Therefore, maintaining TLE water chemistry control is a vital component of overall ethylene plant reliability.

Water chemistry quality control can be complex due to shared ownership and responsibility within an operating organization and with a water chemistry partner. Building a successful partnership relies on having standard, agreed upon quality control practices, effective collaboration, and clear communication between teams.

This presentation intends to discuss, with a case study, how cross organizational troubleshooting can be leveraged to improve TLE reliability through tightened chemistry control and efficient, collaborative response to upset events.

158d Learnings from Cold Box Core Exchanger Replacement on Existing Unit

William Butler, Robert Witt

Beaumont Chemical Plant required replacement of a cold core exchanger due to a long-term fouling issue. The fouling caused high DP and poor heat transfer leading to lost recovery of valued molecules. Graphs of this will be included. This is a brazed aluminum heat exchanger (BAHX) inside of a perlite filled cold box.

The first attempt to purchase a new core used a non-OEM exchanger supplier but the exchanger had to be disposed due to manufacturing issues. The site then utilized another vendor and successfully manufacture the new core. There were multiple learnings due to the fact that neither vendor provided an identical replacement in kind for a 30+ year old exchanger. Details will be provided on some of the technical quality checks and design evaluations.

Installation was successful without issues. A general overview of the process, typical hold points, inspections, leak testing, and findings will be provided. Discussion of the perlite changeout typical process and safety concerns will be included.

Base improvements were made during the outage to prevent future fouling. The paper will discuss some standard cold box exchanger protection elements.

The old exchanger was dismantled by cutting in a fabrication shop to determine the root cause of the fouling. Multiple cuts were made on the old exchanger over the course of three days. This was a unique experience and pictures will be provided. The shop work helped identify particulates as the source of the exchanger performance issues. This method of troubleshooting may have a low probability of success depending on the choice of cuts on the exchanger pieces. Learnings from the process will be provided. The cutting required a lot of senior technical people observing and discussing the issues to find the fouling in the cut exchanger pieces.

Post start-up performance on the new core exchanger was very good and returned the unit back to design performance. Graphs of the performance change will be provided.

158e Feed Flexible Furnaces: What Are the Advantages, Constraints, and Considerations?

Huynh Pham

In the ethylene industry, cracking furnaces are typically designed for dedicated feedstocks. The designs of the radiant coil in the firebox, as well as the coils of the convection section, are catered to the feedstock to be cracked. Furnace designs have become flexible enough to fully crack a gaseous feedstock or a liquid feedstock.

Because of the constantly varying costs of feedstocks, ethylene plant owners may purchase an array of gaseous and liquid feedstocks. Sometimes this is a usage of products from the owner's refineries or the import of liquefied petroleum gas (LPG).

Grassroot designs of today's plants require furnace counts with large ethylene production capacities to meet their needs. Distribution among these large capacity furnaces becomes difficult, when typically the furnace loading is of one type of feedstock. In a revamping design, increased cracking results in increased recycle cracking. However, the recycle flow is not substantial enough to load one of the new higher capacity furnaces. Such was the case at YNCC plant #2 where two (2) new feed flexible 192U were designed and installed at site.

The YNCC#2 plant feedstocks ranged from Heavy Naphtha to light gas feedstock, including the recycle ethane. The furnace design allowed for simultaneous cracking of the liquid feedstock and the gaseous recycle ethane in the same furnace. The design encompasses the practice of independent cell operation (ICO) where different feedstocks can be cracked in the different cells, to which one of the cells can also be in decoking mode – independent cell decoking (ICD). Even further, when needed, the recycle ethane can be cracked in just a quarter of the furnace by using Quad-cracking, where only half of one cell is cracking a particular feed. With all the considerations to be feed-flexible, the furnace was also designed with the Gas Turbine Integration (GTI). The furnace convection section heat integration design allows the furnace burners to be capable of utilizing ambient air as well as gas turbine exhaust as the combustion air source.

This presentation will discuss the considerations that went into the design of the 192U feed-flexible furnace while incorporating the use of GT exhaust and address the furnace availability concerns when the other recycle gas furnaces are decoking.

158f Procedural Automation – Best Operator Always Available Reliably

Stephan Hazzam, Parth Shah

Ramping and switchovers during frequent transitions in petrochemical plants involve manual field and panel operator activities. Normally the ramping of setpoints and turning of automatic valves are executed manually based on standard operation procedures. These manual operations if not performed properly can have an impact to downstream process units and lead to off-spec products, less production time, lower energy consumption and unnecessary GHG emissions. In worst case they can even trigger process upsets.

An automation of these activities by the Automatic Sequence Transition module in an integrated Advanced Process Control System leads to more consistent and reliable operation. The variations during cracking furnace transitions (e.g., standby to cracking or decoking) due to less experienced operators can be mitigated by help of this kind of automation to achieve optimized olefin production.

This paper explains the potential of this solution, architecture, and challenges especially on a brownfield projects and lessons learned.

SPRING24
+20TH GCPS
A Joint AIChE and CCPS Meeting

Paper #81a

PROCESS GAS COMPRESSORS IN THE ETHYLENE PLANT:

Fouling 101

Dave Smith
Senior Marketing Manager – CPI
Nalco Water, an Ecolab Company
mdsmith@ecolab.com

Prepared for Presentation at
American Institute of Chemical Engineers
2024 Spring Meeting and 20th Global Congress on Process Safety
New Orleans, LA
March 24-28, 2024

PROCESS GAS COMPRESSORS IN THE ETHYLENE PLANT

Fouling 101

Dave Smith
Senior Marketing Manager – CPI
Nalco Water, an Ecolab Company
mdsmith@ecolab.com

Keywords: ethylene, compressor, fouling, polymerization, PGC, CGC

Abstract

The Process Gas Compressor (also known as the Cracked Gas or Charge Gas Compressor) is a key unit operation in the production of ethylene via the steam cracking process. It is a complex and expensive piece of equipment, and its operation often determines the run length of the unit – a compressor malfunction can result in the entire unit being shut down. Declining performance of the compressor can negatively impact energy consumption and carbon emissions, product yield, and throughput, can cause damage to the machine, or unplanned unit outage. To prevent these costs, ethylene producers have implemented a number of mitigation measures.

The purpose of this paper is to provide a primer on the various fouling mechanisms, mitigation of fouling, and monitoring of the compressor performance to minimize the impacts of fouling in the compressor and interstage coolers, thus improving reliability and profitability of the manufacturer.

1 Introduction

Ethylene is a critical product of the petrochemical industry. As a reactive monomer, it serves as a building block for many products, such as polyethylene, as well as intermediates such as ethylene dichloride, ethylene oxide, and others. The production process for ethylene also yields valuable co-products such as propylene, butadiene, pygas, and others that serve as key intermediates as well. Global ethylene production totaled 183.5 million tonnes in 2022[1].

A key unit operation of ethylene production is the process gas compressor (PGC), also known as a charge or cracked gas compressor (CGC). As is often the case with equipment used in reactive monomer processing, the process gas compressor and the associated piping and equipment are prone to fouling that can result in serious economic and operational consequences for the ethylene manufacturer.

2 Fouling in the Process Gas Compressor

2.1 Consequences of Fouling

Due to mechanical, operational, and chemical innovations, chemical manufacturers have been able to increase turnaround cycles from 1-2 years to as high as 8-10 years[2], with the compressor often serving as the critical factor in achieving turnaround-to-turnaround performance. Compressor fouling has also been listed as one of the top three reasons for unplanned unit shutdowns[3], and even if a full shutdown is not required, compressor fouling can cause a unit trip.

There are many other serious negative economic impacts on the ethylene unit attributable to compressor fouling. As foulant material builds up on the compressor wheels, the efficiency of the machine decreases, requiring increased energy inputs to maintain the same performance. Decreased efficiency requires increased rotor speed to maintain the required discharge pressure, which results in an increase in steam consumption for a turbine-driven compressor; in addition to cost, this can also result in increased greenhouse gas emissions, or increased electricity load. When the RPM limit is reached, the operator must then either reduce throughput or increase suction pressure to meet the discharge pressure requirement. Increasing the suction pressure puts backpressure on the cracking furnaces, lowering selectivity towards ethylene[4].

Another impact of compressor fouling is potential damage to the machine. If the fouling on a wheel is unbalanced, it can cause increased compressor vibrations. Even if the vibrations do not reach a limit that will trip the machine, they can still do physical damage to the compressor. In an extreme case, it could result in loss of containment and discharge of flammable process gas to the atmosphere.

Fouling in the compressor is not limited to the wheels or flow paths themselves. Foulant can also build up in the labyrinth seals, creating an easier path for backflow. This also reduces the efficiency of the machine. Fouling can also be observed in the inlet guide vanes, balance line, and can even build up in the discharge piping.

Fouling can also impact the performance of the interstage coolers (also known as aftercoolers) that cool the compressed gas between compression stages. Fouled interstage coolers provide less cooling; the resulting higher temperatures increase the fouling severity of subsequent stages, as explained below. As the interstage cooler fouls, the differential pressure (ΔP) across the exchanger increases, which increases the discharge pressure of the preceding stage and increases energy consumption. The increased ΔP consequences also can mirror the impacts of efficiency loss explained above.

Due to the time and cost constraints of an ethylene unit turnaround, often the discharge piping and balance line are not inspected, and thus accumulated material is not removed. As the piping cools and contracts, material that had adhered to the surface can loosen. As the unit starts up, that material can then slough off and flow downstream to the interstage cooler, resulting in a high start-of-run ΔP.

What happens in the compressor system can impact other areas of the plant as well. Downstream of the interstage coolers, the cooled process stream goes through a knockout drum, where liquids are removed. Water from these drums typically is routed to the quench water system. Foulant in these water streams can cause emulsions and fouling in the quench area. Because knocked-out hydrocarbons often feed to a condensate stripper, foulant material can impact the operation of that column as well.

In summary, the compressor operation has a wide-reaching impact on the entire ethylene unit.

Figure 1. Examples of fouling in compressors and aftercoolers.

2.2 Mechanisms of Fouling

Fouling in the compressor and aftercoolers can be caused by inorganic or organic material, although organic-based fouling is the most widely consequential.

2.2.1 Inorganic Fouling

Inorganic fouling is most commonly caused by carryover from the caustic tower, which leaves behind salts when the aqueous phase flashes. This typically does not occur in compressors that use attemperation water to lower discharge temperatures (a practice discussed in more detail below); however, while water addition is common in more recent crackers, not all plants engage in this practice. The other, less common source of inorganic fouling is the deposition of corrosion by-products from the organic acids produced in the ethylene furnace.

2.2.2 Organic Fouling

The mechanism most commonly responsible for compressor fouling is organic in nature: polymer is formed from reactive monomers present in the cracked gas such as butadiene, isoprene, styrene, and vinyl acetylene[5]. These monomers polymerize, forming increasingly longer chains that until they become insoluble and begin depositing on equipment, piping, exchanger tubes, etc.; this polymer begins to foul the equipment and impede performance.

The main reaction mechanisms for organic fouling are free radical polymerization, Diels-Alder condensation, and thermal dehydrogenation[6].

In **free radical polymerization**, there are two different potential initiation steps. In one, a reactive monomer undergoes thermal initiation to form a carbon-centered radical. These radicals are unstable molecules that then react with other monomers, building an increasingly longer-chain radical molecule. In the other possible initiation step, peroxides decompose to peroxy radicals, which then also react with reactive monomers to form carbon-based radical molecules. The chain-building reactions between radicals and monomers is known as propagation.

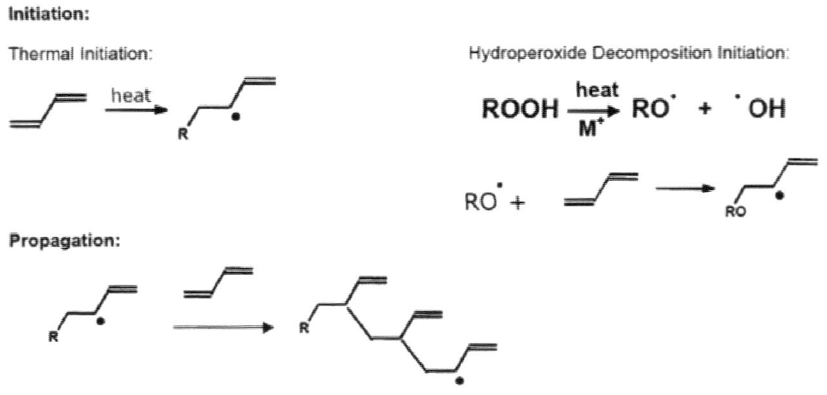

Figure 2. Free radical polymerization reaction steps

In **Diels-Alder condensation**, smaller unsaturated molecules react to form ringed molecules with double bonds that continue to react with each other to form cyclopentadienes, polynuclear aromatics, and tar-like materials that become heavier until they are insoluble and begin to deposit.

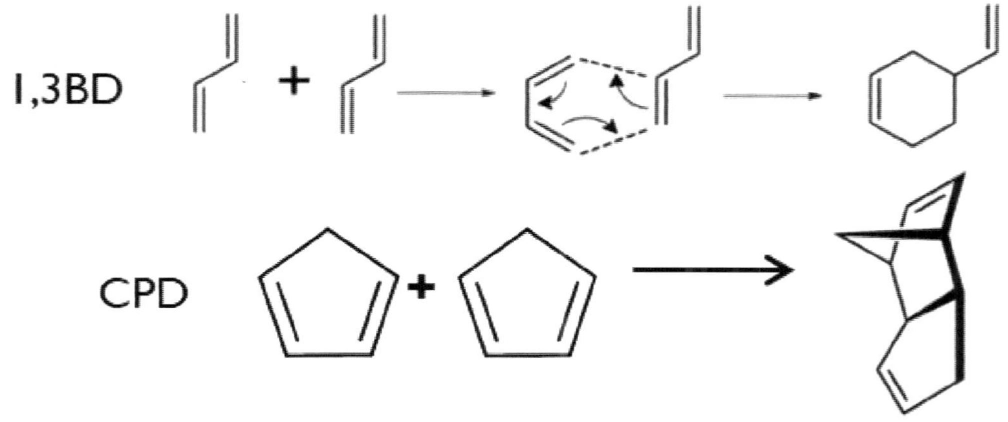

Figure 3. Diels-Alder condensation reaction steps

Over time, polymer formed by either of the above mechanisms can **thermally degrade**, dehydrogenating to form a hard, coke-like material that can be especially detrimental to equipment.

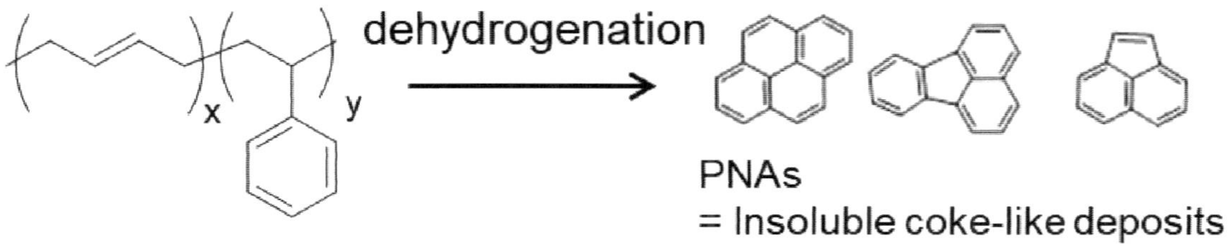

Figure 4. Thermal dehydrogenation reaction step.

2.3 Fouling Mitigation

Because of the serious consequences of compressor fouling, the ethylene industry has developed multiple methods for mitigating fouling, comprising mechanical, operational, and chemical aspects of ethylene production.

2.3.1 Mechanical Methods

Mechanical methods tend to involve either pre-construction design changes and innovations or procedures taken during a turnaround. These methods include:

- Installation of the piping and valving necessary to allow pulling an interstage cooler out of service for cleaning while keeping the unit operating.
- Increased fastidiousness in inspecting and cleaning discharge and kickback piping, knockout drum mesh pads, interstage coolers, and related equipment to maximize performance at startup.
- Installing of a 5-stage machine instead of a 4-stage machine adds capital cost, but doing so reduces the fouling severity by reducing discharge temperatures.
- Adding a booster compression stage; this action is taken to increase throughput, but the effect can be reduced fouling severity.
- Increasing the size of the interstage coolers, or using a lower-temperature cooling medium, e.g., propylene, instead of cooling water can lower discharge temperatures.
- Upgrading to improved and more extensive monitoring systems helps detect fouling earlier to allow for a more proactive response.
- Application of a coating to the compressor internals can help impede the adherence of foulant to the equipment[7].

2.3.2 Operational Methods

Attemperation Water

Temperature plays a significant role in polymerization: as the temperature increases, the reaction rate of peroxide decomposition and styrene polymerization both increase in a non-linear manner, thus increasing the fouling severity potential.

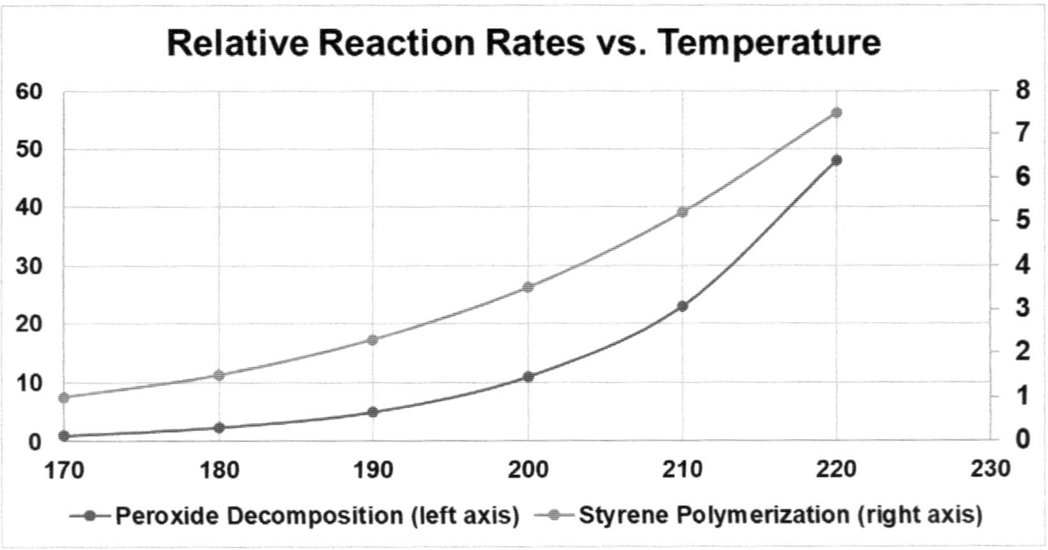

Figure 5. Relative reaction rates vs. temperature for hydroperoxides and styrene[8,9]

Thus, lowering the discharge temperature reduces the polymerization reaction rate. To lower the discharge temperature, many plants inject attemperation water into the compressor. As the water vaporizes, it absorbs heat of compression, resulting in a lower discharge temperature; the water also reduces the volume and the molecular weight of the process gas.

While water addition can remove salts that have deposited, it is important to note that **attemperation water is not wash water**. In fact, water has the opposite effect: polymer is insoluble in water, and a Nalco Water survey has shown that incidents of interstage cooler fouling have increased over the past several years as use of attemperation water has increased. While some of that increase can be attributed to longer turnaround cycles, cracking lighter feedslates, and exchanger design, the decreased solubility due to water addition is a logical factor.

There are caveats to water addition:

- Only deaerated water should be used for attemperation, as oxygen will increase the fouling severity as addressed above. Boiler feed water is the most common source.
- Minimize the water use, as excess water can cause erosion/corrosion.
- Calibrate the flow meters periodically to ensure that the water injection volume is accurate.
- When starting up a new water injection system, monitor the water from the knockout drums to note pH and iron to monitor for potential corrosion and acidic species in the water.

Wash Oil

Compressor wash oil acts as a solvent for low molecular weight polymers and oligomers and provides physical scouring of the surface of the compressor. While the preferred method is to inject it directly to the wheels, water can be added to the suction of the machine; however, if doing so, it is recommended to use an injection quill for good dispersion and to reduce the risk of erosion/corrosion.

Wash oil can be added continuously, intermittently, or both. The results from a Nalco Water survey of North American ethylene manufacturers are shown in Figure 6.

Figure 6. North America compressor survey on wash oil application.

Wash oil may be an internal stream or purchased from a commercial source, but not all wash oils are equal: they can vary widely in properties and quality. In *A Survey of Compressor Wash Oils*[10], the author outlines the characteristics to look for when choosing a wash oil and the associated ASTM test methods:

- **Solvency**. A high solvency for styrene, butadiene, and Diels-Alder polymers is key. (ASTM D-1319, ASTM D-1133)
- **Initial boiling point above 200ºC (392ºF)**. The wash oil is ineffective if it flashes and could actually be a detriment if it leaves behind solids. (ASTM D-2887)
- **Final boiling point below 200ºC (392ºF)**. Wash oil with heavy tails could leave behind residue. (ASTM D-2887)
- **Low solids**. A distilled product with existent gums less than 25mg/100mL is preferred.
- **Low reactive monomer concentration**. The Bromine number should be below 5g Br_2/100g. (ASTM 1319)
- **Density**. A wash oil with a density too close to water could cause emulsions. (ASTM 4052)

Note that when cracking lighter feedslates, fewer high-solvency aromatics are produced in the furnace[11], so the fouling severity increases. This is especially true for mixed crackers moving toward gas feed. In North America, the shift to shale gas feedstocks also resulted in a change in the characteristics and availability of the wash oil on the market[12].

Recycle Streams

Some ethylene plants bring recycle streams from other processing units into their quench system or compressor, either continuously or episodically, e.g., recovered gas from a refinery or polymers unit. Often, the precise composition of that gas is unknown, but can contain deleterious components for the compressor such as oxygenates and reactive monomers. Many plants have moved away from this practice and have rerouted these streams to other areas.

2.3.3 Chemical Methods

Based on various mechanical and operational practices and their impact on fouling in the compressor, Nalco Water developed a matrix for determining fouling severity. Stressors include discharge temperature, lighter feedstocks, recycle stream, and departure from design, while wash oil usage and quality decrease fouling severity.

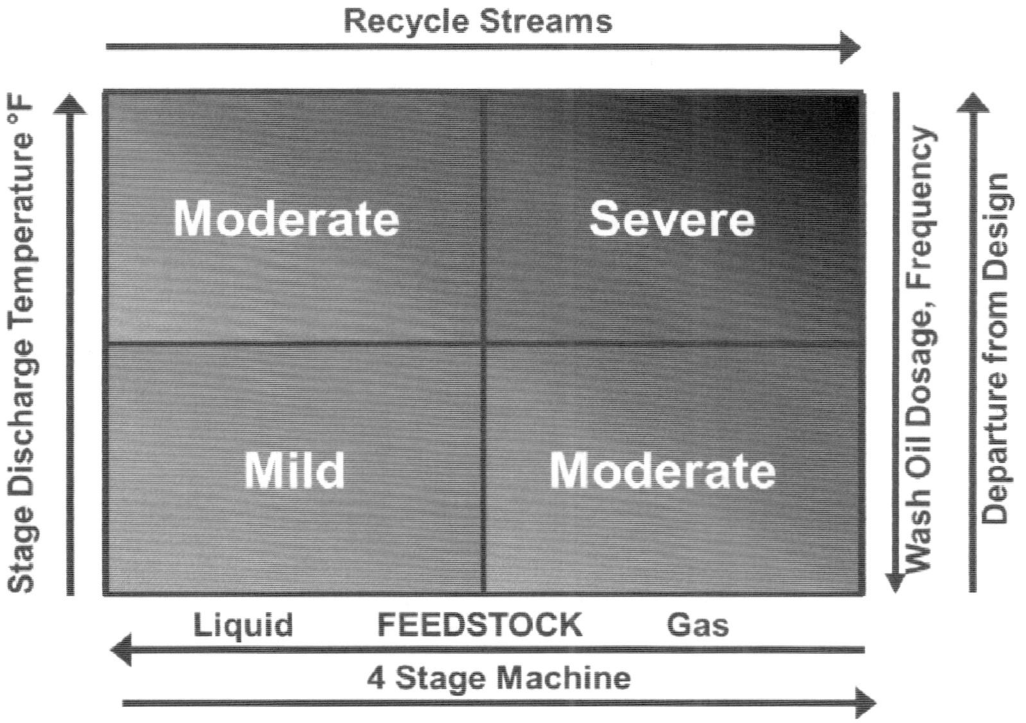

Figure 7. Compressor fouling severity factors

In some moderate or high severity systems, mechanical and operational adjustments still do not provide performance that meets plant reliability goals. In those cases, chemical additives known as antifoulants may be able to provide the solution.

Various antifoulants have different functionalities.

Inhibitors react to terminate polymerization reactions. Various inhibitors work by "scavenging" either the carbon-centered free radicals (antipolymerants) or the peroxy radicals (antioxidants). In Figure 8, the reaction termination mechanisms are shown alongside relative reaction rates.

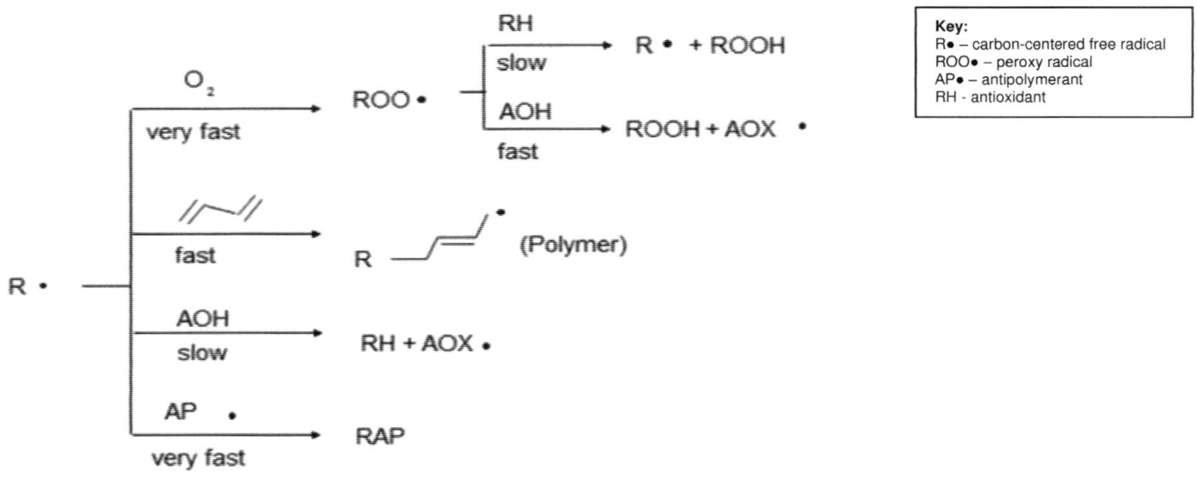

Figure 8. Reaction termination mechanisms of polymerization inhibitors

Dispersants are surface active agents, or surfactants, that do not inhibit polymerization reactions; rather, they disperse polymers and oligomers to prevent deposition and can sometimes even have a cleaning effect, depending on the nature and the tenaciousness of the foulant material[13].

Metal deactivators prevent iron oxide-catalyzed formation of peroxy and carbon free radicals.

Corrosion inhibitors minimize fouling from corrosion by-products and minimize the formation of catalytic iron oxide.

Antifoams can prevent carryover from the caustic tower, a potential source for inorganic salt fouling.

Often, the best approach is to implement an antifoulant program that is a combination of various components. Using the Fouling Matrix can help guide the selection of the appropriate antifoulant program for a specific compressor.

2.4 Compressor Monitoring

Because of the scale of economic and operational impacts of compressor fouling, extensive monitoring is a requirement. A thorough compressor monitoring program should include the following:

- **Polytropic efficiency**. To account for changes in gas composition (e.g., due to feedstocks or a recycle stream) and throughput, the recommended practice is to compare calculated efficiency to design efficiency using the compressor design curves.
- **Vibrations**. Note that vibrations can be an indication of fouling but could also result from non-fouling-related issues, such as a mechanical problem or poor lube or seal oil quality.

- **Interstage cooler ΔP**. To accommodate for changes in throughput, the ΔP should be flow-corrected.
- **Knockout drum pH and iron**. These parameters will help detect corrosion or caustic carryover.
- **Throughput**.
- **Wash oil quality and application.**
- **Suction and discharge temperatures and pressures.** These are used to calculate efficiency but can also be indicators of other changes in operation that are meaningful.
- **Attemperation water flow**. Minimize water injection to ensure an absence of free water.
- **Recycle stream flow.**
- **Equipment inspections**. During every turnaround, the compressor and interstage coolers should be inspected for fouling, corrosion, and mechanical damage, as should the discharge piping and balance line.

Proper monitoring is impossible without accurate data, so periodic checks of the instrumentation is advisable, e.g., a single gauge pressure check in the field to verify pressure data or using an IR device to confirm temperatures. Attemperation water flow meters are notoriously inaccurate, so they should be checked on a regular basis and recalibrated when appropriate.

Polytropic efficiency is a key performance indicator, so calculating it accurately is essential to a successful monitoring program. An accurate efficiency calculation requires gas composition and must be corrected for attemperation water addition – the purpose of the water is to lower the discharge temperature, so using the observed discharge temperature will provide an artificially high calculated efficiency. The ASME steam tables provide the enthalpy data needed to calculate the correct discharge temperature to use in the efficiency calculation.

Polytropic efficiency is calculated as follows[14]:

$$\textbf{Polytropic efficiency} = \frac{\ln (P_d/P_s)}{\ln (T_d/T_s)} * \frac{(k-1)}{k} * 100\%$$

Figure 9. Polytropic efficiency equation

Key:

- T_s and T_d are the suction and discharge temperatures. Remember to use **absolute** temperature (Rankine = ºF +460; Kelvin = ºC +273).
- P_s and P_d are the stage suction and discharge pressures, also in absolute (psia = psig + 14.7; kPaa = kPag + 101.3).

- k is the ratio of specific heat at constant pressure (C_p) to specific heat at constant volume (C_v)

Again, polytropic efficiency is not accurately calculated without considering the attemperation water and gas composition.

Polytropic efficiency trends are a key part of monitoring, but they can still be misleading in determining if a change in efficiency is due to fouling or simply a function of a change in rotor speed, volumetric flow, or gas composition. For example, for the compressor represented by the performance curves in Figure 9, an increase in volumetric flow at 4701 rpm (the leftward-most curve) from 55,000 ACFM to 70,000 ACFM would result in a drop in efficiency from approximately 79% to 72.5%, simply by nature of the compressor design. That is why comparison of calculated efficiency to design is the most accurate method for monitoring efficiency.

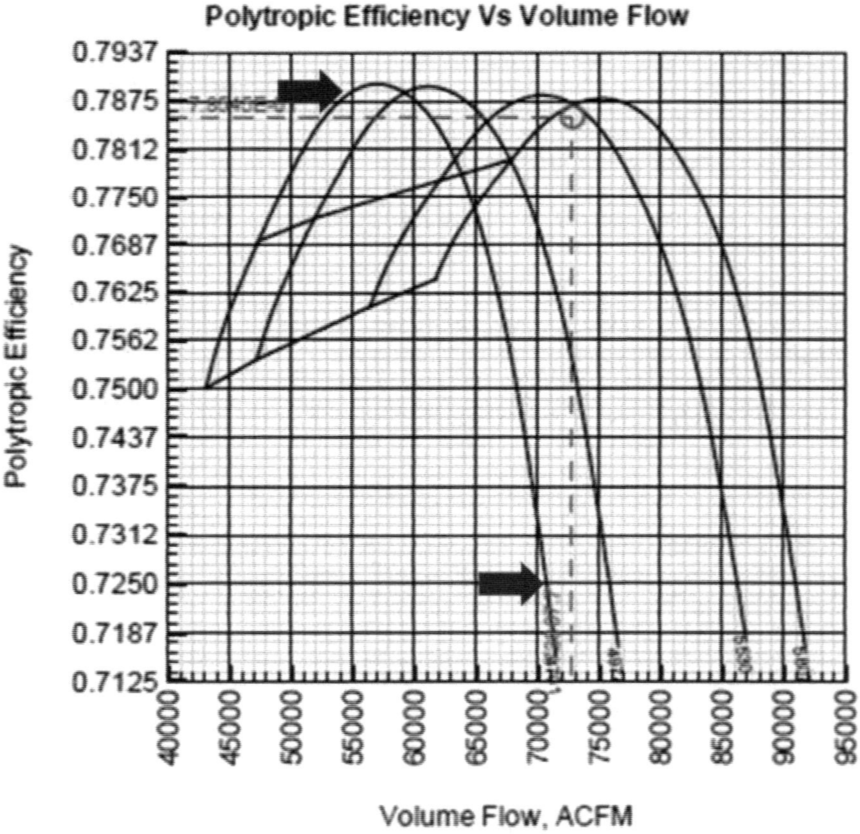

Figure 10. Performance curve for a process gas compressor

Sometimes, however, gas composition is not available. In this case, there is a less ideal option that can still demonstrate performance trends. It is known as the M-Factor, which is similar to the efficiency calculation above but without the k value. Note that the temperature and pressure ratios are swapped between numerator and denominator, meaning that as fouling increases, the M-factor increases.

$$\text{M-Factor} = \frac{\ln (P_d/P_s)}{\ln (T_d/T_s)} * 100\%$$

Figure 11. M-factor equation

Key: same as Figure 9.

One other thing to consider when evaluating compressor performance is seasonal change. For example, depending on the climate, there could be significant differences in performance due to cooling water temperature, and some refineries have seasonal operating modes that could impact wash oil characteristics. For that reason, it is important to look at the entire run of the compressor and, if possible, compare it to previous runs under the same conditions.

3 Summary

Ethylene is a critical product in the petrochemical industry, and the process gas compressor system is critical to the operation of an ethylene plant. Fouling in the compressor can cause significant negative economic impacts due to lost production, unit trip, unplanned shutdown, energy consumption, equipment damage, and overall reliability, as well as impacting safety and environmental issues.

To mitigate the impacts of fouling, ethylene manufacturers have implemented a variety of mechanical, operational, and chemical solutions that have helped to improve performance and reliability of the compressor and increase cycle time between turnarounds.

4 References

[1] Bescond A, Smith K. Chemical economics handbook – ethylene. S&P Global Commodity Insights, 2023.

[2] Smith D. The evolution of light ends fraction fouling control: from mechanism to mitigation, three decades of dramatic improvement. American Institute of Chemical Engineers Spring National Meeting, Ethylene Producers Conference, 2017. Paper #53am.

[3] Metzler R, Woodside L. Olefins plant cracked gas compressor fouling tutorial. American Institute of Chemical Engineers Spring National Meeting, Ethylene Producers Conference, 2019.

[4] Snider S. Ethylene plant cracker gas compressor fouling. American Institute of Chemical Engineers Spring National Meeting, Ethylene Producers Conference, 2006. Paper TA2002.

[5] Khanlari A, Rossana D, Arnst T. Mitigating fouling in cracked gas compressors. American Institute of Chemical Engineers Spring National Meeting, Ethylene Producers Conference, 2018.

[6] De Haan S, Sullivan B. Overview of cracked gas compressor fouling: theories and practices. American Institute of Chemical Engineers Spring National Meeting, Ethylene Producers Conference, 2001. Paper 76b.

[7] Dugas J, Neidhardt R. Compressor antifoulant coating experience. American Institute of Chemical Engineers Spring National Meeting, Ethylene Producers Conference, 2001. Paper 76c.

[8] Brandrup J, Immergut E. *Polymer Handbook Fourth Edition.* John Wiley & Sons; 1999.

[9] Boundy R, Boyer R. *Styrene, its polymers, copolymers, and derivatives.* Reinhold; 1952.

[10] Smith D. A survey of compressor wash oils. American Institute of Chemical Engineers Spring National Meeting, Ethylene Producers Conference, 2023.

[11] Imran M, Pickett T. Ethane cracking in existing liquid Crackers – effects on CGC, cold fractionation, and refrigeration. American Institute of Chemical Engineers Spring National Meeting, Ethylene Producers Conference, 2016.

[12] Hata S, Horiba J, Snider S. Process gas compressor fouling & washing mitigation: a better understanding of fouling phenomenon & compressor design techniques. American Institute of Chemical Engineers Spring National Meeting, Ethylene Producers Conference, 2010. Paper T4 175313.

[13] Khanlari A., Rossana D., Arnst T.

[14] Snider S.

81b CGC Fouling Control for Six Year Run Length Achievement in C2-C3 Feed Gas Cracker

Joice Boll, Abd Halim Bin Sulaiman, Mohd Zakiamani Mohd Nor, Kelvin Chong Chee An

This is a case history of collaboration between PETRONAS Chemical Olefins and Dorf Ketal that improved Cracked Gas Compressor system fouling control in a C2-C3 gas cracker feed to achieve a six-year run length. The collaboration identified root causes of historical fouling problems in this system. Improved understanding of root causes resulted in changes in wash oil & wash water injection, antifoulant chemistry, injection systems and monitoring. A combination of these actions and disciplined execution led to this success.

The paper will review the history of the fouling control program, past and present to demonstrate the correlation between program design and run length results.

SPRING24
+20TH GCPS
A Joint AIChE and CCPS Meeting

Review of Coating Technologies for Centrifugal Compressors

Derrick Bauer
Ebara Elliott Energy
901 North Fourth Street
Jeannette, PA 15644
dbauer@elliott-turbo.com

Prepared for Presentation at
American Institute of Chemical Engineers
2024 Spring Meeting and 20th Global Congress on Process Safety
New Orleans, LA
March 24-28, 2024

Review of Coating Technologies for Centrifugal Compressors

Derrick Bauer
Ebara Elliott Energy
901 North Fourth Street
Jeannette, PA 15644

Keywords: coatings, fouling, anti-fouling, corrosion

Abstract

Coatings have been applied to centrifugal compressors for minimizing the amount of foulant build-up that occurs during operation. The oldest coatings for this application rely on corrosion-resistant undercoat with an organic topcoat system. While these worked effectively for a while, the limitation of these coatings has been caused by a breakdown of the coating, either by mechanical means or by chemical interaction with the environment. Another type of coating that has been applied is electroless nickel which offers greater durability in harsh environments which have chemical and liquid injections. In order to properly evaluate coatings for various applications, testing to confirm that the coating will resist fouling in various application needs to be performed. The durability of the coating can also be tested through standard testing methods, and the resistance of the coating to chemical attack and permeation will need to be considered.

1 Introduction

The turbomachinery industry for petrochemical processing and ethylene production has roots that trace back many years, and this has developed tried and true designs which are manufactured from well-established materials. Given the success of the centrifugal compressors in these applications, customer have come to expect reliability and longevity from their equipment. The challenge that is faced by operators and OEMs is how to increase the operating time between shutdowns while allowing the compressor to maintain optimal operating efficiency during the entire operating time. One major issue that can limit the time between planned outages for a compressor is fouling which is the term used to describe the buildup of solids on the compressor's internal surfaces. This is a process-dependent issue that can have different factors. In hydrocarbon services, fouling can be caused by a polymerization reaction intrinsic to the compression process. Foulant build limits flow through the compressor and can alter the aerodynamics, reduce efficiency over time, increase vibration and can result in an unplanned shutdown in severe cases. Many variables affect the occurrence of fouling, and the exact process is not always fully understood. These factors make predicting the amount of fouling expected with a centrifugal compressor impossible to determine with any amount of certainty. The focus needs to be on how to best minimize the amount of foulant depositing on the surfaces of the compressor.

Coatings have been applied to both the rotating and stationary components within a centrifugal compressor to help maintain this operating efficiency by both preventing the adhesion of foulant from the process gas and providing corrosion protection to the base metal of the components. As there is increasingly more demanding applications for centrifugal compressors, the development of new coatings with improved properties is always necessary. This includes testing for not only the anti-fouling properties of the coating, but also considering the chemical resistance and durability of the coating.

The first applications of anti-fouling coatings in a centrifugal compressors occurred in the 1970s. This coating was a PTFE (abbreviation for polytetrafluoroethylene) coating applied directly the diaphragms of the compressor. In 1995, Chow, McMordie and Weigand reported on a PTFE based coating system that would eventually become what is essentially the industry standard fouling resistant coating [1]. Similarly, in 2000, Dugas [2] presented a case study of a coating in operation. Besides the non-stick properties of PTFE system in the topcoat of the coating system, there is very little data available showing what testing was performed to test the coating system prior to placing it into service. Despite fouling resistant coatings being in use for more than 40 years in centrifugal compressor applications, there is very little reference in literature regarding upfront laboratory testing to determine the suitability to centrifugal compressors.

2 Background

Fouling, or the buildup of deposits, presents an interesting problem for centrifugal compressors. There are a wide array of methods in which fouling can occur. The most simple would be the carryover of rust exfoliated from piping or equipment of the compressor. Adding a little more complication, solids or liquids resulting from the upstream process can be entrained in the gas stream and carried over into the compressor. These types of externally generated foulants tend to deposit on the early stages of the compressor. Figure 1 shows the flow passage of the 1st sate impeller in a compressor operating in a close loop compressing propane. Solid particulate in the loop, most likely present in the system at start up, has deposited on the impeller. Figure 2 is showing an extreme example of polymerization in a charge gas compressor. The occurrence and amount of this type of fouling depends on a number of conditions. Temperature and pressure are obviously important and a general rule of thumb is that the polymerization will occur at temperatures above 90 C.

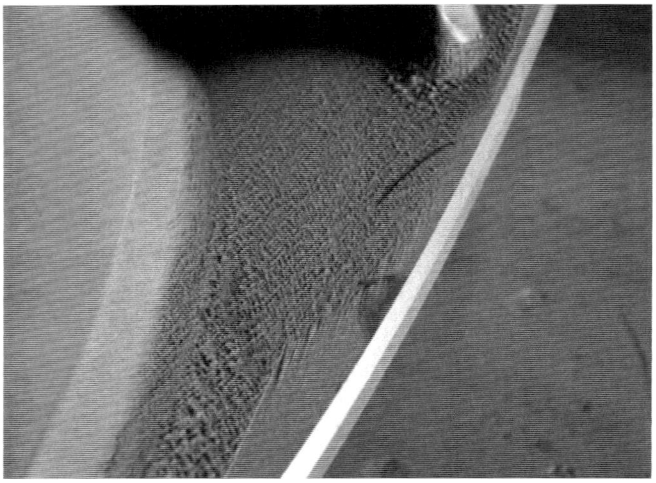

Figure 1: Fouling in the Inlet of an Impeller in Propane Service [3]

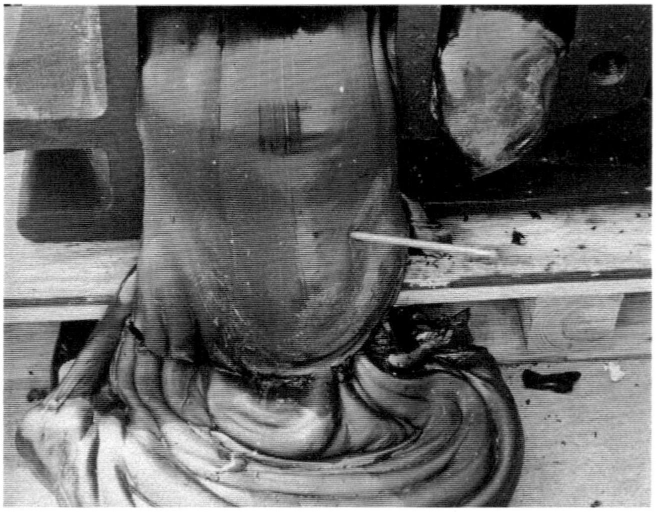

Figure 2: Fouling in the Flow Passage of Diaphragms in a Charge Gas Compressor [4]

The effect the foulant has on the compressor also depends on a number of factors and can be variable. For example, the fouling in Figure 1 is on a relatively large impeller, so it had very little if any effect on the operation of the compressor. Conversely, fouling on a relatively small impeller with tight flowpaths will have a great influence on the performance. Polymerization type fouling shows that the fouling may adhere to inlet guide vanes, impellers and diaphragms within a centrifugal compressor. There are certain locations that will tend to collect foulant more than others.

The methods to deal with foulant in a centrifugal compressor are more limited and more difficult than simply installing a filter. Options include process control to avoid conditions that cause fouling, chemical injection to either physically or chemically remove the foulant, and coatings to alter the flowpath surface to prevent foulant from adhering. All three have their advantages and drawbacks. Given that process control and chemical injection require intimate knowledge of the entire process in order to properly evaluate, the implication of the required change are best left to the user or licensor of the process. The OEMs of the centrifugal compressor can focus on the coatings which should be chemically inert to the process.

3 Testing of a Coating

The primary objective of a coating in a centrifugal compressor is to provide anti-fouling properties. There is a lot expected of a coating as it has to work for preventing the fouling, it has to provide corrosion resistance to the base metal and not react with the process environment, and it has to be sufficiently durable to withstand the operating condition of a centrifugal compressor. The types of foulant, the chemical compatibility and the amounts of liquids and solids in the process gas can range very widely. Even though there are many variables, the testing of the coatings can be evaluated to the following considerations.

3.1 Wettability

The ability to resist the buildup of a deposit beyond the capability of the base materials is obviously the property that a fouling resistant coating should possess. There are two possible routes that can be taken for a coating to perform this function. Nonstick cookware is an easy example of a fouling resistant or nonstick coating. The most well-known non-stick coating is fluoropolymer material commonly referred to as Teflon or polytetrafluoroethylene (PTFE). Coatings for this type of application have low surface energies [5]. When a liquid of a higher surface energy is in contact with the coating, the liquid is more attracted to itself than the coating. This gives rise to wetting angles greater than 90 degrees and the liquid tends to bead up on the surface and can easily roll off or be repelled away. This is called a hydrophobic coating if the liquid in question in water and oleophobic if this is an oil.

Considering hydrophobic coatings, it makes sense that the best fouling resistant coatings should have the highest wetting angles possible. Materials are termed 'superhydrophic' when the contact angle of a droplet exceeds 150 degrees [5]. Figure 3 shows a superhydrophobic coating at very low tilt angles. Tilt angle, or roll-off angle, is the measure of the angle from horizontal at which a drop begins to move. Low angles are desirable in a fouling resistant coating. In this type of evaluation, a superhydrophobic coating would be superior.

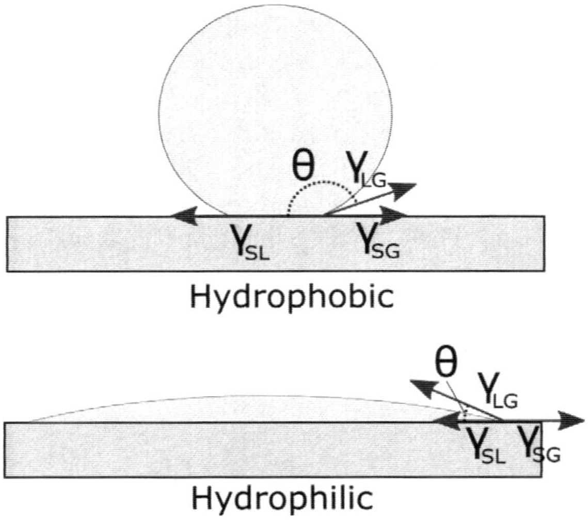

Figure 3: Schematic Showing the Difference between Hydrophobic and Hydrophilic Surfaces

3.2 Hydrocarbon Fouling

Testing the wetting angle against water can provide a general guide to the suitability of a coating, but this needs to be tested against a simulated hydrocarbon foulant. The first thing that is needed to is to have a simulated foulant that can be used for testing purposes. The composition of a simulated hydrotest foulant is shown below in Table 1.

Table 1: Composition of a Simulated Hydrocarbon Foulant [6]

Simulated Foulant	
50 mL	Corn Oil
2 g	Carbon Lampblack
0.4 g	Fumed Silica
10 g	Iron Oxide

Initial testing of the fouling resistance of a coating was reported by Wang [6] and others in that this foulant was applied to coated panels. The panels were scrubbed with a wet abrasive pad and the number of cycles needed to remove the foulant were recorded. While this provides a ranking of the coatings in terms of their performance, there is nothing that should be scrubbing the flowpath of a centrifugal compressor during operation.

The next approach for testing the anti-fouling coatings is to spray this simulated foulant at coated panels and measure the amount of deposit which sticks during a certain period of time. There are no known industrial standards to follow for this testing, and there are many variations in terms of the exact foulant formula, temperature, spin speed, testing time that can be used. An example of a testing rig is shown in Figure 4 along with a schematic of the setup. This provides a ranking system of how likely a foulant is to stick to the surface which is coated. A control sample of a bare carbon steel in a machined condition should be used for a reference, and the improvement due to the presence of a coating can hopefully be measured. Examples of coating coupons that were evaluated in the apparatus shown in Figure 4 are displayed in Figure 5 [7]. Within Figure 5, Coating C had the best performance as very little foulant remained adhered to the surface during testing.

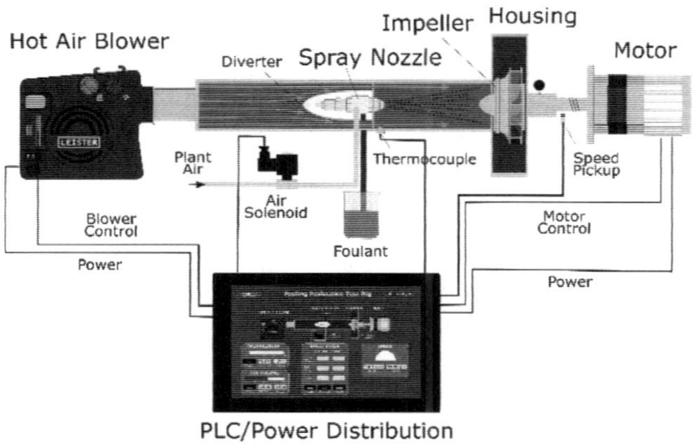

Figure 4A: Schematic of a Hydrocarbon Fouling Test Apparatus

Figure 4B: Picture of Actual Hydrocarbon Fouling Testing Apparatus

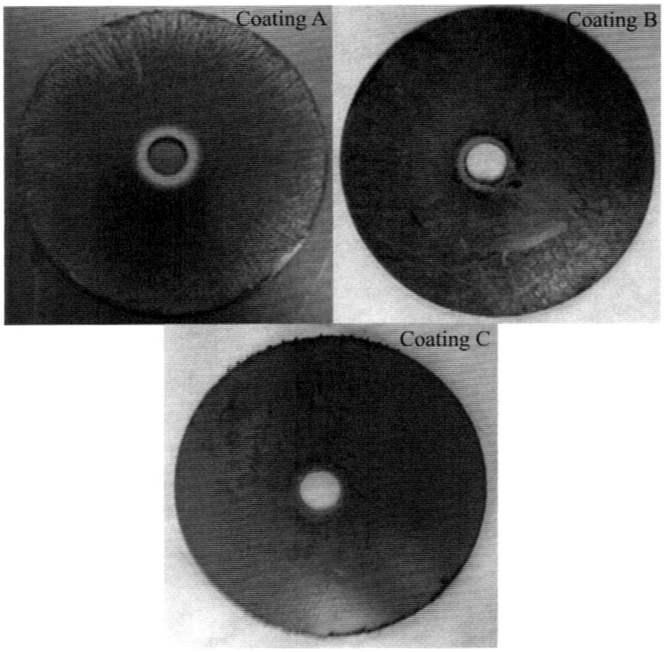

Figure 5: Examples of Various Coating Samples after Hydrocarbon Fouling Testing

3.3 Wear Resistance

The most common reason for the failure of a fouling resistant coating is related to wear, or more specifically, erosion. Figure 6 displays a compressor rotor that had been taken out of service where the coating had experienced significant wear damage. Given that is often desirable to have a 5 to 7 year operational period between shutdowns, it is very desirable to maximize the wear life of the coating.

Figure 6: Fouling Coating on Centrifugal Compressor Rotor Showing Wear Damage

Organic topcoat coatings are typically utilized for their anti-fouling properties, and these organic coatings are significantly softer than the metals they are applied to. Despite best efforts to prevent solid particulate form entering the compressor, it is inevitable in most cases that some solids will be entrained in the process gas. With respect to the coating, the particles act as blast media and can quickly remove the coating. Liquids are also frequently encountered, with many compressors injecting chemicals as part of the process or in conjunction with the coating to prevent fouling. Purposely injected liquid should be atomized to minimize damage to the compressor. Even the process gas itself can cause wear if it is directed at the coating with sufficient pressure.

Designing wear resistance into a coating is difficult with respect to organic coatings in fouling applications. Fluoropolymers are typically used in the coating system due to their fouling resistance and broad chemical resistance. In the ideal case, they are used as a pure, solid layer since the best takes advantage of their properties. Unfortunately, fluoropolymers are not high strength materials and are expected to exhibit higher wear rates than desired. To improve the wear resistance, there are two routes that can be taken. The fluoropolymers be used as a filler in different organic materials such as PAI (polyamide imide) or epoxy which may be more wear resistant. They may also be used as the matrix material with a hard particle filler such as a silicon carbide. While both of these have the potential to increase the wear resistance, it is often at the expense of the fouling resistance and in some cases can influence the corrosion resistance of the coating. The non-fluoropolymer material on the surface provides anchor points for the foulant and the interfaces between the two materials can provide short circuit paths with respect to permeation.

Wear testing of these types of coatings has primarily focused on wear due to solids. One reason for this is that applications for coating more commonly encounter solid particulate under conditions that may cause wear than liquid droplets. Coatings for pots and pans for example are more critically damaged by utensils than they are during washing. Automotive coatings are also more prone to damage from dust, road debris and solids such as hail rather than rain. Typically, whirling arm tests are used to measure the wear resistance against liquid droplets. This testing has proven to be too severe for organic topcoat coatings as they do not often survive to the first inspection level. This is a main reason why liquid injections should be atomized when being introduced to a compressor.

ASTM D968 [8] outlines a test method for evaluating the abrasion resistance of organic coatings when being impinged by an abrasive. This is a relatively simple test and is used as a standard method for comparing the wear resistance of coatings, and schematic of the testing apparatus is shown in Figure 7. The ASTM D968 specification specifies the dimensions for the both the testing apparatus as well as specifics for the abrasive particles including the composition, size and shape of the particles. The testing continues when the coating is worn away and the base metal is exposed, and the results are reported in terms of liters of abrasive per mil (0.001 inches) of coating. It is strange that the reporting combines metric and imperial units, but this helps to provide a visualization of the wear resistance of the coating.

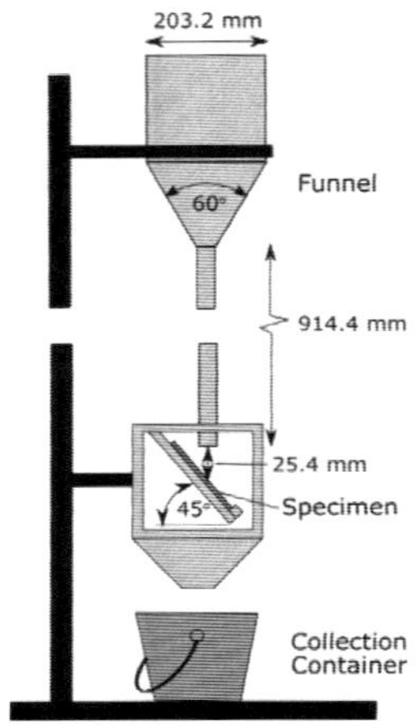

Figure 7: Schematic of the falling sand testing apparatus

3.4 Corrosion Resistance

When considering fouling, it is easy to understand the effect that it can have on the performance and behavior of the compressor during operation. It is a little more difficult to see the damage that it can cause in the form of corrosion until the internal components can be visually inspected. The vast majority of compressors are required to meet NACE MR0103 [9] and/or NACE MR0175 [10] as there can be some level of hydrogen sulfide present in the process gas. It is stated in these specifications that coatings are meant to enhance corrosion protection, not to be stand-alone protection against corrosion or stress-corrosion cracking. The base materials of the compressor components must be resistant to the environment and the coatings acts as enhancement. Internal static components within a compressor are often designed with a corrosion allowance while the critical, highly stressed rotating components are often made from stainless steel or even nickel-based alloys. The presence of the foulant on the surface alters the local environment and can promote corrosion, even in otherwise corrosion resistant materials. This can be very detrimental to impellers which are dynamically loaded. Pitting corrosion can occur underneath the deposits which can act as initiation sites for fatigue cracking.

In theory, the fact that the base material should be resistant to the process gas should relax the requirements of the coating. Basically the coating would only need to be able to withstand the environment without a care for the protection of the substrate. Data from the coating manufacturer would be able to provide all of the data necessary to make this assessment, however, this is not the case. As an example, many organic coatings do not perform well in the presence of amines or benzene, even though the coating material itself can be chemically inert in this environment. An example of this is shown by the attack on the coated diaphragm in Figure 8 [6]. It was report that the coating was being dissolved during operation of the compressor. Unfortunately, without the coating acting as a barrier, it could act much like the foulant does in terms of changing the local environment and promoting corrosion. The permeation resistance is another factor that should be considered a prime property for a corrosion resistant coating even when it does not lead to substrate corrosion. An example of this is shown in Figure 9 which displays blistering in a coating that was attributed to benzene in the process gas permeating the coating. As the pressure decreased during a shutdown, the benzene pressure trapped under the coating was higher than the pressure on the outside of the coating. The benzene gas tried to escape faster than diffusion through the coating would allow and resulted in the blistering that is shown. This is very similar explosive decompression that is sometimes found in O-rings. While this had no effect of the corrosion of the base metal, the organic topcoat was removed and fouling can occur.

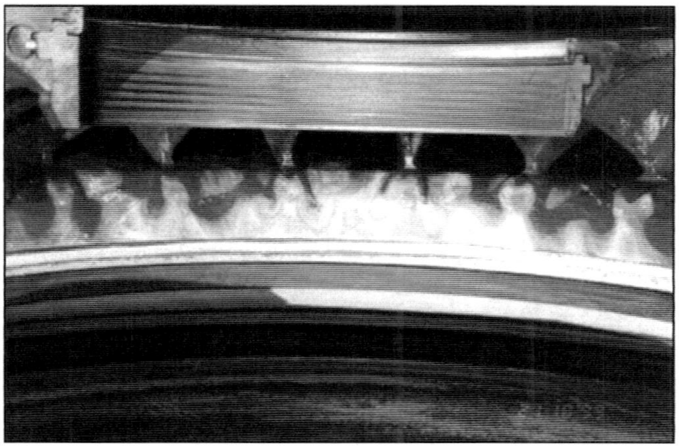

Figure 8: Example of a Chemical Attack on an Organic Topcoat Coating

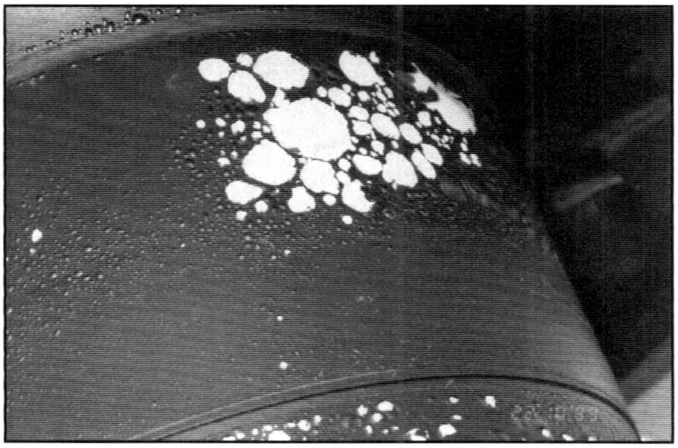

Figure 9: Blistering of an Organic Topcoat Coating [6]

3.4.1 Salt Spray Testing

One of the most common corrosion tests performed on organic coatings is a salt spray test per ASTM B117 [11]. This is a relatively simple test in which coated steel panels are placed in a salt fog chamber and periodically visually evaluated. This process is continued until there is a visual failure or a specific time period is reached. The panels may also have scribe lines through the coating applied prior to exposure which allow for a method in which a breach in the coating is exposed to the salt atmosphere. While this test is relatively easy to run and is relatively inexpensive, there is only limited value that can be gained from it. The reporting of the data is somewhat vague in that it will only state that a sample lasts some number of hours in a salt spray test or that it failed after a certain number hours. Even when is failure is reported, the criteria of exactly how the coating failed is not normally listed. An example of coated test panels from a salt spray test is shown in Figure 10.

Figure 10: Example of results from a salt spray test with scribed specimens [12]

3.4.2 Electrochemical Impedance Spectroscopy

A different type of a method which allows for more variations in the test environment along with a numerical measure of the corrosion resistance of a coating is through Electrochemical Impedance Spectroscopy, initialed EIS testing. ISO 16773-1 defines EIS as an "electrochemical technique which allows the impedance spectrum of an electrochemical system to be recorded as a function of the frequency of the applied signal, and the spectrum thus obtained to be analyzed by transfer function analysis." [13] Basically, electrochemical systems can act similar to an electronic circuit and the measurement of the electrical resistivity of the system can provide numerical data to evaluate the effectiveness of a coating. A schematic of an EIS testing apparatus is given in Figure 11.

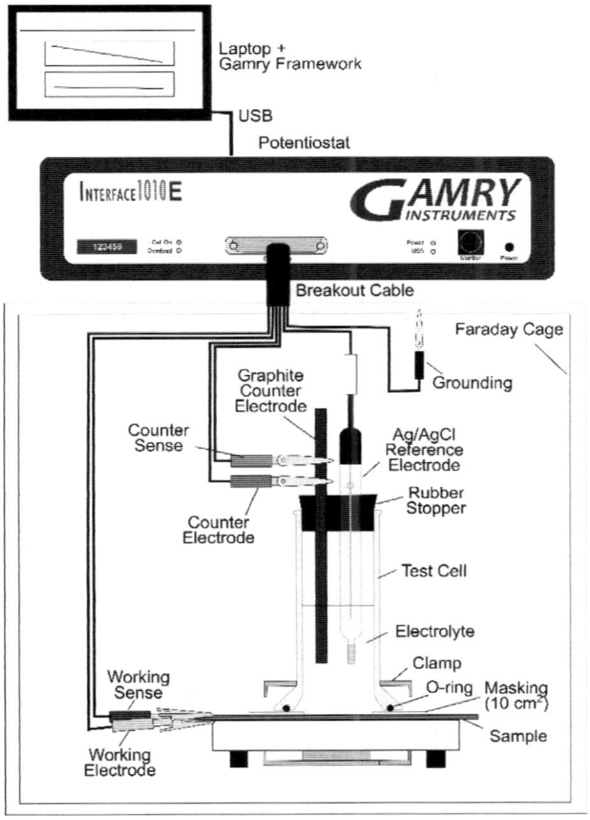

Figure 11: Schematic of an Electrochemical Impedance Spectroscopy Apparatus [7]

The Impedance of the system can be calculated from measurement of the voltage and the current. The impedance will decrease as the frequency increases, and the results are displayed as a diagonal line. The impedance can be measured over a range of frequencies at different time intervals to see how the system changes over time. A coating will likely initially act as a perfect barrier coating, and changes over time will show that a corrosion cell is being setup. If the coating remains as a perfect barrier coating, there will be no changes over time. Examples of this are shown in Figure 12. The coating showing good results actually has seven different lines all on top of each other, which is showing that no corrosion cell is being set up over a 2,000 hour time interval. In the example of a failed coating, you can clearly see how the damage of the coating has changed the results over time.

The benefits of an EIS test is that the coating substrate, coating and the corrosive solution are all variable. Testing solutions containing ammonium chloride are given in Figure 12, however, any process gas constituent or additive can be included in this test. Along with providing the ability to test a coating's effectiveness to specific test environments, the results over time can be shown numerically.

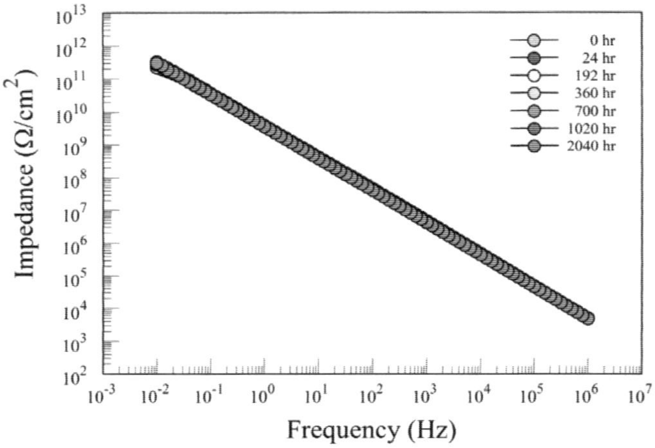

Figure 12A: EIS Testing Results Showing no Corrosion After 2,000+ Hours

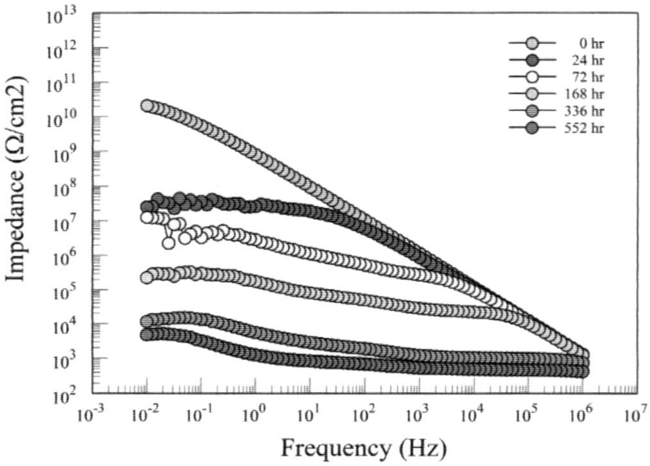

Figure 12B: EIS Testing Results Showing Corrosive Attack on a Coated Sample

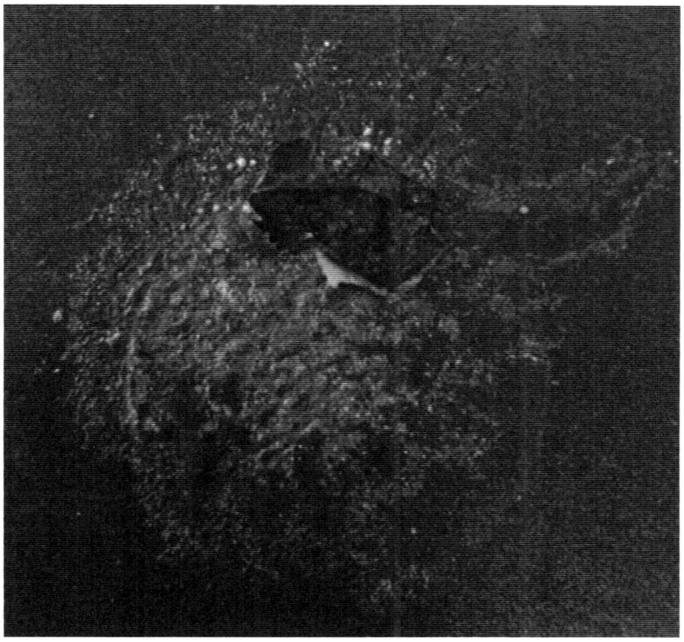

Figure 12C: Permeation and Corrosion Attack during EIS Testing from Figure 12B

4 Types of Coatings

The most commonly utilized coating systems for anti-fouling are organic topcoat coatings and electroless nickel coatings.

4.1 Organic Topcoat Coatings

As previously mentioned, organic topcoat coatings have been utilized in centrifugal compressors since the 1970s. This multi-layer coating system contains an aluminum-filled chromate-phosphate base coat which is applied for corrosion protection as the base layers of the coating. The first pre-requisite of an anti-fouling coating is that is must provide corrosion protection against the service environment. A bond coat layer may be necessary on top of this coating, and then the organic topcoat is applied. There are many variations to these coatings, but they are all are applied with the same structure. An example of this type of coating system is provided in Figure 13.

Figure 13A: Picture of Centrifugal Compressor Rotor with Organic Topcoat System

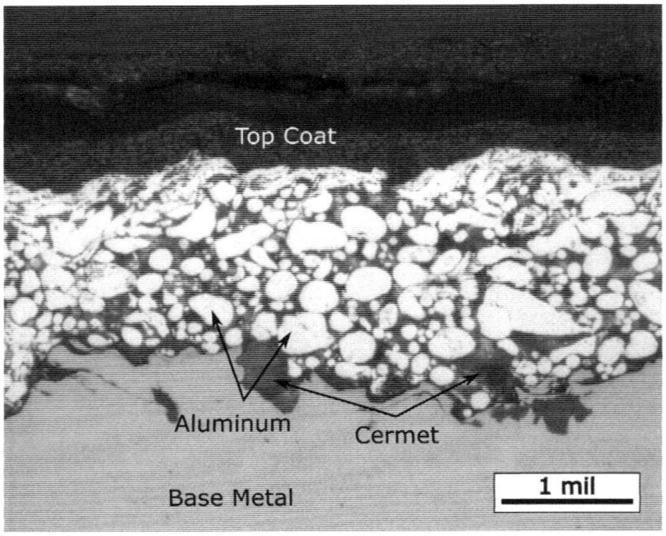

Figure 13B: Cross-Sectional Photomicrograph of Organic Topcoat System

These coatings are usually applied through a 'spray and bake' application. Each layer is applied through a spray which is usually a manual operation. Coating suppliers have their own developed techniques to reach inside of the sometimes complicated internal flowpaths of the impellers or diaphragms. Each layer is thermally cured at temperatures that can range from 500 to 700 degrees F. Diaphragms are easier to handle for this application as they can be placed in a curing oven. The rotors are usually coated as an assembly, and the curing operation requires the rotor to either be cured in a vertical position or to rotate horizontally in a clamshell oven during the entire curing cycle. This type of furnace is shown in Figure 14. The rotation is usually slow at approximately 3 rpm. The heat rate and the cooling rate must be carefully controlled so that the impeller shrink fit is not altered. After the coating process is complete, it is recommended that a high-speed balance operation is performed. There will usually be some amount of grinding corrections that are required, and the areas that are ground for these balance corrections should be outside of the impeller flowpath. A touch-up paint is usually applied to these ground areas for cosmetic reasons, the touch-up paint will not have the same anti-fouling properties as the main coating.

Figure 14: Rotor during Coating Application in Clamshell Furnace

4.2 Ceramic Topcoat Coatings

These coatings have been utilized in applications where organic topcoat coatings have not been able to withstand the operating conditions of centrifugal compressors, specifically in charge-gas compressor applications. These coatings rely on the same base coat system. Rather than applying an organic top-coat, the topcoat is simply a ceramic sealer. While the anti-fouling properties may not be as good as organic top-coat systems, these have been used as a trade for a little bit of efficiency to gain durability. The smooth surface of the un-reactive ceramic topcoat layer provides some benefit in terms of anti-fouling properties in comparison the exposed base metal, however, the anti-fouling benefits are lower than that of a fluoropolymer topcoat.

These coatings are often applied in the same 'spray and bake' method as the organic coatings previously mentioned.

Figure 15: Picture of Rotor with a Ceramic Topcoat Anti-Fouling Coating

4.3 Electroless Nickel Coatings

Electroless nickel has been used on centrifugal compressor components in applications where maximum coating durability is required. Electroless nickel is a nickel-phosphorus alloy with an amorphous structure meaning that that atomic arrangement contains no long-range order. Because of this amorphous structure, there are no grain boundaries in the coating which adds to the enhanced corrosion protection. Electroless nickel is chemically inert in many environments. The electroless nickel coating has a hardness between 40 and 45 HRC which makes this harder than the base metal of the impeller. If the injections in the compressor are able to cause erosion damage on the impeller, the electroless nickel coating will be able to withstand these conditions. The coating has a dull, metallic look which does not make it the visually most appealing coating as shown in Figure 15A.

The application of the electroless nickel is more complicated than the spray-and bake methods utilized for the other coating systems. The application process involves a series of cleaning tanks which remove all organic particulate from the surface followed by immersion in a plating tank. The solution in the tank contains a Nickel compound which reacts with the surface of the metal and deposits the nickel-phosphorus alloy. The benefit of this application method is that the electroless nickel will conform to any geometry as the component is completely submerged. A photomicrograph of the electroless nickel coating on a steel substrate is shown in Figure 15B. The drawback to this coating application is that the compressor rotor has to be unstacked so that each impeller can be coated individually. The coating tanks are not large enough to handle the coating of a fully assembled centrifugal compressor rotor.

Figure 15A: Picture of Impeller with Electroless Nickel Coating

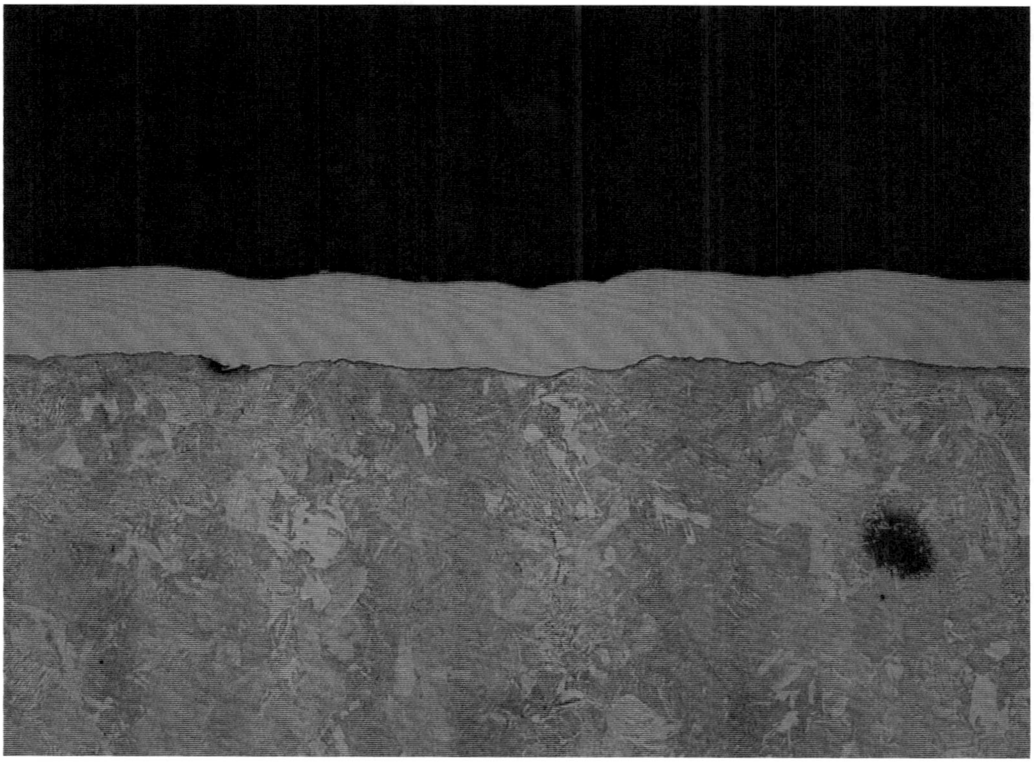

Figure 15B: Cross-Sectional Photomicrograph of Electroless Nickel Coating

4.3.1 Case Study of Electroless Nickel Coating

A major chemical plant had been utilizing an organic topcocat coating system for anti-fouling resistance in a charge gas compressor. The coating had suffered deterioration two times during a four year operational period as shown in Figure 16, and analysis indicated that this deterioration was related to the heavy washing injection which contained aggressive chemical additives along with steam cleaning operations. The washing and steam cleaning operations could not be removed as they were essential to the operation of the compressor. In order to withstand these operating conditions, the compressor rotors and the diaphragms were coated with electroless nickel during a planned turnaround.

Figure 16: Picture of Organic Topcoat Coating System after 18 Month Service in a Charge Gas Compressor [6]

The compressor internals with the electroless nickel ran successfully for a 5-year operational period without an unplanned outage. The rotor vibration had been monitored at a much lower level than the previous run periods as shown in the Figure 17. This improvement is attributed to the endurance of the electroless nickel coating against the heavy washing and chemical injections that had previously affected the organic topcoat coating system. The rotor was only taken out of service after this 5-year run as the compressor was being replaced with a new machine, and the customer requested the same electroless nickel coating be applied to the new rotor and diaphragms. The electroless nickel rotor was evaluated after the operation period in Figure 18 and the coating remained intact after the entire after the entire operating period.

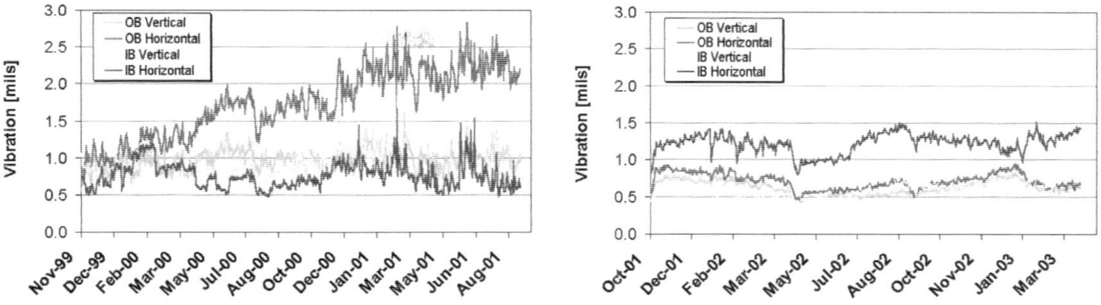

Figure 17: Vibration Data of Operation with Organic Topcoat Coating (left) vs Electroless Nickel Coating (Right) [6]

Figure 18: Electroless Nickel Coating after 5-year Operational Period

5 Conclusion

Coating have been used effectively for anti-fouling resistance in centrifugal compressors for many years. While there are documented service operations which show that the coatings help the compressor maintain efficiency and prevent fouling, the service life of the coatings have been the limiting factor. The most common types of coatings are corrosion resistant undercoats with an organic topcoat which is usually a Teflon (PTFE) or some variation or derivative of this. In order to gain durability, an option is to replace the organic topcoat with a ceramic topcoat. Electroless nickel coatings have been used in conditions where there has been severe liquid injections or chemical additives. The testing of the coating for the compressor operation should consider the effectiveness of the coating to resistant foulant buildup, ensure that the coating will not experience chemical attack or permeation during surface, and testing of the durability of the coating to evaluate how durable it will be during service.

6 References

[1] R. Chow, B. McMordie and R. Wiegend, “Performance Maintenance of Centrifugal Compressors Through the Use of Coatings to Reduce Hydrocarbon Fouling,” Elliott Company, Jeannette, PA, 1995

[2] J. Dugas, “Compressor Antifouling Coating Experience,” in *29th Turbomachinery Symposium*, Houston, TX, 2000.

[3] N. White, S. Laney and C. Zorzi, “RCFA for Recurring Impeller Failures in a 4.7 Mtpa LNG Train Propane Compressor,” in *Proceeding of the First Middle East Turbomachinery Symposium*, Doha, Qatar, 2011.

[4] S. Snider, "Ethlyene Plant Cracker Gas Compressor Fouling," in *18th Ethylene Producers Conference*, Houston, TX, 2006

[5] K. Law, "Definitions for Hydrophilicity, Hydrophobicity, and Superhydrophobicity: Getting the Basics Right," J. Phys. Chem. Lett., vol. 5, pp. 686-688, 2014.

[6] W. Wang, P. Dowson and A. Baha, "Development Of Antifouling And Corrosion Resistant Coatings For Petrochemical Compressors," in *32nd Turbomachinery and Pump Symposium*, Houston, TX, 2003.

[7] S. Laney, “Consideration for the Evaluation of Fouling Resistant Coatings for Centrifugal Compressors,” in *Proceedings of ASME Turbo Expo 2021*, Virtual, Online, 2021.

[8] ASTM D968 - Standard Test Method for Abrasion Resistance of Organic Coatings by Falling Abrasive, West Conshohocken, PA: ASTM International, 2017.

[9] ANSI/NACE MR0103/ISO 17945 - Petroleum, Petrochemical and Natural Gas Industries -- Metallic Materials Resistant to Sulfide Stress Cracking in Corrosive Petroleum Refining Environments, Houston, TX: NACE International, 2015.

[10] ANSI/NACE MR0175/ISO 15156 Materials for use in H2S Containing Environments in Oil and Gas Production - Part 3, Houston, TX: NACE International, 2015.

[11] ASTM B117 - Standard Practice for Operating Salt Spray (Fog) Apparatus, West Conshohocken, PA: ASTM International, 2019.

[12] ASTM D173 - Standard Test Method for Evaluating Degree of Blistering of Paints, West Conshohocken, PA: ASTM International, 2017.

[13] ISO 16773-1 - Electrochemical impedance spectroscopy (EIS) on coated and uncoated metallic specimens — Part 1: Terms and definitions, International Organization for Standards, 2016.

81d Combining Advanced Digital Monitoring and Chemical Applications to Maximize Cracked Gas Compressor Performance

Sara Bruscoli, Domenico Berterame, Chris May, Lisheng Xu, Mehdi Teimoori

Objectives/Scope: The global petrochemical market demand for ethylene is growing steadily. While operators are striving to maintain maximum throughput, expand margins, and decarbonize their operations, equipment fouling and unplanned maintenance is constantly pushing plants in the opposite direction. With plants extending the time between turnarounds, these avoidable situations are hurting the ethylene industry's bottom-line more than in the past.

Methods, Procedures, Process: Most ethylene plants have limited capabilities in predicting charge gas compressor failures resulting from either fouling or mechanical issues. This jeopardizes the reliability of the compressor operation. The purpose of this paper is to provide a case study demonstrating that fouling and unplanned outages involving the cracked gas compressor can be significantly reduced. Baker Hughes developed its next level real time monitoring program by combining its digital assets for remote diagnostics (iCenter) with effective chemical treatment programs to improve compressor performance and minimize the total costs over the 5-10 year run cycle. By using proprietary Baker Hughes monitoring technology, current plant instrumentation, real time plant data, machine efficiency, and key mechanical parameters, the cracked gas compressor can be constantly monitored by experts. This new approach can more accurately predict the real time demand for antifoulant chemistries and provide predictive mechanical failure information. The ability to predict mechanical failures provides the plant with the lead time needed to properly prepare for an unscheduled outage.

Results, Observations, Conclusions: The result is plant availability and efficiency improvement leading to economic gains and carbon footprint reductions. The presented case study demonstrates that on a 1.2 MTA ethylene plant, the fouling associated cost over typical mean time between maintenance (6 years) can be halved (from approx. 60M$ to 30M$) with potential CO2 emission reduction of 50,000 tons. The amount of savings generated by avoiding an unplanned outage due to a mechanical failure would vary depending on specific outage situation and duration.

81e

Large Machine Electric Drives – How LNG experience can be applied to ethylene plants

Martin I Mayer

VP LNG

McDermott

Aivars E. Krumins

VP Commercial – Low Carbon Solutions

McDermott International, Inc.

Sherif Nossair, PE

Senior Principal Project Engineer

McDermott

Prepared for Presentation at the 2024 Spring National Meeting

New Orleans, LA

March 24 - 28, 2024

81e

Large Machine Electric Drives – How LNG experience can be applied to ethylene plants

Martin I Mayer

VP LNG

McDermott

Aivars E. Krumins

VP Commercial – Low Carbon Solutions

McDermott International, Inc.

Sherif Nossair, PE

Senior Principal Project Engineer

Abstract

The decarbonization of ethylene plants via electrification is gaining prominence both for existing facilities as well as for future expansions and grassroots projects. The cracking heaters are the major source of emissions in an ethylene plant and emission reductions can be achieved through the use of air preheat and ultimately through the electrification of the Steam Cracking Heaters. The cracking heaters generate the Super High Pressure Steam that is used to drive the ethylene plant compressors. As a result of lower firing there will be less steam available necessitating switching the compressor drivers from steam turbines to electric drivers. While electric drivers have been applied for the smaller machines, such as the ethylene refrigeration compressor and, at times, for the other compressors, such as the charge gas, propylene, ethylene refrigeration or similar compressors, they have not been utilized to the size and scale associated with world scale crackers.

McDermott has unparalleled experience utilizing large variable speed electric drivers in LNG facilities of up to 75 MW. We will discuss the experience associated with the design, start-up and operations of such electrically driven compressors and then relate this to the implication of implementing similar electric drive systems in an ethylene plant.

Ethylene Plant Electrification

The Petrochemical Industry is one of the largest industrial sectors in terms of Greenhouse Gas Sources as highlighted in the Figure below:

Figure 4-1: Industrial Processes and Product Use Sector Greenhouse Gas Sources

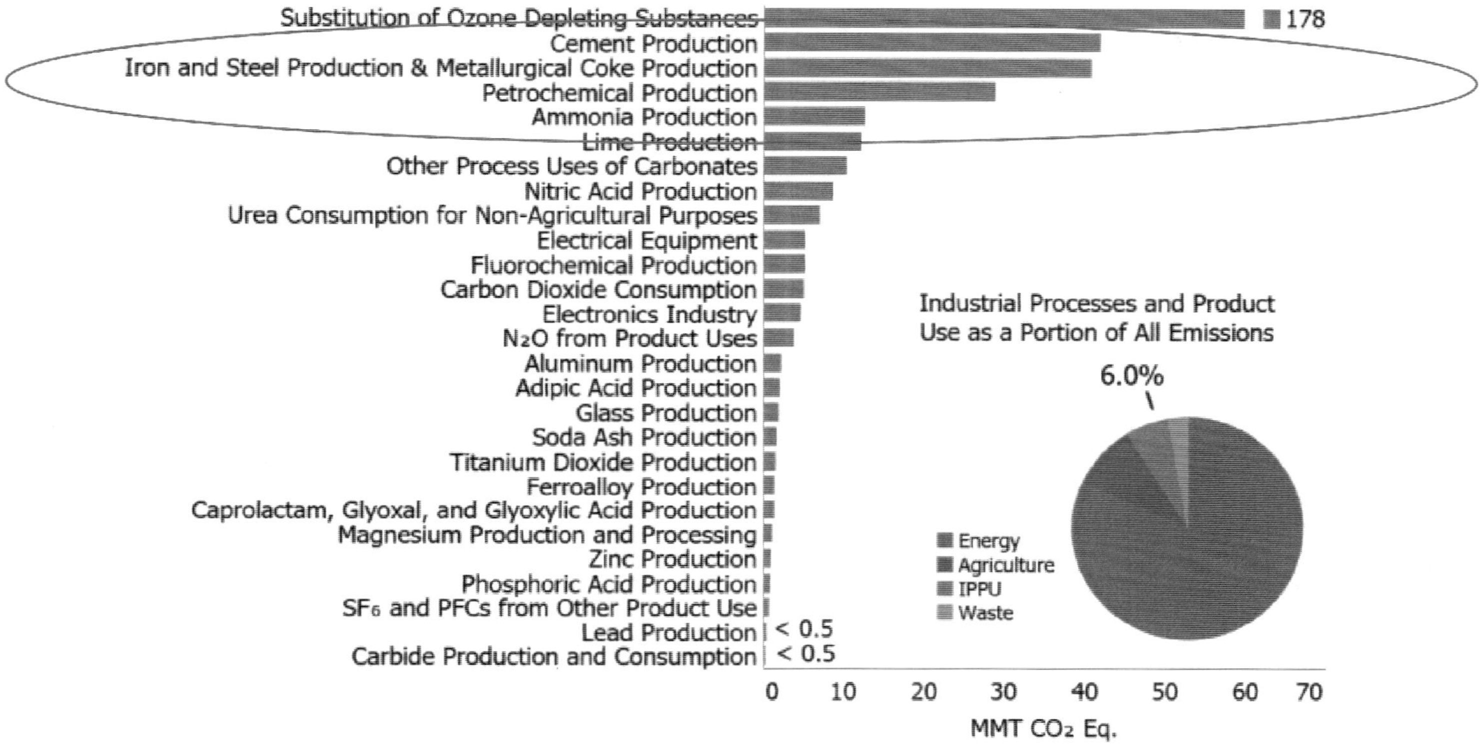

Source: US EPA's Inventory of US GHG Emissions & Sinks: 1990 – 2022 - Industrial Processes and Product Use (DRAFT)

Two other industries, steel and cement production are also large emissions sources. Consider the volume of steel and concrete used to build an ethylene plant and the embedded carbon within. The ethylene industry has a big impact on the emissions.

At the 35th Ethylene Producers Conference in 2023, Steve Sloame, NexantECA, presented that in 2022, there were 371 ethylene assets globally with an approximate installed capacity of 214.2 million tons of ethylene with over 250+ million tons CO_{2eq} annual emissions. That means there is 1.17+ tons CO_{2eq} / ton of ethylene produced! This is not news to anyone familiar with this industry. Ethylene licensors and operators have always looked to save energy to improve operating margins. The challenge has been that energy efficiency improvements, doesn't always meet techno-economic returns required to proceed with the investment. Today there is still an emphasis to save energy, however, it is from the perspective of limiting greenhouse gas emissions which in turn changes the results of the aforementioned techno-economic returns.

There have been multiple presentations made at past Ethylene Producers Conferences describing different strategies for reducing an ethylene plant's carbon footprint. One of these strategies is via electrification. This is gaining prominence both for existing facilities as well as for future expansions and grassroots projects. The cracking heaters are the major source of emissions in an ethylene plant and emission reductions can be achieved through the use of air preheat and ultimately through the electrification of the Steam Cracking Heaters. The cracking heaters generate the Super High Pressure Steam that is used to drive the ethylene plant compressors. As a result of lower firing there will be less steam available necessitating switching the compressor drivers from steam turbines to electric drivers. While electric drivers have been applied for the smaller machines, such as the ethylene refrigeration compressor and, at times, for the other compressors, such as the charge gas, propylene, ethylene refrigeration or similar compressors, they have not been utilized to the size and scale associated with world scale crackers.

This paper is not about the process impacts as this is has been covered very well by other presentations during this and past Ethylene Producers Conferences. This paper is about actual implementation, McDermott's experience with LNG where we've implemented electric drives in a big way. We've been all phases – its design, installation, commissioning, start-up as well as the warranty period. Consequently, if you decide to install an large electric drive on your compressor, that you'll have some guidance as to how to get this right. This will be done through two case studies, one for a very large LNG facility and the second case study for an ethylene plant.

LNG Case Study

Hardware Layout

Simple stated, Liquified Natural Gas (LNG) is produced by purifying natural gas and super-cooling it to -260°F / -160°C to turn it into a liquid. McDermott has demonstrated a true and successful experience implementing electric drives for the large compressors for an LNG project in Texas. This was a $14B project!. The Front End Engineering Development (FEED) was started in 2011, followed by the Engineering, Procurement and Construction for the three LNG trains. The LNG trains were started up in sequence from late 2019 until the last train was started up in mid-2020. The three trains combined, produce 15.6 MTA LNG. Each train had 3 x 75 MW electric driver motors, so with three trains you have nine drivers. There was also a tenth driver that served as a spare. The services for each of the trains were propane refrigeration, low pressure mixed refrigeration and high pressure mixed refrigeration. Gas turbines are typically used to drive these compressors. Additionally, LNG plants have a power island with gas turbine generators. So in essence, there are a lot of opportunities for emissions reductions.

The project is enjoying additional benefits as a result of eliminating the compressors' gas turbine drivers. The site has low pressure energy storage so you need to deal with the boil-off gas. Also, there is no power island for this project. The client elected to maintain a gas turbine generator (GTG) for the upstream pretreatment facility. This allowed the plant to maximize its overall efficiency by recycling the boil-off gas to the GTG and producing supplemental electrical power for the site.

The compressor and driver are located on a concrete table top, just like for an ethylene plant as shown in sketch below.

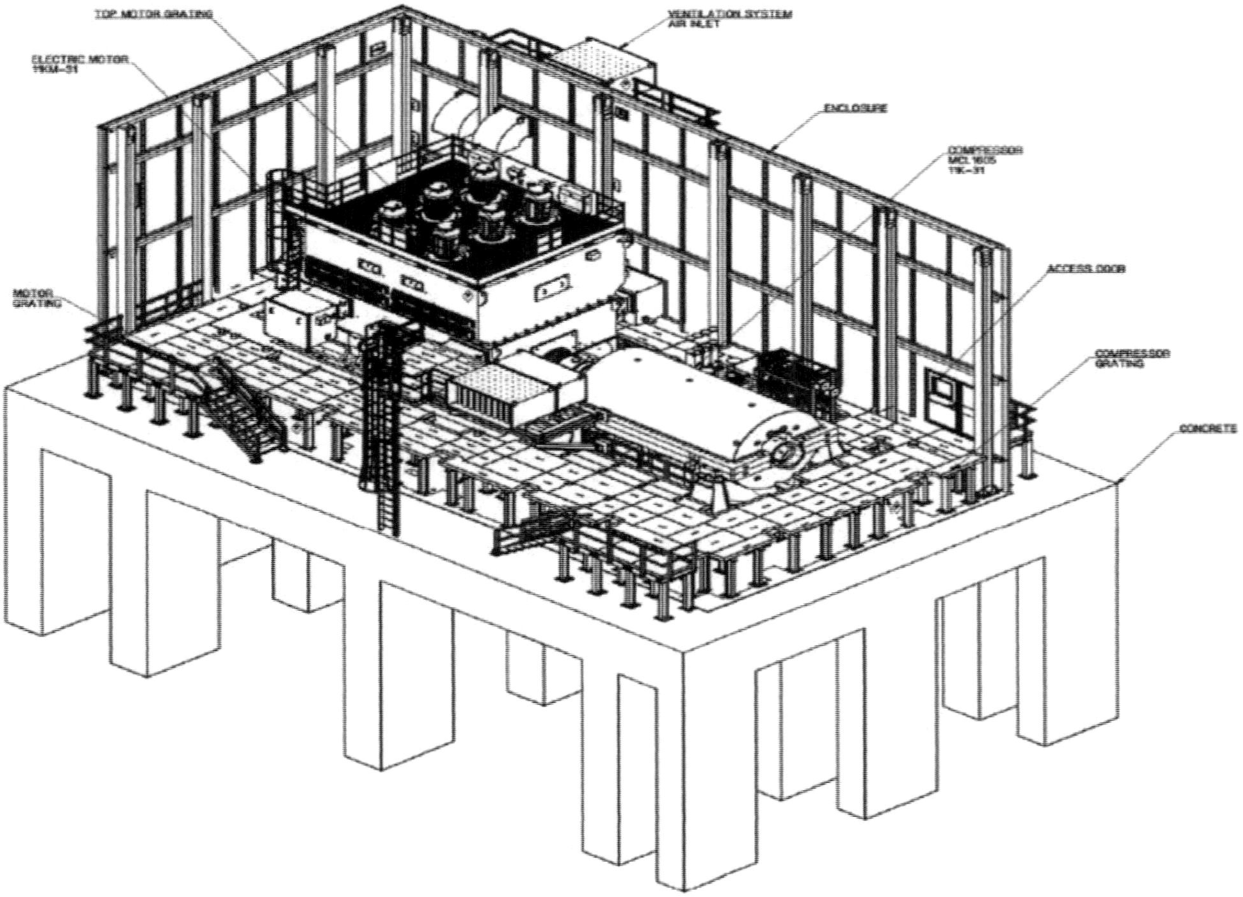

There's a direct drive motor to the two pole 3000 RPM compressor. Normally, you'd see a gear reducer to connect to the compressor. While this was considered during the FEED, the conclusion was that while eliminating the gear box creates some inefficiencies, increasing the operating cost (OPEX), the lower capital cost (CAPEX) was deemed more beneficial in the long run.

The E-house includes a ventilation system, fire suppression and deluge system and noise insultation.

It is a down nozzle, direct drive gas compressor from LCI 75 MW electrical motor. The motor is being cooled by water which is referred to as a Totally Enclosed Water to Air Cooling or TEWAC design. The motor has a six cooling fans on the top of it which cool down the motor enclosure.

The noise enclosure around the compressor is the largest by any extent around any machinery. Besides its noise reduction function, it also provides the ventilation necessary to keep the compressor motor area inside classified as Class 1, Div 2. The amount of ventilation provided in and out allows you maintain a Class 1, Div 2 in case of any kind of gas leakage. The result is that all the instrumentation inside the enclosure is also classified as Class 1, Div 2.

The enclosure was also designed with a water suppression system and a firefighting system in case of any fire within it.

There are four key benefits with the electric drive motors:

- The VFD provides flexible operation from 80 to 105% of the speed providing flexibility in meeting the performance and production demands during the course of operations.

- If there is any trip and/or emergency shutdown, there is no need to vent the trapped refrigerant because you can contain and readily restart from settle out pressure using the VFD. This also prevents venting and subsequent flaring of any valuable gases, thereby both reducing OPEX and GHG emissions during the plant's operations.

- Typically, motors have better mean times than steam or gas turbines for inspection and maintenance requirements. Consequently, maintenance planning is fixed around the compressors which typically have maintenance every 5 years. A motor only requires a major overhaul every 10 to 12 years. The motors do not require any major, extensive overhaul as you may have with steam turbines. You don't have to open the steam turbine upper casing and change the seals, inspect the buckets and all of the other typical maintenance activities. The motor by its electrical design is much better than a turbine in terms of its maintenance requirements.

  During the FEED, the owner added a 10th motor and transofmer in case of a catastrophic failure. This is discussed later.

- The use of the motor drives reduces the carbon emissions by at least 75% and, as was described earlier, that 25% is attributable to the GTG set installed for the pretreatment system for meeting its power generation needs.

Project Electrical System

The electrical system single line diagram for each 75 MW string is shown below.

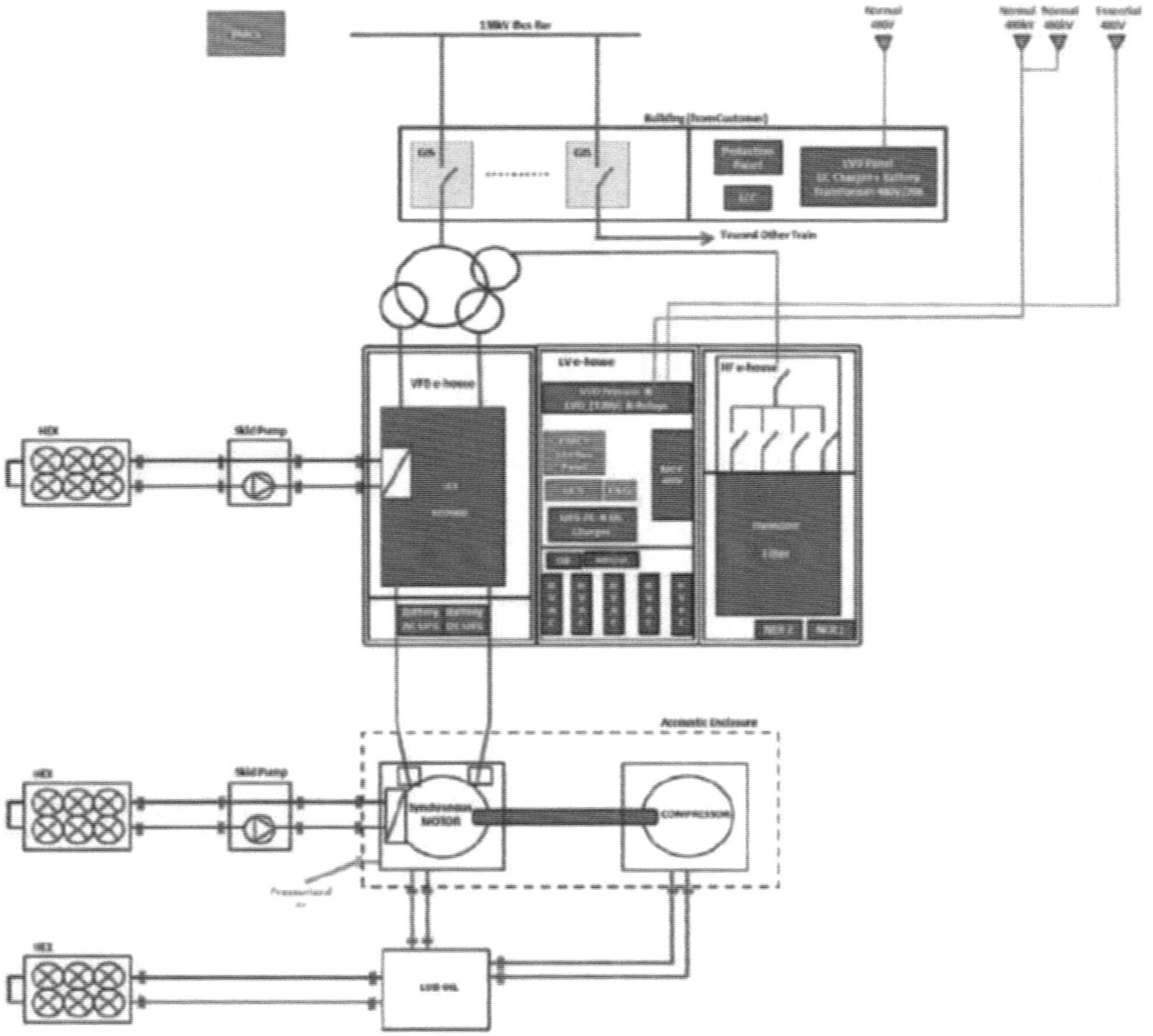

Starting from the top, you have an incoming 138 kV power supply from the local utility company coming into the Gas Insulated Switchgear (GIS), which is the, the main disconnect for each string. This feeds the isolation transformer, which is between the GIS Switchgear and the E-house enclosure which generates some undesirable harmonics that are discussed later.

The E-house has 3 partitions which, going from left to right, serve the following functions:

- The one on the left is the VFD itself that is being cooled by a closed cooling water system.

- The one in the middle has the low voltage and main controls for the string

- The right hand side has the harmonic filters. By its nature, this kind of VFD which is feeding the motor is generating undesirable harmonics which are fluctuations in the voltage going back to the utility. You have to filter those harmonics from varying currents, voltage, etc. because, otherwise, you are sending them back to the utility.

Below the e-house, you see the VFD cables feeding the motors.

Overall, the system has cooling systems for the variable drives and the motors. The common lube oil system at the very bottom is being cooled by air coolers.

LNG Electrification Lessons Learned

We had many lessons learned from this project, but we can group them by the following six major areas:

- There are a lot of interfaces during the course of developing the design and installation for one of these systems. There needs to be a good interface between the owner, the supplier and the contractor. The supplier's lead engineers were invited to sit in our office becoming embedded within our project execution team for the first year. This was a huge benefit because it enabled a much more effective review of the initial documentation and all the associated additional reviews. The supplier's engineers were then recalled for the last six months to close out the project. This was critical for finalizing all the documentation, to ensure that all the documentation was in good order and avoided unnecessary recycling documents back to the supplier.

- The motors solo runs are the major milestones for the project. There were two tests made. The first was at the factory where three of the nine strings were tested. The initial motor runs were made at the factory before connecting to the compressor. The VFD, controls and motors were all assembled together and then commissioned. This made it possible for the motors solo run where the full load and full speed of the string were reached thereby passing the string test. The strings were then decommissioned, packaged and shipped to the site, where it was reassembled and the same procedure was repeated with the site conditions and the site incoming utilities. The motors solo run requires a significant effort including about a year of planning efforts prior to the solo run itself.

- There is an early need of construction and commissioning early in the project to identify the red flag reviews

- Early identification of the inspection team is important. For the first train, there were a number of punch list items that carried over from the factory to the site. However, these lessons were then applied to Trains 2 and 3. It is important to have a well-trained and efficient inspection team prepared and ready ahead of the inspection windows. Clear, effective communications are required between the source inspection and engineering teams because there are parts and pieces in different factories and different countries. Your team needs to be ready and available when the inspection window opens for inspection and testing.

- As noted earlier, the client elected to purchase a tenth motor and a tenth transformer. Consequently, a large warehouse is required for the #10 motor and transformer which you don't typically see for steam turbine applications because you would only require minimum spares. However, a large warehouse is needed in this case because of the size of the motor and transformer as well as the ventilation needed and the trained personnel to maintain the spare. This becomes an operating cost that needs to be considered by the client. This is a major commitment because you have to have this in place for about 25 years.

- An important consideration that is applicable to the propane machine, and may be applicable to propylene machines, is when they are operating at refrigeration conditions and they trip. The compressor rotor would face what is called rotor bowing phenomena because the thermal distribution around the rotor wouldn't be equal. Typically, the compressor flange to the bottom side would be at a different temperature from the compressor casing to the top side which can create a kind of unbalanced thermal distribution around the rotor. If the rotor suddenly stops rather than go into a slow rotation, you would see the rotor bowing phenomena. You need to keep slow rolling the rotor until you clear the emergency trip to avoid this and restart again. For turbines, you have turning gears, but here you don't have that so we have to tweak the design for the VFD in order to slow roll the compressor rotor in case of an emergency trip.

Ethylene Plant Case Study

This case study is associated with an ethylene plant 33% capacity expansion executed in 2012. This represents a similar story to what was done for the LNG facility, but on a very small scale. Two new compressors with motor drives were installed as part of this expansion. One was an 8.6 MW induction motor driving a charge gas booster compressor and the second was a 14 MW induction motor driving an enhanced binary refrigeration compressor. The two motors were a Totally Enclosed Air to Air Cooling or TEAAC design,. There is no need for a cooling water system to cool down the motor because of its small size. If you have a 90 MW or even a 45 MW machine, you will require a cooling water design for the motor. Otherwise, it's design is similar to the LNG plant with the transformer, VFD, motor and compressor. The one difference in this was that the E-house was done by the EPC contractor and the VFD was by the supplier which required a lot of interfaces and which you had to continually manage, unlike the LNG project where the entire system was by the supplier. Interface management is always a critical element of complex projects, installations.

Summary

The use of large VFD machines is proven for large scale LNG facilities. If renewable power is available from hydro, nuclear, wind or solar sources, you would then have or be approaching a net zero design. The implementation process, lessons learned for the LNG plant can be readily applied for an ethylene plant. The key to success is proper planning from the beginning, during the FEED and integrating the efforts of all the parties involved- the supplier, the owner and contractor, and within those groups early involvement of the all functions, engineering, procurement, inspection, construction and commissioning.

SPRING24 +20TH GCPS
A Joint AIChE and CCPS Meeting

Analysis of Steam Turbine Replacement with Electric Driver in Existing Ethylene Plants to Decarbonize

Yuta Okumura
Mitsubishi Heavy Industries Compressor Corporation
4-6-22, Kan-on-shin-machi, Nishi-ku, Hiroshima 733-8553 Japan
yuta.okumura.rs@mhi.com

Keywords: Ethylene, Electrification, Decarbonization, Footprint Replacement, Energy transition

Prepared for Presentation at
American Institute of Chemical Engineers
2024 Spring Meeting and 20th Global Congress on Process Safety
New Orleans, LA
March 24-28, 2024

Analysis of Steam Turbine Replacement with Electric Driver in Existing Ethylene Plants to Decarbonize

Yuta Okumura
Mitsubishi Heavy Industries Compressor Corporation
4-6-22, Kan-on-shin-machi, Nishi-ku, Hiroshima 733-8553 Japan
yuta.okumura.rs@mhi.com

Abstract

The reduction of carbon dioxide emissions, electrification, improved fuel utilization and overall energy transition have become urgent corporate mandates and plant owners are now considering feasible solutions to revamp and/or replace steam turbine drivers in order to minimize steam consumption.

Mitsubishi Heavy Industries Compressor Corporation (MCO) is now focusing on the technical and commercial viability of steam turbine replacements with electric driver, from the viewpoint of a steam turbine manufacturer, in order to support end user's path to lower carbon petrochemical plants. As a steam turbine OEM MCO has been collaborating with process licensors to optimize plant's steam balance with the overarching objective for decarbonization while keeping the turbomachinery highly reliable.

MCO has many experiences replacing both compressors and steam turbines over the last several decades in order to increase plant capacity and/or improve energy efficiency. Additionally, MCO has vast experience providing large complex electric driven compressor trains with electric driver. As a result of these experiences, we can provide single point technical guidance and support operators on the best path forward when reviewing their existing plant install base.

In this paper, MCO introduces analysis result of various aspects for steam turbine replacement with electric driver from plant owner's viewpoints.

1 Introduction

1.1 World Trend for decarbonization

Decarbonization efforts are being actively promoted toward the carbon neutrality by 2050. The global decarbonization market is anticipated to grow at a high rate to reduce the carbon dioxide emissions of industrial facilities and the transition to Net-Zero carbon economy has become a key challenge for industry.

As the government commitment, reduction of CO2 emissions must be push forward with Net-Zero carbon society. In implementing various decarbonization initiatives, improvement of plant utilization and steam consumption are one of the solutions to reduce CO2 emissions by revamp/replacement the existing steam turbine trains with electrification. This plant electrification is proceeded rapidly in the future.

1.2 Decarbonization by Revamp/Replacement

Plant owner focuses on the revamp/replacement of compressor train as part of decarbonization efforts for existing plant. Because it can be contributed to the reduce CO2 emissions by improving energy efficiency through modernization, upgrading plant facilities to improve reliability, increasing plant capacity, and electrification.

The demands are basically achievable by Internal Revamp or Footprint Replacement. MCO has rich experiences of compressor revamp/replacement services which have been accumulated as know-how for many years.

1.3 The potential of Electrification for CO2 reduction

The major source of direct CO2 emissions in an ethylene plant is from firing fuel gas in the cracking furnaces and auxiliary steam boilers. The benefit of electrical drivers instead of steam turbine are lower CO2 emissions, lower maintenance costs and simple operation. Furthermore, since electric drivers can be operated without requiring steam raising and warming of turbines, the electrification allows end users faster start-up and turnarounds, with a potential to reduce overall flaring during these operations.

This study is mainly focused on replacement the steam turbine with electric driver in existing ethylene plant.

2 Technical Feature for Revamp/Replacement

Before the detailed study of electrification, the general practice of revamp/replacement for compressor and steam turbine is introduced.

2.1 Approach for Revamp/Replacement of Existing Compressor

When footprint replacement of existing compressor train is executed, there are four kinds of the reasons as follows and **Figure 2.1**.

1. Capacity expansion (To increase plant capacity)
2. Energy saving (To improve OPEX)
3. Plant renewal (To improve reliability, Modernization)
4. **Decarbonization** (To improve plant efficiency, Electrification)

As per the customer's demands, we select the best solution by footprint replacement or internal revamp of compressor and steam turbine. Then, in case of upgrading the plant capacity, we have to consider these three factors with plant owner and process licensor as follows and **Figure 2.2**.

1. Process requirement
2. Mechanical limitation
3. Plant turnaround schedule

Plant revamp/replacement study is started from decision of the process requirement by process licensor, and this is most important factor to design the compressor train. If plant owner request to increase the plant capacity, larger machine than existing one would be required. Compressor installation location in existing plant has many mechanical limitations such as size of existing foundation and baseplate, the specification of auxiliary equipment, crane capacity and so on. The relation between the new equipment and the existing facilities should be checked the re-usability of the existing ones as much as possible. In general, the equipment modification work is carried out during the turnaround period, so that the field work to modify the existing equipment has to be completed within this period. In such case, minimum modification to be technical advantage.

Based on the above consideration, there are four steps for the capacity expansion.

- Step-1 : Process adjustment

  Production capacity would be raised less than +10%. The change of suction pressure to handle more flow or sometimes adjust the speed without changing compressor design.

- Step-2 : Revamp (Re-rotoring) with an internal remodeling

  Production capacity-up between 10% and 30%, it would be achieved by Revamp (Re-rotoring) with an internal remodeling which the rotor with impellers and internal parts (diaphragms, etc.) are replaced to new ones by re-using the existing casing.

- Step-3 : Replacement of compressor (Footprint Replacement)

  Production capacity-up more than 30%, it would be achieved by replacement of the compressor train. In this case, existing compressor is removed, and the new units will be installed on the existing foundation (Footprint Replacement). Existing foundation shall be re-used, and existing baseplate is checked the re-usability.

- Step-4 : Establishing additional compressor in parallel

  In this case, simply add new compressor train by keeping existing compressor train as it is in parallel operation.

Considering the applicable solution to achieve the scenario of Step-2 and 3, the size and scope of the modification parts should be minimized to utilize the existing casing or to fit on the existing foundation.

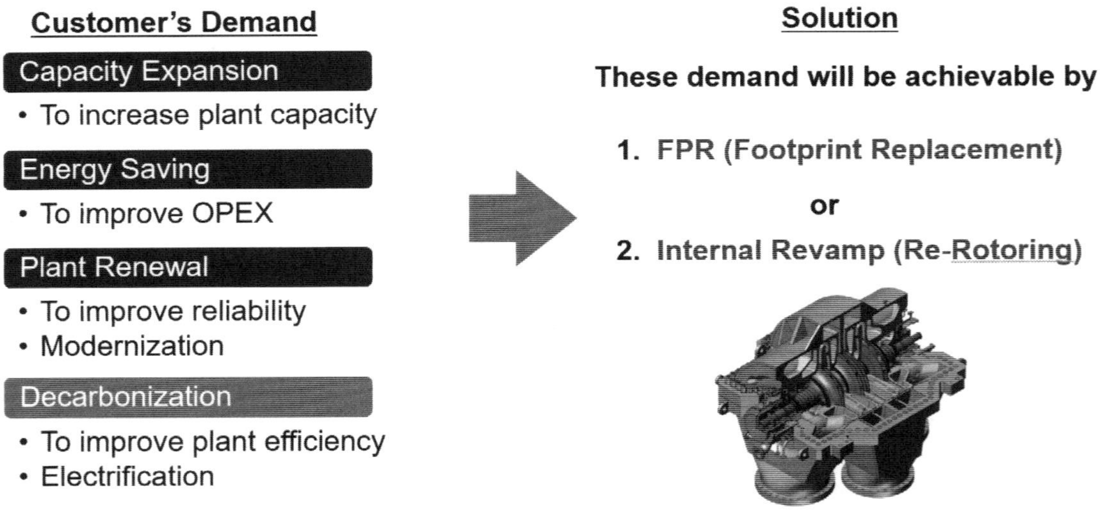

Figure 2.1 Revamp/Replacement demands and solutions

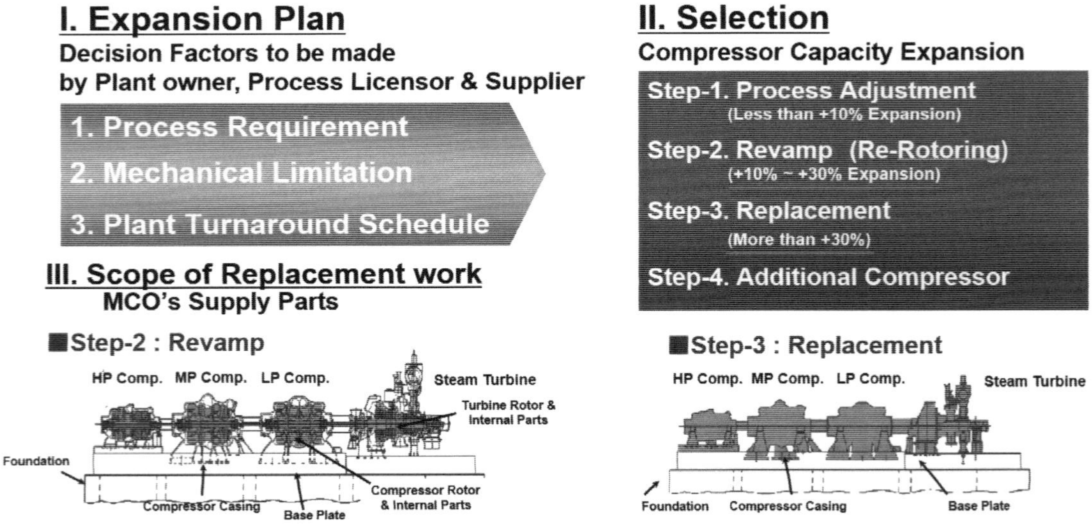

Figure 2.2 Revamp/Replacement plan and selection

2.2 Revamp/Replacement Experience

MCO has rich experiences of compressor and steam turbine revamp/replacement services, and it can be contributed to reduce the CO2 emission by improving the energy efficiency through modernization, reliability improvement and capacity increase.

Figure 2.1 shows the capacity trend of ethylene plant based on MCO experiences. Ethylene plant capacity is increasing year by year as growth the ethylene market size. Typical three replacement experiences are shown in below.

Case-1: **Figure 2.2** shows upgrading history of a certain company's ethylene plant. This original plant capacity is 400 kTA. Then, this plant had carried out 2 times expansion. As the result, 134% capacity upgrading from original plant by repetitive replacement. Capacity expansion, energy savings and plant upgrading have been achieved by footprint replacement of compressor and steam turbine.

Case-2: **Figure 2.3** shows efficiency improvement by replacing the compressor. Modification scopes are replacement the compressor and coupling, and existing baseplate is re-used. Total 26.5 ton/hr of steam consumption can be saved and 66.3 ton/hr of CO2 can be reduced. It contributes to decarbonization.

Case-3: **Figure 2.4** shows capacity expansion with plant electrification project which aims to electrify the ethylene plant by optimizing the steam balance and process conditions to improve the efficiency of the entire plant.

These three experiences are typical case that MCO has know-how and vast experience for both internal revamp and footprint replacement in ethylene plant.

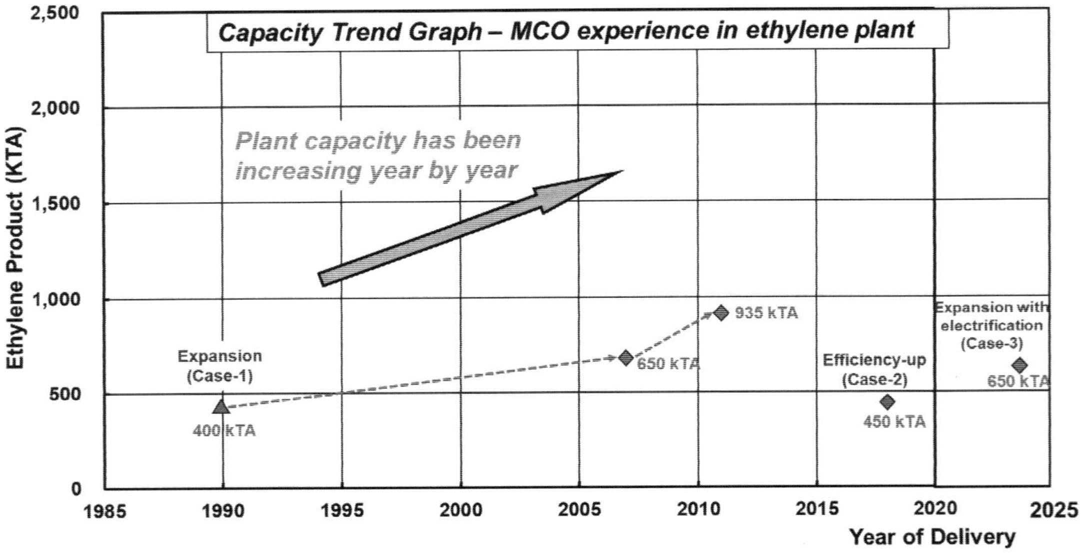

Figure 2.1 Capacity trend of ethylene plant

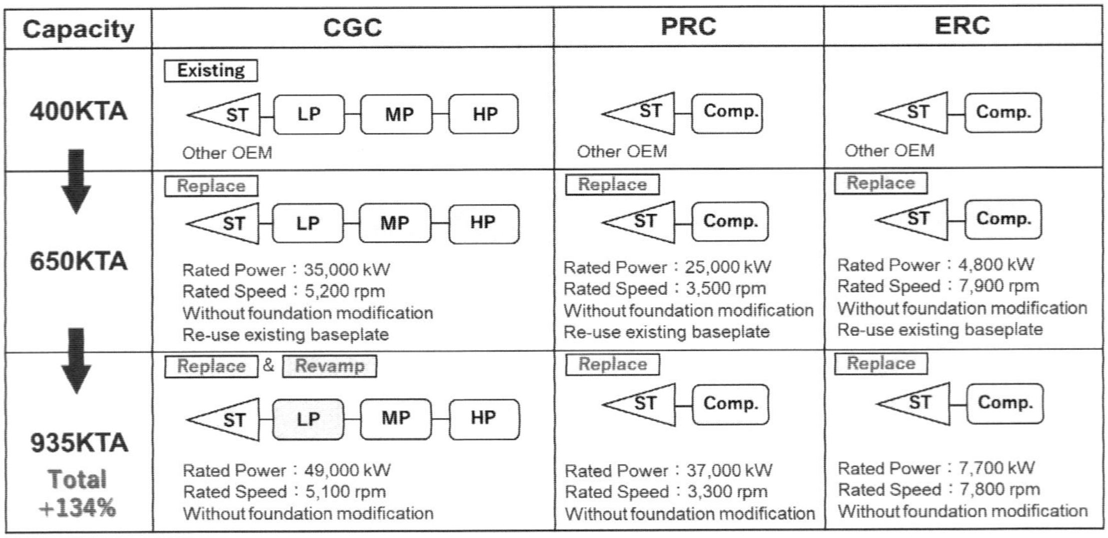

Figure 2.2 Upgrading history of ethylene plant

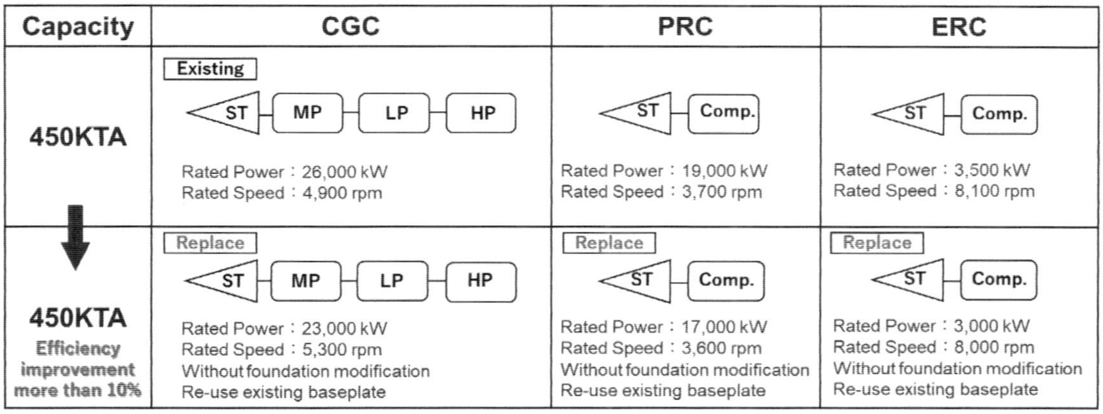

Figure 2.3 Efficiency improvement by replacement of ethylene plant

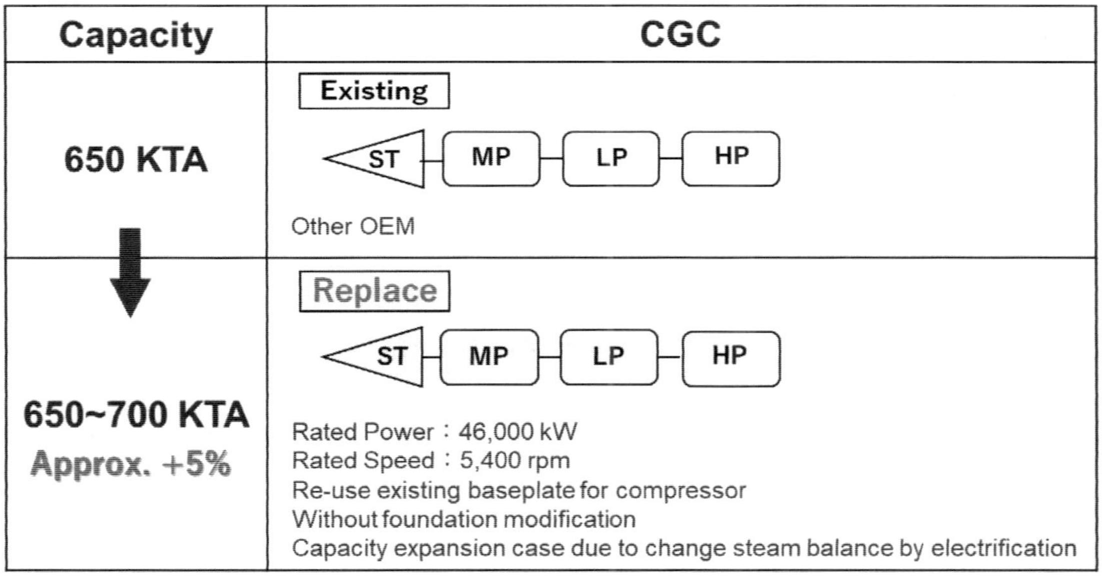

Figure 2.4 Capacity expansion with plant electrification project of ethylene plant

2.3 Revamp/Replacement Technology

Technical advantage for footprint replacement is to minimize field work. Also, successful plant operation can be achieved by following three advantages.

2.2.1 High reliability of compressor performance

The compressor casing has a limit on the maximum volume flow rate. Excessive internal modifications may result in a large discrepancy between the expected and actual performance, especially for larger flow rate due to choking.

In case of both internal revamp and footprint replacement, the compressor design can be optimized from the viewpoint of both fluid dynamics and rotor dynamics. There is also the possibility of further production increases by future internal modifications.

2.2.2 Design flexibility and stable operation

The steam turbine rotor has stable rotor dynamics characteristics due to the short bearing span. The rotor bending during stand-still is small and easy to restart.

Also, compressor has proven design that the swirl canceller and overhang damper contribute to long term continuous operation as stable the rotor behavior.

Stable operation both steam turbine and compressor rotors can be achieved.

2.2.3 Best in class support for field installation work

In case of footprint replacements, we check re-usability of existing baseplate, auxiliary equipment and existing maintenance crane by considering compact design.

Based on our extensive project execution experience with technical know-how, smooth construction and early start-up without any field operation troubles can be achieved. In addition, a skilled technical field advisor can support the field work during the installation and commissioning.

We have the technology the custom design the compressor nozzle to arrange the same size and location to existing machine. It is possible to shorten the modification work period of main process piping at site. Also, we conduct the 3D digital measurement around the existing compressor train in order to prevent the interference between existing items and new equipment. And then, compressor main piping nozzle location can be adjusted with high accuracy.

As considering above, field installation work of compressor train can be minimized.

3 Electrification for Compressor Train

The focus of this study is to find the technical availability to reduce the carbon footprint of ethylene plant by electrification of compressor driver. The following study explores the changes required to achieve decarbonization of existing ethylene plant with electrification and investigate the impact to the plant including CAPEX/OPEX, CO2 emissions and Maintenance/Site work.

3.1 Proposed footprint replacement

Approximately 700 KTA ethylene plant is selected as case study to investigate the availability with electrification on existing plant design. The major compressor train services in the ethylene plant are "Cracked Gas Compressor"(CGC), "Propylene Refrigeration Compressor"(PRC) and "Ethylene Refrigeration Compressor"(ERC). In a typical ethylene plant, steam turbines drive all the compressors.

Table 3.1 shows the compressor train configuration for the typical ethylene plant. For compressor trains, required total power is approximately 72 MW and rated speed is approximately 5,000 rpm of CGC and PRC trains and approximately 6,000 rpm of ERC train. We conduct the electrification case study based on this basic train arrangement.

Table 3.2 shows the train arrangement study of steam turbine replacement with Electric driver. This study examines the case where one driven compressor is connected to a drive machine as in the PRC and ERC train arrangement. The left side of table shows the conventional turbine drive train. The right side shows the electric drive train after the steam turbine replacement, and the two cases are compared. Case-1 shows the train arrangement that the direct drive with high-speed motor is applied. Case-2 shows the train arrangement of conventional motor with gearbox configuration. From these two cases, we study which is suitable design for electrification.

Table 3.3 shows the study results of comparison of two cases from a technical point of view. The technical items for electrification replacement are shown on the left side, and the evaluation results are shown on the right side. The evaluation results show that each case has the following technical characteristics.

For Case-1 :

1. Direct drive with high-speed motor arrangement can minimize installation space.
2. Since the motor weight is heavier than steam turbine, the strength evaluation of existing foundation has to be considered.
3. From a maintenance point of view, the equipment height of the motor is equivalent to the steam turbine, and motor weight is heavier than steam turbine. So that the re-usability of existing crane capacity must be checked.
4. From the viewpoint of equipment design, since there are not many experiences of high-speed motors as compressor train driver, it is necessary to examine the technical design verification in detail and select a proven design.

5. It is important to examine the train torsional resonance analysis to verify the stable operation. Also, as API requirement, the shaft-end and coupling must be changed to high strength design.
6. Oil amount of motor might be less than the steam turbine. The oil console may use the existing equipment after completed the detail design study for each project.

For Case-2 :

1. The motor can be selected by adopting a conventional design with sufficient experiences.
2. Since the gearbox need to be applied, a large installation space is required, and huge foundation modification work such as foundation expansion and overhung baseplate are occurred.
3. Detail technical examination of foundation strength is necessary due to increase the train total weight.
4. If plant owner request to re-use the existing plant design as much as possible, it seems that the train arrangement in this case-2 is inappropriate due to some technical issues.

Therefore, it was concluded that a direct drive with high-speed motor arrangement in Case-1 is suitable for replacing an existing steam turbine in this study.

Next, replacement of steam turbine to electric drive in an actual ethylene plant is examined. **Table 3.4** shows the proposed train arrangement summary for electrification.

For CGC train, direct drive high-speed motor cannot be applied since motor design is out of application range due to high required compressor power with high operating speed. Therefore, it is necessary to install a gearbox to adjust the rotational speed. Therefore, in this train arrangement, the motor is applied a single base plate, and the gearbox and compressor are applied a common base plate.

For PRC and ERC trains, the same train arrangement is assumed. High operating speed with a rated compressor power about 20 MW is required for both the PRC and ERC trains, but direct drive with high-speed motor can be applied since motor design is within the applicable range. There is an opening hole in the existing foundation at steam turbine's nozzle location, and the common baseplate of the motor and compressor is applied to ensure the strength of the motor installation position.

Because of these train arrangement study result, it is expected that the CGC train must require more wider installation space and the train total weight might be increased. That's why, it is difficult to replace to electric driver. On the other hand, PRC and ERC trains have the potential to be replaced to electric driver because it can apply the same train configuration as the turbine drive. From the above, we will focus on the electrification replacement of PRC and ERC.

Table 3.1 Compressor train configuration for the typical ethylene plant

Service	CGC	PRC	ERC
Train Arrangement			
Driver Rated Power [MW]	Approx. 35	Approx. 20	Approx. 17
Total Power [MW]	Approx. 72		
Rated Speed [rpm]	Approx. 5,000	Approx. 5,000	Approx. 6,000

Table 3.2 Train arrangement study of steam turbine replacement with electric driver

Service	Steam Turbine Train	Electric Driver Train
Case-1	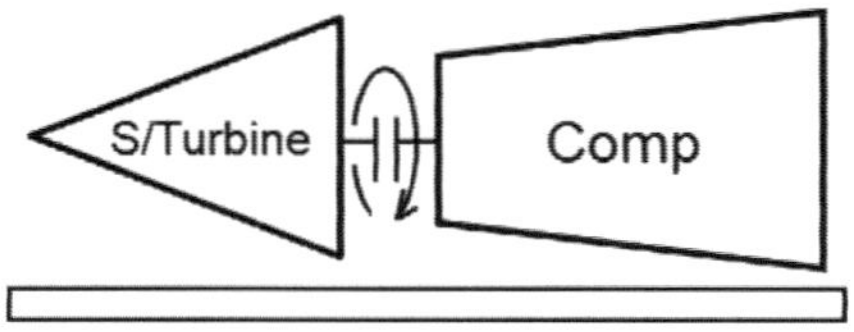	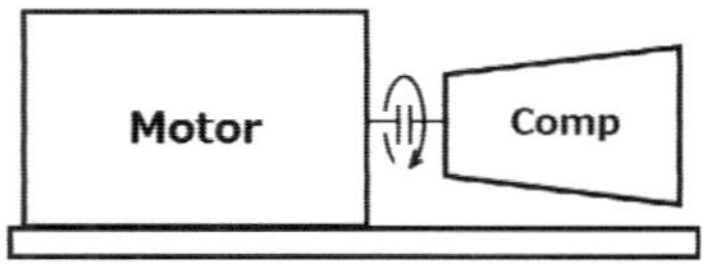 Direct drive with high-speed motor arrangement
Case-2		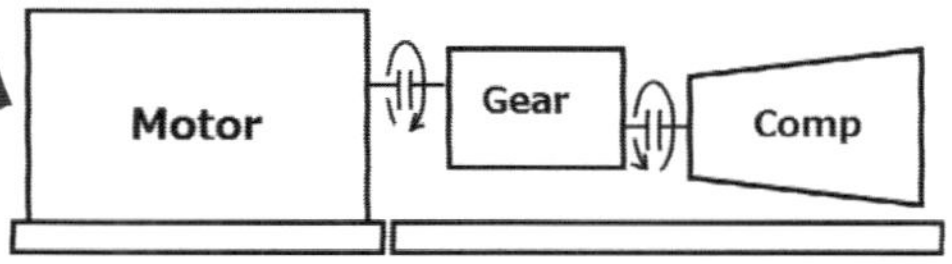 Gearbox arrangement

Table 3.3 Evaluation Result of electric driver train

Technical items	Technical check points	Evaluation Result	
		Case-1 (Direct drive)	**Case-2 (Gearbox arrangement)**
Foundation	Installation space	Same level as S/T train (Good)	Extended space is required (Bad)
	Foundation strength	High-Speed Motor might be heavier than S/T (Moderate)	Motor and Gearbox weight are considered (Bad)
	Modification of opening hole	Foundation modification is required (Even)	
Crane	Crane height	Maximum lift height is same level as S/T train (Even)	
	Crane capacity	High-Speed Motor might be heavier than S/T (Moderate)	Same level as S/T (Good)
Equipment Design	Motor design	Less experience (Bad)	Many experience (Good)
	Train torsional resonance analysis	Same level as direct drive train (Good)	Severer torsional resonance condition than direct drive train (Moderate)
	Coupling and shaft-end design	High strength more than the S/T train is required (Even)	
	Oil console design	Required oil flow is less than S/T train (Good)	Required oil flow is same level as S/T train (Moderate)

Table 3.4 Proposed train arrangement summary for electrification

Service	Train Arrangement Summary
CGC	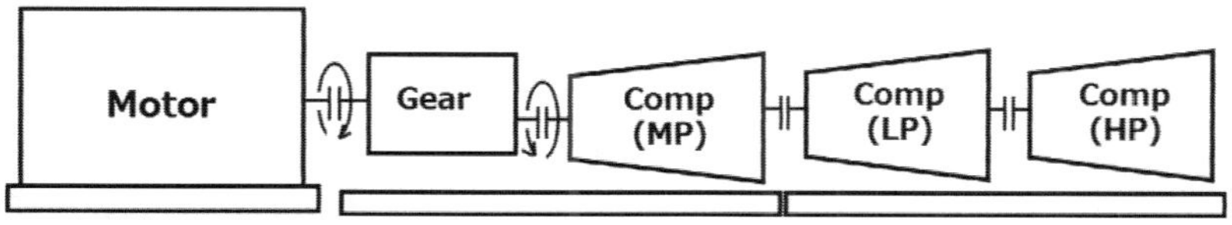
	- Motor and gearbox arrangement - Single Baseplate: Motor - Common Baseplate: Gear and Comp.(MP), Comp.(LP)&Comp.(HP)
PRC & ERC	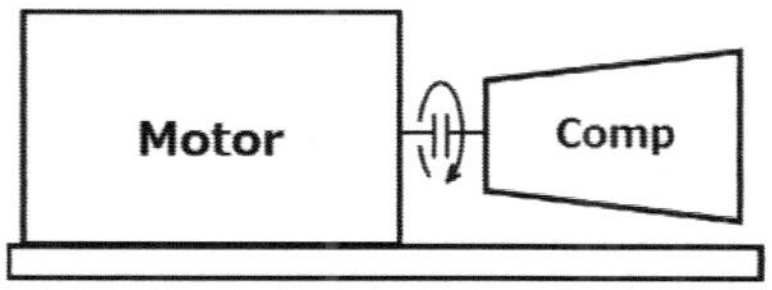
	- Direct drive with high-speed motor - Common Baseplate: Motor and Comp.

3.2 Pros and Cons for Footprint Replacement

As the result of chapter 3.1, Pros and Cons of PRC and ERC trains are compared with replacing to electric driver and re-using existing steam turbine. The evaluation results for each case are shown below.

3.2.1 OPEX and CAPEX evaluation

Cost analysis for electrification is performed for PRC and ERC on typical ethylene plant. Both in terms of Operating Expenditure (OPEX) and Capital Expenditure (CAPEX). Cost consideration results of train arrangement are shown below.

- OPEX considerations:

- Required inlet steam flow is approximately 100 t/hr (PRC), 80 t/hr (ERC)
- High pressure steam price is assumed as $40 per 1 ton.
- Operating time is assumed as 8,000 Hr/Yr.
- Electric price in U.S. is assumed as $0.2 per 1 kWh.

Calculation

Operating cost (Steam) = Required inlet steam flow x steam price x operating time

Operating cost (Electric) = Required electric power x electric price x operating time

Calculation result

Re-using existing steam turbine case.

Operating cost (PRC) = 32.0 million $/Yr (= 100 t/h x $40/t x 8,000 Hr/Yr)

Operating cost (ERC) = 25.6 million $/Yr (= 80 t/h x $40/t x 8,000 Hr/Yr)

Replace to motor case.

Operating cost (PRC) = 32.0 million $/Yr (= 20,000 kW x $0.2/kWh x 8,000 Hr/Yr)

Operating cost (ERC) = 27.2 million $/Yr (= 17,000 kW x $0.2/kWh x 8,000 Hr/Yr)

- OPEX evaluation result:

As the result of above rough calculation, OPEX cost is almost same level between re-using the existing steam turbine and replace to motor. However, if high pressure steam cost can be reduced, operating cost of steam turbine train become lower than motor train. On the other hand, if electric price can be reduced, operating cost of motor become lower than steam turbine. Therefore, in case of plant owner's study for the replacement of existing steam turbine to motor, it is important to optimize the steam balance of existing plant and reduce electric prices.

- CAPEX evaluation result:

PRC and ERC trains require the application to meet various operating cases from process requirements. In that case, applying the Variable Frequency Drive System (VFDS) to motor is one of reasonable solution. So that the total train cost is higher than the steam turbine train. If the supply voltage from the plant facility to the motor is high, a transformer is required that is further increasing the train cost. In addition, it is necessary to increase the capacity of plant electrical facility with additional investment.

From the view of utility consumption, the steam which is using the existing steam turbine will become surplus steam by replacement to motor. It can be used in other plants and/or other driver of compressor train. Therefore, there are more options for optimizing the steam balance of entire plant process and it can be increased the capacity of other compressor trains. These ideas are one of the solutions of decarbonization.

3.2.2 Maintenance and field work schedule

Maintenance and field work schedules are critical factor to the completion of all field work during the plant turnaround period. **Table 3.5** shows the maintenance evaluation compared with motor and steam turbine. Maintenance evaluation result is as follows.

- Maintenance evaluation result:

As evaluation items for maintenance, there are main three factors are considered. The minor inspection assumes that bearings and shaft seals are inspected. In this inspection, it takes three working days every three years for motor. Compared to steam turbines, motor have longer maintenance cycles and shorter maintenance periods.

As for the major inspection such as casing overhaul, the motor takes twelve working days every six years. On the other hand, the steam turbine takes eighteen working days every four years. Both required maintenance period and working days of motor is shorter than steam turbine.

For the operability, the motor can automatically operate simple start-up with one push. On the other hand, conventional design of steam turbine might be required the manual valve operation during start-up, which needs some technical skills and experience of the workers.

Therefore, the motor is superior in maintainability compared with the steam turbine.

- Field work schedule evaluation result:

The plant owner considers the field work period when replacing an existing compressor train. In order to achieve high economic rationality in the commercial operation of the plant, it is necessary to minimize the maintenance period. Plant owners, EPC companies, and compressor OEMs should discuss to reduce the field work period.

Figure 3.1 shows field work schedule for replacement to electric driver of typical ethylene plant as reference schedule. We considered the replacement of existing steam turbine and compressor to motor and new compressor. There are four specific field work items such as preparation work before starting the hot work, hot work for main equipment, piping work around main equipment and installation for auxiliary equipment. **Figure 3.1** shows the required working days for each task.

As the results, the field work is expected to be completed in approximately 35 days from plant shutdown to plant start-up. If existing steam turbines and compressors are to be replaced with new steam turbines and compressors, the field work is expected to be completed in about the same period as for electrification.

Table 3.5 Maintenance evaluation comparing motor and steam turbine

Item	Recommended Maintenance Requirement	
	Motor	**Steam Turbine**
Minor Inspection (Bearing, Shaft Seal Disassembly)	Period: 3 years Workday: 3 days (Good)	Period: 2 years Workday: 10 days (Moderate)
Major Inspection (Casing Overhaul)	Period: 6 years Workday: 12 days (Good)	Period: 4 years Workday: 18 days (Moderate)
Operability	Automatic Startup with One Push Start (Good)	Multiple Operations are required to start-up (Moderate)

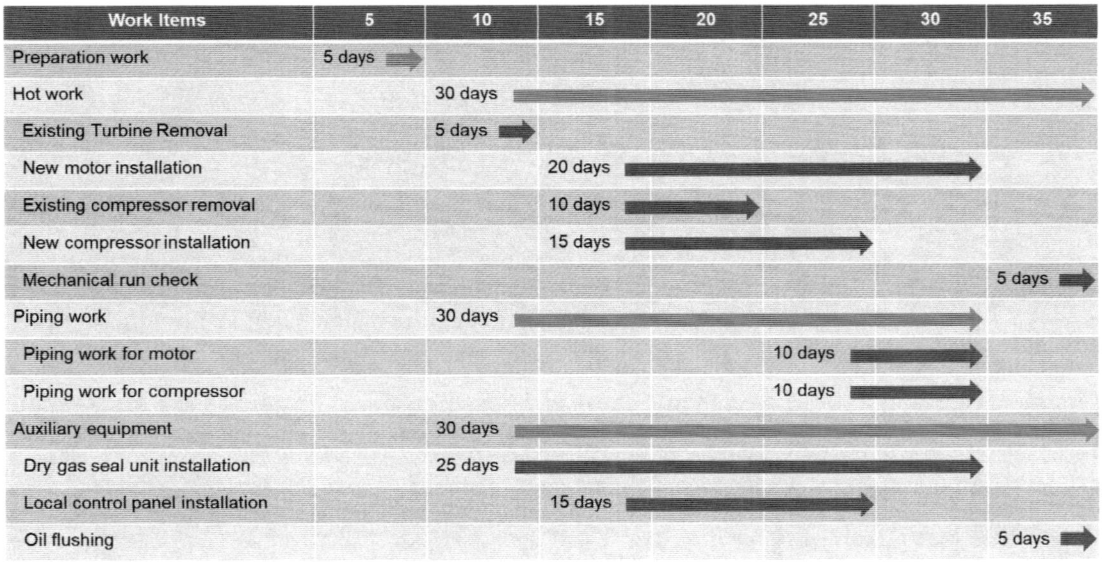

Figure 3.1 Field work schedule for replace to electric driver of typical ethylene plant

3.3 Carbon footprint Evaluation

In the previous chapter, we evaluated electrification replacement of the compressor train and showed that PRC and ERC train can be replaced. Based on the results, this chapter evaluates the impact to electrification on decarbonization.

3.3.1 Quantity of fuel used for boiler

As estimate the energy cost for carbon pricing, one of the main CO2 emissions in ethylene plant is boiler that produces pressure steam. The amount of CO2 emissions can calculate using the boiler efficiency formula as shown in **Figure 3.2**.

Calculation

$$\text{Boiler efficiency } (\eta): \quad = \frac{\mathbf{Q \times (H - h) \times 100}}{\mathbf{(q \times GCV)}}$$

Where **Q** = Quantity of steam generated per hour (kg/hr)
q = Quantity of fuel used per hour (kg/hr)
GCV = Gross calorific value of the fuel (kCal/kg)
H = Enthalpy of steam (kCal/kg)
h = Enthalpy of feed water (kCal/kg)

Figure 3.2 Standard formula to calculate boiler efficiency [1]

Assuming a 100% conversion rate of heat from cracked feedstock into steam, and a standard boiler efficiency of 85% for gas-based boilers [2], the carbon pricing is shown below.

(Note: Although a 100% conversion rate is not possible, it was chosen to get an estimate about gas quantity as this study doesn't go into the details of steam cracker design)

Q = 100 t/hr (PRC), 80 t/hr (ERC)
H = 800 kCal/kg
h = 35 kCal/kg (Room Temp.)
GCV of gas = 12476 kCal/kg
η = 85 %
Quantity of fuel, q = 7.2 t/hr (PRC), 5.8 t/hr (ERC)

3.3.2 Carbon emissions pricing

As per estimates [2], 1 kg of fuel produces nearly 2.5 kg of CO2. Quantity of CO2 reduction is as below.

PRC: Quantity of CO2 reduction = 7.2 t/hr * 2.5 = ~ 18 ton/hr

ERC: Quantity of CO2 reduction = 5.8 t/hr * 2.5 = ~ 15 ton/hr

Total 33 ton/hr of CO2 emissions can be reduced.

For the CO2 emission price, the latest regional greenhouse gas initiative quarterly auction is resulted in a clearing price of $13.9 per ton for CO2 emissions [3].

PRC: CO2 emissions price = 18 t/hr * $13.9/t * 8,000 hr/year = ~2.0 million $/year

ERC: CO2 emissions price = 15 t/hr * $13.9/t * 8,000 hr/year = ~1.7 million $/year

Total 3.7 million $/year of OPEX cost can be reduced.

From the above, electrification has economically advantageous in terms of carbon pricing and it can lead to Net Zero emissions.

4 Summary

The following **Table 4.1** shows the evaluation summary for footprint replacement of electrification based on above study results. Electrification of replacing to electric driver is more effective from the view of OPEX which is considered both power consumption and carbon pricing cost. As plant owner look for opportunities to reduce the carbon footprint for decarbonization, it can be achieved since utility consumption and maintenance are better than re-use to existing steam turbine.

Table 4.1 Evaluation summary for footprint replacement of electrification

Item	Replace to Motor	Re-use to existing Steam Turbine
OPEX (Power consumption)	Good (Considered Carbon Pricing)	Even
CAPEX	Bad (Motor & VFDS)	Even (Re-use existing equipment)
Utility Consumption	Good	Even
Maintenance	Good	Even
Site Work Schedule	Even	Even
Decarbonization	Excellent	Even

(1) High-speed motor arrangement is suitable for replacing an existing steam turbine of ethylene plant. PRC and ERC trains have the potential to be replaced to electric driver because it can apply the same train configuration as the steam turbine drive.

(2) Electrification of replacing to electric driver is more effective compared with re-using the existing steam turbine for decarbonization.

(3) In case of electrification for PRC and ERC, total 33 ton/hr CO2 reduction can be achieved. Also, total 3.7 million $/year of OPEX cost can be reduced. Electrification has economically advantages in terms of carbon pricing and it can lead to Net Zero emissions.

REFERENCES

[1] ENERGY PERFORMANCE ASSESSMENT OF BOILERS
https://www.beeindia.gov.in/sites/default/files/4Ch1.pdf

[2] U.S. Energy Information Administration; Carbon Dioxide Emissions Coefficients
https://www.eia.gov/environment/emissions/co2_vol_mass.php

[3] U.S. Energy Information Administration; Northeast U.S. carbon dioxide emissions prices return to last year's
https://www.eia.gov/todayinenergy/detail.php?id=60982#:~:text=The%20latest%20Regional%20Greenhouse%20Gas,ton%2C%20set%20in%20March%202022.

Near-Misses Temperature Excursions below MAT Identified during AR/BF Studies on Unexpected Equipment.

Giorgio Franceschetti, Thais Helena A. B. de Camargo, Jussara da Rocha Duarte, Leila Lisboa Santos

Braskem has implemented a preventive program to analyze, evaluate and mitigate the risk of auto-refrigeration/brittle fracture in its existing plants and units according to 'API 579 Recommended Practice for Fitness-for-Service, Assessment of Existing Equipment for Brittle Fracture'.

This program includes a level 2 assessment of a complete section of a plant or unit, including some equipment that, in a level 1 assessment, was not considered a low temperature candidate and therefore would not be susceptible to a brittle fracture scenario.

The result of this level 2 analysis, using historical data, identified low temperature excursions close to the MAT caused either by the phenomenon of auto-refrigeration, as by the loss of heat or contamination with lighter HC.

Although these temperature excursions did not result in operation below the MAT, they reached very close values and therefore we decided to include specific recommendations to reduce the frequency of these scenarios reoccurring.

Operation and Maintenance of Safety Instrumented System (SIS)

Rajeev Limaye

Safety Instrumented Function (SIF) implemented in SIS or Burner Management System (BMS) is an important protection layer to bring the process risk to an acceptable level. SIS follows IEC61511 which is a lifecycle standard. The standard covers all aspects in the SIS lifecycle starting from inception, design, engineering, commissioning, maintenance till decommissioning. The operation & maintenance is the longest span in the SIS lifecycle. The owner/operator is responsible for following the maintenance and documentation requirements of IEC61511 to be in compliance with OSHA PSM. Each SIF must be tested at an interval determined during the SIL calculations. This test interval usually matches the Turnaround interval as most SIFs cannot be tested while the plant is running. Many users inadvertently ignore the useful life of devices specified by manufacturers. A device in safety service must be replaced or refurbished at the end of its mission time. Management of Change, documentation, training, validating device failure rates in SIL certificates against the field failure rates, developing decommissioning plan are examples of the tasks every user must perform as part of the SIS maintenance stage of the lifecycle. This paper will make the users aware of all such requirements of IEC61511 to stay in compliance with OSHA PSM.

182a

HUMAN PERFORMANCE IN ETHYLENE UNITS MAKING IT EASIER TO SUCCEED AND HARDER TO FAIL

Ryan Grothe

Process Safety Execution Manager

Chevron Phillips Chemical Company

Tony Amador

Human Performance Safety Specialist

Chevron Phillips Chemical Company

Prepared for Presentation at the 2024 Spring National Meeting

New Orleans, LA., March 24-28th, 2024

HUMAN PERFORMANCE IN ETHYLENE UNITS
MAKING IT EASIER TO SUCCEED AND HARDER TO FAIL

Ryan Grothe

Process Safety Execution Manager

Chevron Phillips Chemical Company

Tony Amador

Human Performance Safety Specialist

Chevron Phillips Chemical Company

Abstract:

Are there tasks in an ethylene unit that require a person to perform perfect 100% of the time to avoid a serious process safety or lost profit consequence? Is it reasonable to expect a human being to never make a mistake?

This engaging presentation will discuss various scenarios that may be more prone to human error when operating or maintaining equipment in an ethylene unit. Learning is required to prevent unacceptable consequences. We must learn from incidents and from everyday work. Leaders must get to know the work for the purpose of creating safeguards. It is the leader's responsibility to make it easier to succeed and harder to fail. We will share scenarios where it may be too easy for a simple mistake to result in a big incident. We will discuss ideas for potential safeguards and methods for identifying them.

The intent of this presentation is to give ethylene producers a different perspective to take back to their facilities that could bring them a step closer to a truly learning culture.

MAIN BODY

Human Performance (HP) is our industry's next frontier toward improved safe and reliable operations, both in manufacturing and non-manufacturing settings. It is not a new program. HP is a leadership mindset shift, a different way of looking at the activities we do every day.

Human error is in some way involved in the causes of nearly all high consequence and high potential consequence incidents. Errors occur in front-line operations and maintenance. They occur in design and fabrication. They occur in engineering, projects, finance, logistics and leadership. Managing the consequences of error through a Human Performance approach is the next step toward a future of zero incidents of unacceptable consequence.

With good systems, high-performing humans focused on operational discipline can achievably work 97-99% error free. Where we have processes that rely on human perfection 100% of the time, we need a safeguard. A simple visual is the bollard or jersey barrier in a parking lot that occasionally saves equipment from impact. HP builds safeguards for the 1-3% of the time that high-performing humans make mistakes.

The single objective of Human Performance is to build safeguards that prevent or mitigate high potential consequence events. The target audience is leaders and staff who wish to make it easier for their colleagues to succeed and harder to fail. Leaders must work with and listen to the frontline personnel in their workplace to understand how the work is actually performed for the purpose of identifying and developing safeguards. Practitioners can use and develop tools to help leaders understand the work and to implement safeguards. Remember to keep in mind, it is about safeguards, not paperwork.

In our industry we still need humans to perform to their full potential and consistently reach 97-99% of error-free actions in the most critical tasks. Human performance can be variable over time due to fatigue, learning, distractions, complacency, etc. Strong operational discipline and performance management systems are still required to be reinforced and continually improved. HP starts with strong operational discipline.

Leaders and staff have been driving for strong operational discipline for years. This has contributed to the increased success of the industry in preventing high potential consequence events. HP is an addition to the toolbox. Leaders are responsible for performance. Leaders set priorities and distribute resources. Leaders can develop and deploy safeguards. Thus, leaders must learn about the work as they listen to the frontline workers with a continuous improvement mindset. Together they can identify the steps or tasks that could result in a high potential incident when a mistake occurs.

Make it easier to succeed and harder to fail. This sentence is a simple representation of the entire HP program. Safeguards make it harder to fail when mistakes occur, and leaders need to understand the work in order to make it easier to succeed.

This presentation at the 2024 AIChE Spring Meeting will bring the Human Performance concept to ethylene units. The basic of concept that humans will make mistakes will be established as fundamental. We will discuss and review some tasks or scenarios in an ethylene unit could result in unacceptable consequences if people do not perform perfect 100% of the time. Where perfection is required and a human can make a mistake, we need a safeguard. Simple. We will then discuss a hierarchy of safeguards that make it easier to succeed or harder to fail and facilitate an exercise on how to use it. Stronger safeguards are desired for less risk tolerant tasks. Lastly, we will reiterate the message to leaders with a call to action and self-reflection. What will you do to drive Human Performance?

References:

1. Material based on Human Performance strategy at Chevron Phillips Chemical Company, LLC.

189b Cold Safety Considerations for C3 Services

Ashley Pedretti

Previous presentations on cold safety have provided an overview of hazard recognition and mitigations, including piping assessment and mitigations, and cold flare system monitoring and sustainment. Further work indicates that scenarios associated with C3 system may be difficult to identify. At normal operating pressures of C3 refrigeration systems and C3 recovery equipment in olefins plants, C3s can condense at near-ambient temperatures, which can introduce cold safety risks during maintenance activities and abnormal operations of systems that normally operate with vapor-phase C3s. Recognizing the risk of brittle fracture in such scenarios may not be intuitive, since liquid hydrocarbon is not expected to be present. However, condensation of C3s can be influenced by factors such as weather conditions, starting temperature of the vessel or catalyst bed, and duration of time that a line is left stagnant. This paper provides learnings on C3-service equipment during clearing, commissioning and abnormal operations, and potential approaches for better recognition and mitigation of these scenarios. In addition, incidents will be presented to serve as examples of scenarios and mitigations to be considered.

189c Safe State: Preparing Ethylene Units for Severe Cold Weather Events

Sarah Skillern

In February of 2021, a multi-day hard freeze weather event impacted most of the United States, including Texas and the Gulf Coast. This event resulted in widespread utility and facility outages, and significant impact to the ethylene industry. Industry impacts included an increase in process safety events, first aid and injuries, environmental exceedances, and impacts to production. This paper will explore one company's response to this event, and subsequent decision to implement a corporate-wide cold weather directive. It will also present as a case-study the actions taken by one facility to incorporate this directive into a unit-specific approach for a Gulf Coast ethylene unit.

In December of 2022, another cold weather event gave the company and site an opportunity to put these new weather policies into practice, this paper will compare the response and learnings from this subsequent event.

The intent of this paper is to give ethylene producers some tools and best practices to take back to facilities, to better prepare ethylene units for severe freezing weather events.

189d Evaluating Runaway Reaction Hazards of Cracked Gas in Zeolite Media

Soham Dutta, Andrew Athon, Jonathan Lunn, Veronica Li

Solid zeolites are widely used in hydrocarbon plants for drying and adsorption applications. The tunability of zeolite properties by varying them chemical composition and structure makes them particularly useful to a diverse range of unit operations. These same characteristics, however, also render upon them catalytic properties that may lead to un-desired reactions such as hydrogenation and oligomerization in these unit operations. As a result, dryers and adsorbers have well-defined operating envelopes that must be adhered to for safe operation. These operating envelopes typically fix the chemical composition, temperature, and pressure of the feed to the zeolite-containing packed bed. In order to define these envelopes, temperature, followed by chemical composition is typically used as the guiding criteria while the effect of pressure is not always considered.

In this study, the authors reveal the importance of considering pressure while assessing potential for thermal runaway as a result of un-desired reactions and ways in which well-designed tests can provide realistic assessment of un-desired reaction rates. Using the case of cracked gas dryers, we will discuss results of calorimetry experiments that reveal the effect of pressure on un-desired reaction rates under adiabatic conditions. Combined with thermokinetic modeling, such studies can reveal induction time of un-desired reactions under various conditions. Using these studies, we will highlight shortcomings of existing calorimetry techniques that under certain conditions, may lead to over-conservative un-desired reaction rates and thus constrain operational flexibility. Lastly, we will discuss calorimetry tests that rely on pressure scanning techniques that can allow users to overcome such shortcomings. These results will then be contextualized by connecting how they can be used to make realistic estimates of runaway potential in zeolite media under various unit operations and modes.

189e Reactor Exotherms Due to Hydrogen Ingress

Mark Nierode, Selma Kwok, Keng-fai Kuan, Yong-Lai Ng

Hydrotreating reactors are commonplace throughout the steam cracking recovery process. These reactors are required to meet product specifications and, in some cases, to satisfy downstream unit operational requirements. These recovery sequences often include C2, C3, and C4 acetylene converter reactors. In addition, there are additional hydrotreaters for diolefin/olefin saturation (e.g. C4, SCN) as well as desulfurization. Most of these reactors require controlled hydrogen injection to regulate the reaction, with the exception of front end acetylene converters where hydrogen is present in excess.

The reactants in these hydrotreaters are primarily hydrogen and acetylene(s) and to some degree olefins/diolefins. All of these hydrotreating reactions are exothermic and these reactors have the potential to runaway due to the increasing reaction rate with increasing temperature. A runaway is usually coincident with the onset of unselective reactions which are favored by high temperature, reactives concentration, and/or residence time. An example of high reactives concentration would be having too high an injection of hydrogen, e.g. a control failure. Another concern is maldistribution which can lead to high residence time. Therefore, these reactor systems often include trip systems based on detection of high temperature or residence time (e.g. low flow limits). The action of these trip systems stops the reaction by removing one or both of the reactants (e.g. isolate/depressure or remove hydrogen injection).

This presentation will share examples where hydrogen continued to these reactors despite the trip systems activating. These examples illustrate the potential for hydrogen to be introduced unintentionally into the reactor vessels. This presentation will also include potential options to mitigate these examples or similar scenarios from happening in the future.

189f Incident Learning: Popcorn Polymer in Butadiene Extraction Unit

Ruben Vandewijer, Peter Loecker

Popcorn Polymer (PCP) is a well-known industry safety and reliability risk in Butadiene Extraction Units, Steam Cracker Units (primarily light-ends systems), as well in logistic facilities. The formation of highly cross-linked polybutadiene (polyBD), often termed "popcorn polymer" is hazardous as it could lead to losing equipment integrity, explosion, pyrophoric behavior or by taking safety critical equipment out of function.

Recently, one of the Shell Butadiene Extraction Unit experienced plastic deformation of two condensers within a couple of months, caused by popcorn polymer formation. Previously, it was assumed that the risk of PCP formation was relatively low at these locations. This paper will explore the potential causes that are likely to have caused the PCP initiation in the condensers and counter measures taken by the company to prevent such a PCP growth incident in the future.

AIChE Paper Number: 120a

Powering the Transition to Net-Zero: Update on Technology Demonstration of Electric Cracking Furnaces and Future Integration Options into Petrochemical Sites

Tobias Sinn
Process Design Engineer
Sustainable Hydrocarbon Solutions - Linde Engineering

Martin Hofstaetter
Process Design Engineer
Sustainable Hydrocarbon Solutions - Linde Engineering

Rainer Kemper
Technology Advisor
Sustainable Hydrocarbon Solutions - Linde Engineering

Gunther Kracker
Head of Product Management & Cost Estimation
Sustainable Hydrocarbon Solutions - Linde Engineering

Prepared for Presentation at the 2024 AIChE Spring National Meeting
36th Ethylene Producers Conference
Session: Ethylene Plant Technology Fundamentals
New Orleans, LA., March 24-28, 2024

Powering the Transition to Net-Zero: Update on Technology Demonstration of Electric Cracking Furnaces and Future Integration Options into Petrochemical Sites

Tobias Sinn
Process Design Engineer
Sustainable Hydrocarbon Solutions - Linde Engineering
Martin Hofstaetter
Process Design Engineer
Sustainable Hydrocarbon Solutions - Linde Engineering
Rainer Kemper
Technology Advisor
Sustainable Hydrocarbon Solutions - Linde Engineering
Gunther Kracker
Head of Product Management & Cost Estimation
Sustainable Hydrocarbon Solutions - Linde Engineering

Abstract

To meet Paris Climate Agreement targets and limit global warming, energy- and emission-intensive industries must transform their production assets. Olefin technology licensors and operating companies need carbon emission neutral plants to produce safely, reliably, and sustainably. Cracking furnaces are steam cracker plants' largest emitters and significantly impact the petrochemical industry total emissions. Electrification is key to reducing carbon emissions from cracking furnaces. By converting to electric heating using renewable power, flue gas emissions can be avoided. This disruptive innovation in the petrochemical industry can reduce overall CO2 emissions by 95% or more, and thereby significantly contribute to meet sustainability goals.

In 2021, the consortium partners BASF, SABIC and LINDE announced a collaboration to jointly develop electric cracking furnaces. In 2022, construction works began for a multi-megawatt demonstration plant at BASF in Ludwigshafen (Germany), targeting the demonstration of two different electric heating technologies at industrial scale. Mechanical completion of the demonstration plant was achieved in January 2024, this is followed by the commissioning and test phase. A dedicated test program was established and is planned to be executed throughout 2024 and beyond.

The conference contribution provides information on the two electric heating technologies developed as well as the characteristics of the demonstration plant. Main benefits and learnings from the EPC-phase and the demo plant operation will be outlined. As the plant is completely integrated into the commercial production of the existing steam cracker the operational experience gained will be highly representative for the operation of future commercial-size electric furnaces.

The authors will furthermore share their detailed view on how electric cracking furnaces (STARBRIDGE™) are integrated into existing and new petrochemical sites. Conventionally fired cracking furnaces are highly integrated units, preheating feedstock and delivering the cracked gas and energy streams to the separation section.

Replacing fired furnaces with electrical units or adding electric cracking furnace to a given site will have a significant impact on the whole energy balance of the plant. Showcase examples highlight tailored electric cracking furnace integration solutions, developed for site-specific constraints. An outlook will be given on how such integration can be designed today to ensure a smooth implementation of electric cracking furnace projects at commercial scale.

Introduction and motivation

The petrochemical industry contributes approx. 5% to the global CO_2 emissions [1], with steam cracking processes responsible for a significant share of this output. Steam crackers generate as much as 25% of the greenhouse gas emissions of the European chemical industry and hundreds of millions of tons per year worldwide [2, 3, 4]. In light of this, innovative technology must be developed and rapidly made accessible to the industry in order to counteract the negative impact of emissions and achieve greenhouse gas neutrality.

Upwards of 90% of CO_2 emissions in modern steam cracker plants are caused by the high energy requirements of the process [3]. The reaction heat necessary for the endothermic conversion in the cracking furnaces is supplied at high temperatures (the coil or reaction tube outlet has an operating temperature of roughly 850°C and a maximum tube metal temperature of up to 1100 °C). The additional energy necessary for the work-up section, which separates the raw cracked gas into commercial "high value chemicals (HVCs)," is also significant.

The main energy supply for the entire plant currently comes from burning the methane fraction by-product and natural gas in the cracking furnaces, there also the necessary utility steam is generated from waste heat. A substantial portion of a steam cracker plant's scope 1 GHG emissions are released into the atmosphere via the furnace stacks. The electrification of cracking furnaces represents an opportune starting point for significantly reducing GHG emissions and achieving sustainability goals set out in the 2015 Paris Climate Agreement.

The differentiation of the furnace electrification from other steam cracker decarbonization methods, including flue gas CO_2 capture and blue hydrogen firing (pre-CCS), is based on its CO_2 avoidance character. Other decarbonization methods depend on CO_2 sequestration, which requires CO_2 outlets and subsequent storage and/or utilization. As a further distinction, electric furnaces are particularly suited to locations with high availability of renewable power supply, potentially enabling in the long-term low energy import costs without additional operating expenses for CO_2 capture and handling. Hence, there is a great incentive to develop technically and commercially mature solutions for the electrification of cracking furnaces.

In the first part of this paper the corresponding technology and the demonstration activities by the joint collaboration partners BASF, SABIC and LINDE are presented. Two alternative electrical heating concepts for steam cracker furnaces have been developed and are currently being evaluated in a large-scale demonstration plant.

The second part highlights the main differences in the heat duties of an electrically heated cracking furnace compared to those of a conventional fired furnace. Said difference may significantly impact a plant’s energy balance but - at the same time - reveals large potential for increasing the overall plant efficiency. Selected heat integration schemes, from less complex to the more intricate, enable the adaptation of the e-furnace periphery to project specific boundary constraints and are explained in detail. In this context, special equipment designs for high temperature heat exchange from reactor effluent to reactor feed are disclosed, providing the reader with specific details on the technical design.

Technology demonstration

For the core part of the electric furnace, the partners BASF, SABIC and LINDE have developed two alternative technical solutions, whose principles are sketched in Figure 1. Both heating concepts of the STARBRIDGE™ technology are based on resistive heating i.e. with electric power supply to the resistive heating material via solid electrical conductors (unlike inductive heating). The difference between the two concepts lies in the location of the resistive heating material within the furnace.

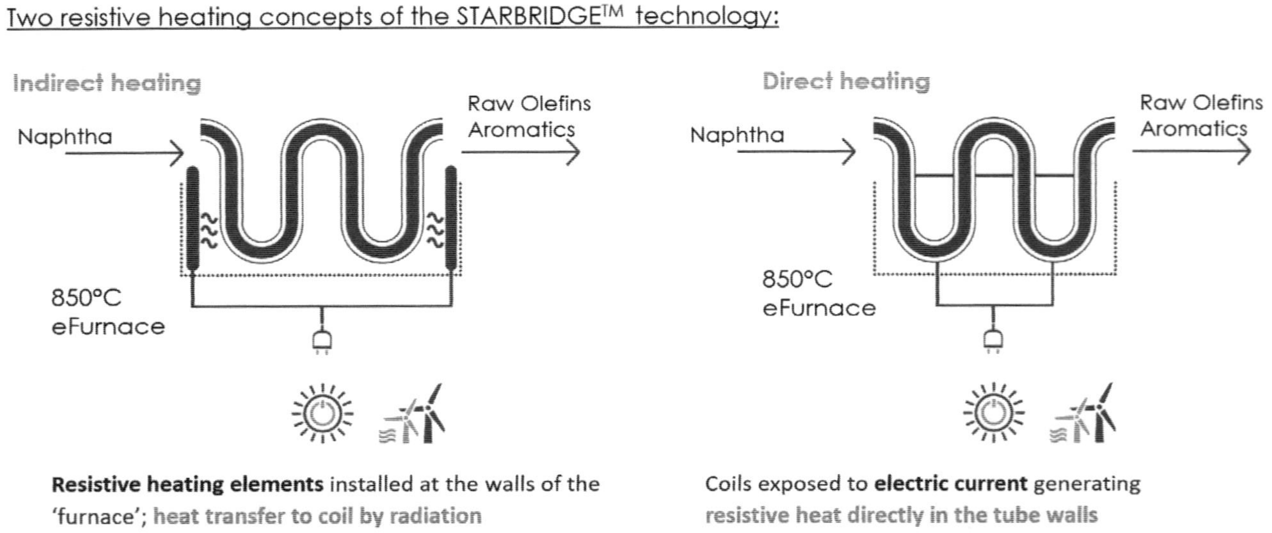

Figure 1: Principal sketches of the developed indirect (left) and direct (right) heating concepts for electrified steam cracker furnaces.

In a furnace with "indirect heating", an electric current is applied to distinct heating elements that are brought to the necessary temperature by Joule heating. The supply of the thermal energy required in the process tubes is subsequently achieved by mostly radiative heat transfer from the heating elements and the insulated walls of the coilbox.

In a furnace with "direct heating", an electric current is applied directly to the process tubes/cracking coils inside the reactor. Joule heating inside the tube wall, caused by the electric resistivity of the cracking coil material, provides the required thermal energy for the chemical reaction.

Extensive pilot testing activities and numerical simulations with different calculation approaches were undertaken to confirm the feasibility of both heating technologies and to establish a sound knowledge and data basis for the design of a demonstration plant and future commercial plants. The techno-economic analysis conducted by the partners showed that both heating concepts have their own advantages depending on the specific use case, while offering comparable energy efficiency/power consumption. This is the reason why both approaches are demonstrated alongside each other in one demonstration plant. A 3D model snapshot is shown in Figure 2, indicating also with faded colours the surrounding elements of the existing steam cracker into which the demonstration plant is integrated.

Figure 2: 3D model snapshot of electric furnace demonstration plant at BASF Ludwigshafen

The demonstration plant at the BASF Verbund site in Ludwigshafen (Germany) has the main goal to demonstrate continuous olefin production using electric energy as a heat source and to serve as a reference for an industrial-scale plant. The plant is currently in commissioning/test operation and will be operated under industrial conditions with the purpose to demonstrate the reliability of key components, like materials of construction and custom-made components, for use in this type of high-temperature reactors. Mechanical completion of the demonstration plant was achieved in January 2024. The test operation will last for several years, and the plant will be accessible for reference visits.

The demonstration plant has the following key features:

- Both heating technologies (direct and indirect) will be demonstrated in separate coilboxes.
- In total, up to 6 MW of renewable power can be drawn from the 6kV grid and used for preheating the feed and sustaining the cracking reaction.
- In total 4 tons per hour of naphtha feed can be processed, and the cooled cracked gas is being transferred to the existing BASF steam cracker for the commercial production of HVCs.
- Cracking coils in both coilboxes are of commercial dimensions.

The indicated capacity and the layout of the demo unit were selected to enable a reliable upscaling. Major equipment is already close to commercial-size dimensions, or the scaling factor is below 10, which is considered in the industry as factor not implying significant upscaling risks.

The demonstration plant has been designed with maximum flexibility in mind. It can run the two radiant boxes in parallel or operating one furnace, while the other may be in standby, decoking or shutdown mode. The decoking process is carried out similarly to the traditional fired cracking furnaces, utilizing a mixture of steam and air to burn coke. Figure 3 provides a functional sketch of the demonstration plant's architecture.

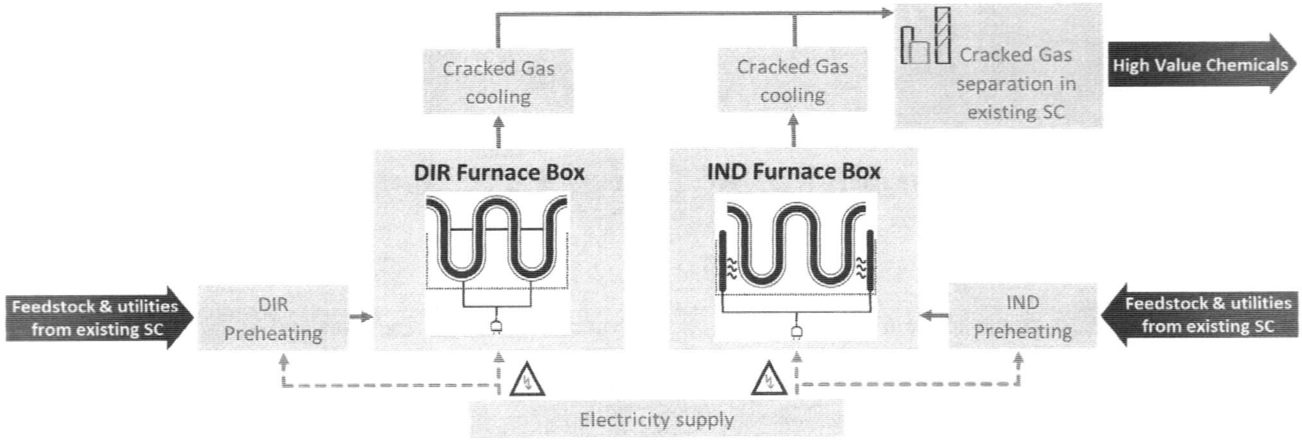

Figure 3: Functional sketch of demonstration plant architecture

The integration of the demonstration plant into the infrastructure of one of the existing BASF Ludwigshafen steam crackers is a key feature. But it also means that the demonstration plant EPC execution as well as the plant operation must comply with the strict regulations and

permitting processes of a commercial production facility. Consequently, all common industry-standard design reviews have been conducted, documented, and followed up, based on best practice from conventional process plant execution projects.
In addition, the standard procedures were supplemented by a newly devised multi-disciplinary review format, namely an "electric HAZOP", which was conducted with support from notified body experts.
This review specifically focussed on the novel aspects of the implemented technologies and their innate, high share of electric components. Furthermore, the demonstration plant did undergo the complete certification process as for a commercial unit.

In Figure 4 some major fields of knowledge gain from the demonstration plant are illustrated, showing that the EPC execution as well as the operation of the demonstration plant are providing a wealth of knowledge and information that are invaluable in bringing the STARBRIDGE™ technology to full commercial deployment.

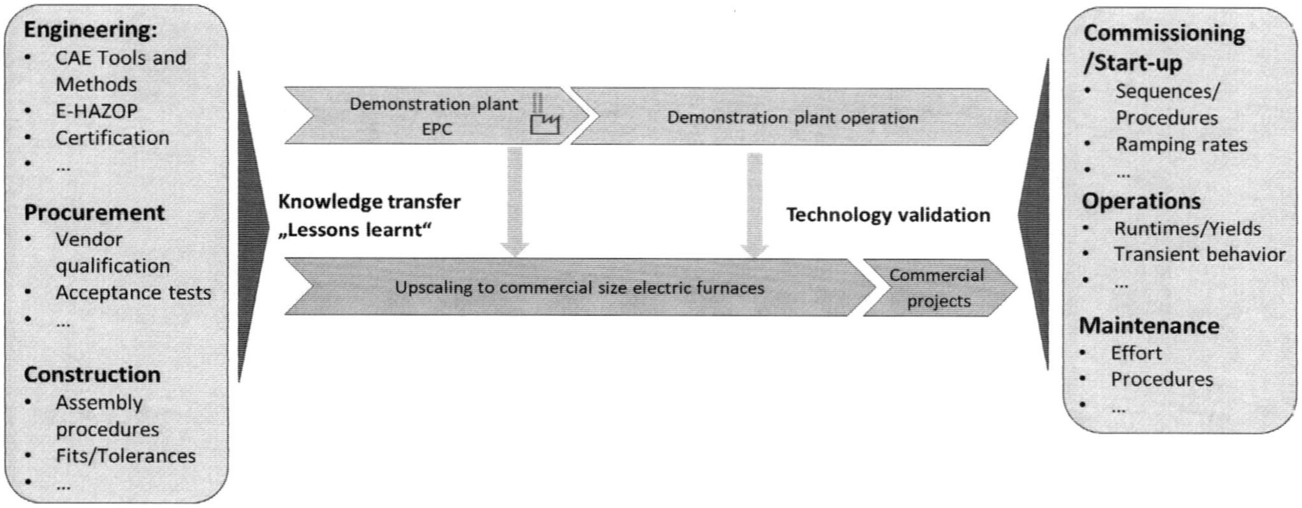

Figure 4: Benefits and learnings from demonstration plant

To ensure the successful implementation of the novel heating technologies, the companies have devised a comprehensive testing program. Starting in 2024, the program spans several years and encompasses a wide range of objectives. At its core, the program seeks to prove the effectiveness of the two heating concepts. This effort will be supported by simulation models further validated by experimental results generated during the program.

In addition to proving the effectiveness of the heating concepts, the program will also generate valuable operational experience. Company personnel will gain expertise in the operation and maintenance of the heating systems. This experience will be extremely useful in scaling the technology to commercial application. Moreover, the testing program will identify ways to optimize and further improve the heating technologies as part of the scale-up.

Impact of transition from fired to electric furnaces

Conventional fired furnaces form the heart of a steam cracking plant, not only because they host the cracking reaction, but also because of their central role in the plant's energy system. Large quantities of heat are recovered from the furnace flue gas and cracked gas streams leaving the cracking coils. This heat is used to preheat furnace feed streams and to generate high pressure utility steam, which is later used in turbines and heat exchangers for mechanical work and heat supply in the downstream separation section.

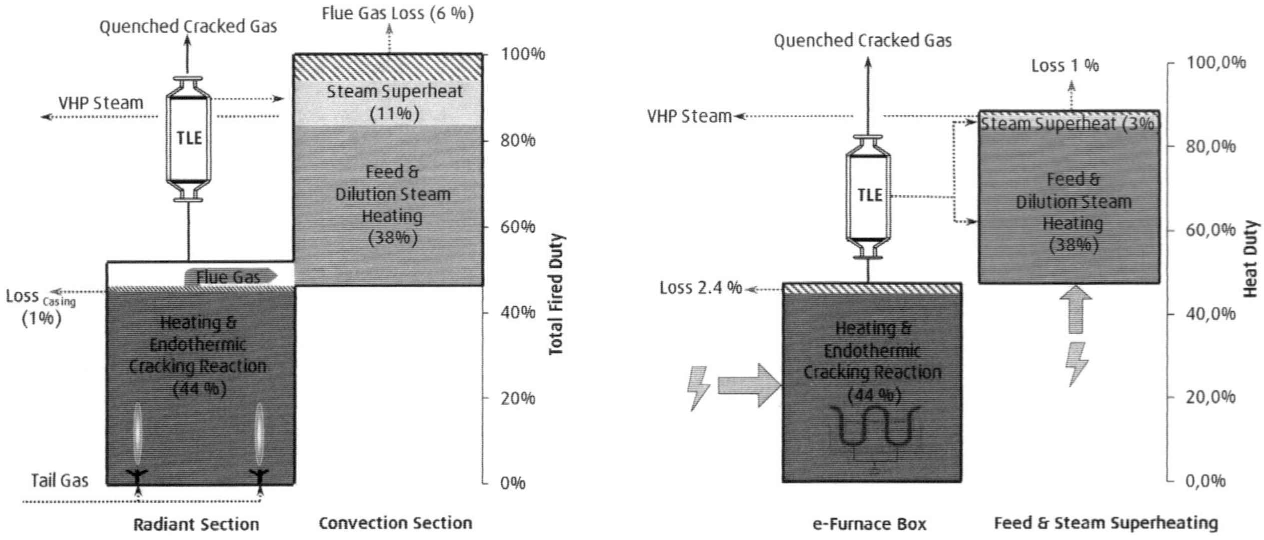

Figure 5: Comparison of heat duties required in a conventional fired furnace and electrically heated furnace including heat integration unit

Figure 5 schematically illustrates on the left side the division of a fired cracking furnace into two main sections, a radiant section and a convection section. In the radiant section, heat from combustion of fuel gas is absorbed by the cracking coils, which heat up the process feed to temperatures where cracking of long-chained hydrocarbon molecules into short-chained product molecules occurs. Typical inlet temperatures for the process gas into the coils (XOT) are between 550 and 750 °C, and the outlet temperatures of the cracked gas coming from the coils are typically in the range between 800 and 900 °C. Typically, 35% to 50% of the total firing duty is transferred to the coils in the radiant section, whereas the residual heat must be recovered in the convection section.

The convection zone of a fired cracking furnace is located above the radiant zone and consists of various tube bundles that cross the flue gas duct. The main purpose of the convection zone is to recover as much energy as possible from the hot flue gas coming from the radiant zone. As mentioned earlier, this heat is used for process heating such as preheating of boiler feed water and/or hydrocarbon feeds, (partial) vaporization of liquid hydrocarbon feeds (with or without process steam injection) and superheating of process steam and high-pressure utility steam. In convection sections of low-emission furnaces with reduced carbon dioxide footprint, parts of the energy recovered from flue gas are used in addition directly or indirectly to preheat the combustion air.

Taking radiant and convection zone together, modern fired cracking furnace may make use of 90 to 95% of the overall fired duty (based on the lower heating value of the combusted fuel). Consequently, the flue gas is typically cooled down to temperature levels between 120°C and 140 °C before leaving the convection section and being released to the atmosphere via stack.

The cracked gas heat is recovered in a quench section located downstream of the radiant section along the main process route. This is achieved by utilizing one or more quench exchangers to rapidly cool the process gas below a critical temperature level. It effectively halts the cracking reactions while simultaneously cooling the gas for downstream processing, as well as recovering and utilizing its sensible heat for additional energy usage. The recovered heat in the quench section is typically used for vaporizing (super-)high-pressure boiler feed water (of a typical pressure range between 60 and 125 bar absolute pressure), and for preheating the same boiler feed water, before it is fed to a steam drum. The generated high-purity (super-)high-pressure steam is then superheated in the convection zone against flue gas, and from there distributed to the central steam system of the cracker plant, providing heat and power used in heat exchangers and steam turbines.

As explained above, an electrically heated cracking furnace heat generates heat for the cracking reaction by joule heating either directly in cracking coils or indirectly in heating elements, radiating heat to the cracking coils. Moreover, the e-furnace is characterized by the fact that at least 95% of the consumed electrical energy is transferred to the process fluid in form of heat, eliminating the need for any waste heat recovery from vent streams or the like. Consequently, heat for feed & dilution steam heating, for boiler feed water heating, for high pressure steam superheating needs to be supplied in a different way. Thus, to ensure full functionality, an e-furnace must be coupled with a heat integration unit replacing the former convection zone and providing the required heating. With the absence of flue gas, the system inherent heat sources are limited to cracked gas heat only. Electrical power can be sourced to support the system if required.

In the new design, the operation of the preheating section can be adjusted far more independent from the reaction section since it is no longer constrained by the available flue gas heat. The use of supplementary electrical power for preheating purposes enables the decoupling of the preheating and reaction sections, even if cracked gas heat is integrated. (Specific details on this topic will be discussed in the following sections, concerning the heat integration designs). This advantageous feature, combined with the operational flexibility of the coil heating, represents a significant advantage of the e-furnace technology, allowing optimization of furnace run length, product yields and feed flexibility to the fullest extent.

Figure 5 indicates a decrease in required heat duty for an e-furnace with a heat integration unit compared to a fired cracking furnace with the same capacity. It is evident that significant heat loss to the atmosphere resulting from hot flue gas is eliminated, which leads to a higher overall efficiency of the e-furnace box and heat integration unit. The losses occur mainly through heat losses in exchangers and in the electrical infrastructure, particularly transformer losses. In comparison to conventional fired cracking furnace, the total losses in the e-furnace & heat integration unit are substantially lower.

It must be noted that the overall energy surplus from the furnace unit to the plant in form of superheated high pressure steam decreases in the e-furnace case, as part of the cracked gas heat is utilized (directly or indirectly) for feed & dilution steam heating purposes.

In an earlier publication of Linde at the EPC, an overview was presented on overall integration scenarios of commercial-size electric cracking furnaces into grassroot and existing plants [5]. The subsequent sections aims to provide additional insights into potential configurations of heat integration units with a focus on relevant boundary constraints but also on equipment design.

Design aspects for an electrical furnace and heat integration system

Having explored the differences between a fired and a electrically heated furnace in the preceding section, the next part of the paper delves deeper into design aspects and the larger implications on balancing-of-plant engineering.

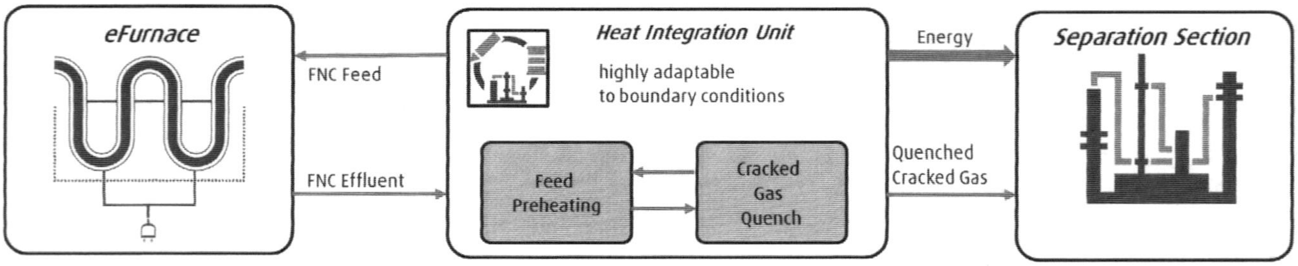

Figure 6: Sketch of an electrified furnace system with heat integration system. Both heating technologies are suitable for the entire range of relevant feedstocks

Aspects like the overall plant energy balance, the furnace configuration and finally, the feed type and flexibility are determining the design of a heat integration system for preheating feed and quenching the cracked gas.

The degree of cracked gas heat integration for preheating the feed and generating an energy surplus to the separation section must align with the plant's overall energy balance, which considers site utility conditions and availability and guarantees the supply of energy to process heat consumers and/or turbine drives. In case of grassroot projects, these central systems are likely to be very different than in existing plants, with high degrees of electrification throughout the plant and reduced utility steam circulation. In case of plant revamps, the changes in energy balance for a single fired furnace replacement may still be within the plant's operational margins whereas multiple furnace replacements may induce the need for simultaneous modifications in the separation section of the plant. The lack of excessive heat from a convection section and the availability of electrical power from renewable sources increase the incentive to utilize energy recovered from cracked gas cooling only for process heating instead of powering pumps and compressors. The superior efficiency of electrically powered drives for rotating equipment, typically over 90%, compared to conventional condensing steam turbines, typically only 30 to 40%, encourages the adoption of such systems to further improve the overall plant efficiency.

The quantity and dimensions of process equipment rely not only on the complexity of the heat integration system but also on the overall furnace configuration, which covers feed capacity, numbers of e-furnaces installed and desired operational flexibility. Other factors to be considered include availability and maintenance requirements, all of which must be taken into account when determining the installation effort and plot space needed.

As for the feed flexibility, both electric heating technologies are equally suitable for grassroot and revamp applications as add-ons or replacements. This also includes emergent, sustainable feedstock alternatives such as bio-feedstock and plastic waste pyrolysis oil, which enable reductions in scope 3 GHG emissions. This high feedstock flexibility results from the fact that either type of electrically heated coilbox, i.e. with direct or indirect heating, can be combined unrestrictedly with multiple layouts/topologies of furnace heat integration. The customization of an electric furnace to a given feedstock scenario therefore mostly takes place at the periphery of the coilbox, in particular in the heat integration system.

Figure 7 illustrates various feed specific boiling curves – starting with full range naphtha in the lower temperature range up to heavy hydrocarbon feedstocks such as AGO or chemically recycled plastic in the high boiling range. Analogous to different convection section designs between fired furnaces for liquid or gas feed, an e-furnace feed preheating system must account for such feedstock characteristics. To be more specific: While the temperature level of 100 bar high-pressure steam is sufficient for complete evaporation and superheating of naphtha, an AGO based feedstock requires much higher temperatures and therefore other methods and possibly different heat sources.

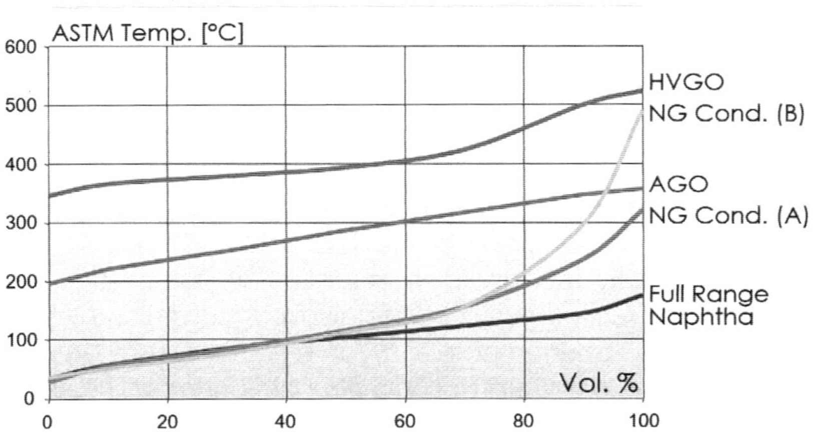

Figure 7: Typical boiling curves for liquid cracker feedstock

At the same time, a feed-specific coking mechanism limits the operating temperature range of the cracked gas quench to not sacrifice valuable furnace run length between decoking intervals. As can be seen in Figure 8, the sensitivity of furnace & quench run length as a function of quench outlet temperature is much higher for heavy feedstock than for lighter feedstocks. This is particularly due to the mechanism of increased condensation coking due to the higher concentration of very heavy cracked gas components originating from cracking high boiling feedstock [6]. Minimum run length requirements could therefore limit the extent of a high-temperature cracked gas heat integration for certain heavy feedstocks.

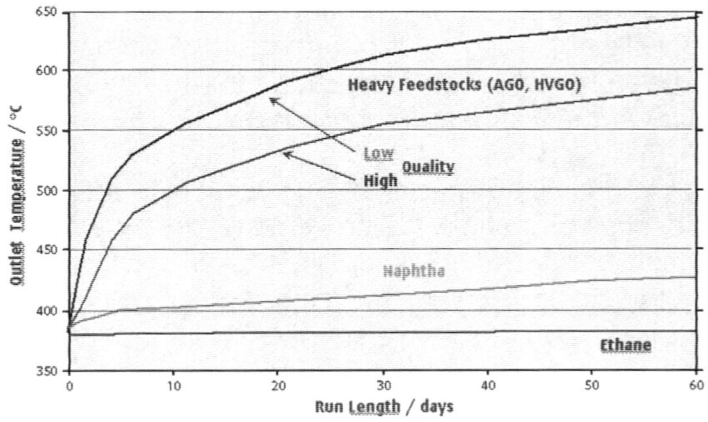

Figure 8: Furnace run length as a function of TLE outlet temperature.

Selected design approaches for different levels of cracked gas heat integration

The following sections will offer a high-level summary of three distinct approaches to designing a heat integration unit, using Naphtha as a feedstock example. The first approach with a conventional cracked gas heat recovery fits inside a cracker with a high steam demand of the separation section, e.g. as in a conventional Naphtha cracker. The second approach with a more advanced heat integration is suitable for a cracker integration with a moderate separation section steam demand, e.g. a Naphtha cracker with partial electrified compressor drives. The third approach represents an intense heat integration scheme and is suitable for a low or no separation section steam demand, e.g. a Naphtha cracker with full compressor drive electrification. As will be demonstrated, achieving higher steam surplus rates comes at the cost of increased electric power imports, ultimately resulting in a design that resembles that of a "concealed boiler."

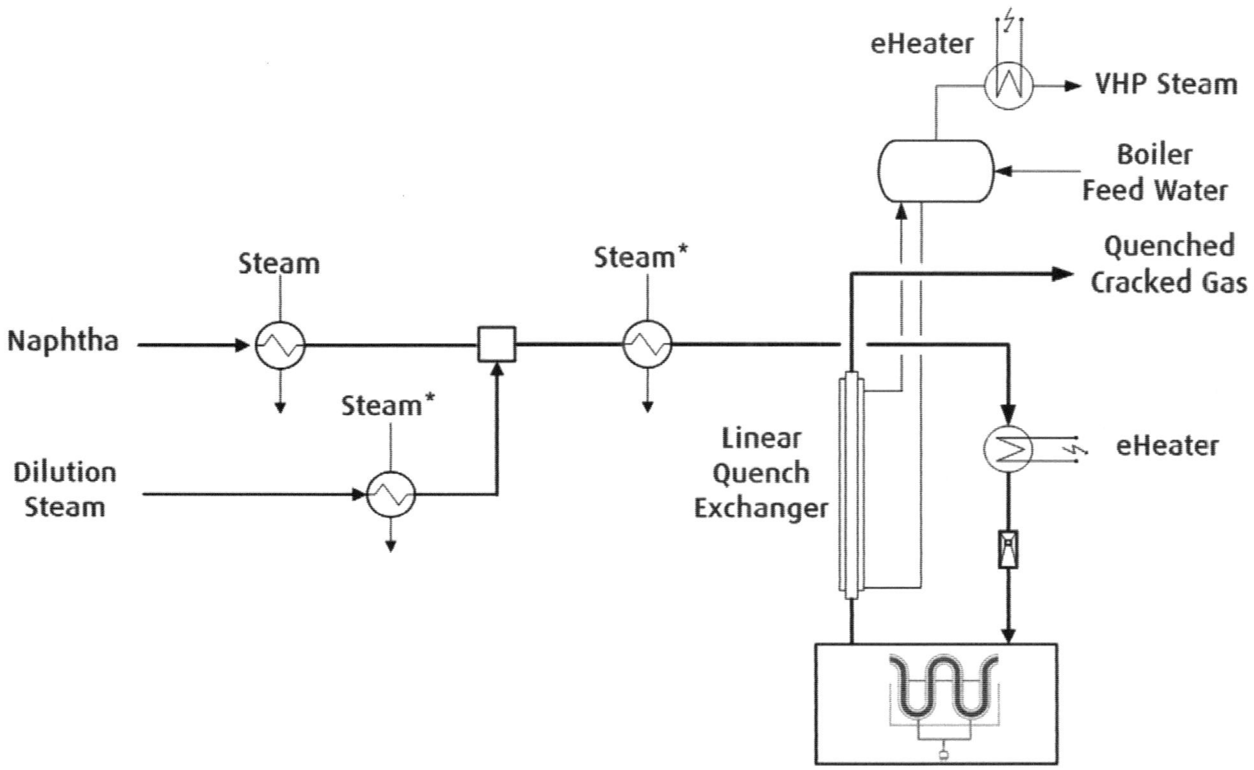

Figure 9: Conventional cracked gas quench for low cracked gas heat integration

A heat recovery scheme in Figure 9 illustrates a conventionally designed cracked gas quench section with one single quench stage comprising a linear quench exchanger cooler or conventional quench cooler. In this quench exchanger VHP steam is raised on the shell side and routed to a steam drum. The same steam, either taken directly saturated from the steam drum or extracted from a turbine, can then be used to vaporize and further superheat the hydrocarbon feedstock.

This allows to achieve process temperatures up to 300°C just by steam heating, which corresponds to approx. 40% of the feed preheating duty. At the same time, 45% of the high temperature heat of the furnace effluent is thus transferred to the furnace feed.

The remaining preheating duty from 300°C up to coil inlet temperature (~600°C) is covered in this example by electrical heating in an electrical heater. Thereby, electrical heating is applied in an efficient way at those elevated temperatures that exceed conventional heating media such as steam or thermal oil.

55% of the high temperature cracked gas heat is available as a net surplus of VHP steam carrying energy for the rest of the plant. The steam surplus accounts for approx. 0.5 tons of VHP steam per ton of naphtha feed, whereas the specific power demand of coilbox and heat integration unit accounts to 1.26 MWh/ton_naphtha.

Conventional quench exchangers and TEMA type exchangers are applied in this scheme. The electrical heater is placed in a service where fouling due to vaporizing hydrocarbons can be omitted and designed to avoid pre-cracking of feedstock molecules.

One important aspect of the e-furnace technology is the challenge of superheating the produced high-pressure steam. If and to which extent superheating of produced steam is required depends mainly on the purpose of steam use, whether for steam turbine drive or for process heat consumers. With the absence of hot flue gas, one of the few remaining options is the electrical superheating. However, it must be highlighted that the use of such steam in steam turbines is not recommended from a thermodynamic perspective as the turbine hardly produces more shaft power than previously spent as electrical duty for the superheating of the steam. Thus, it is far more efficient to directly electrify the compressor drive and utilize the cracked gas for process heating to the maximum possible extent.

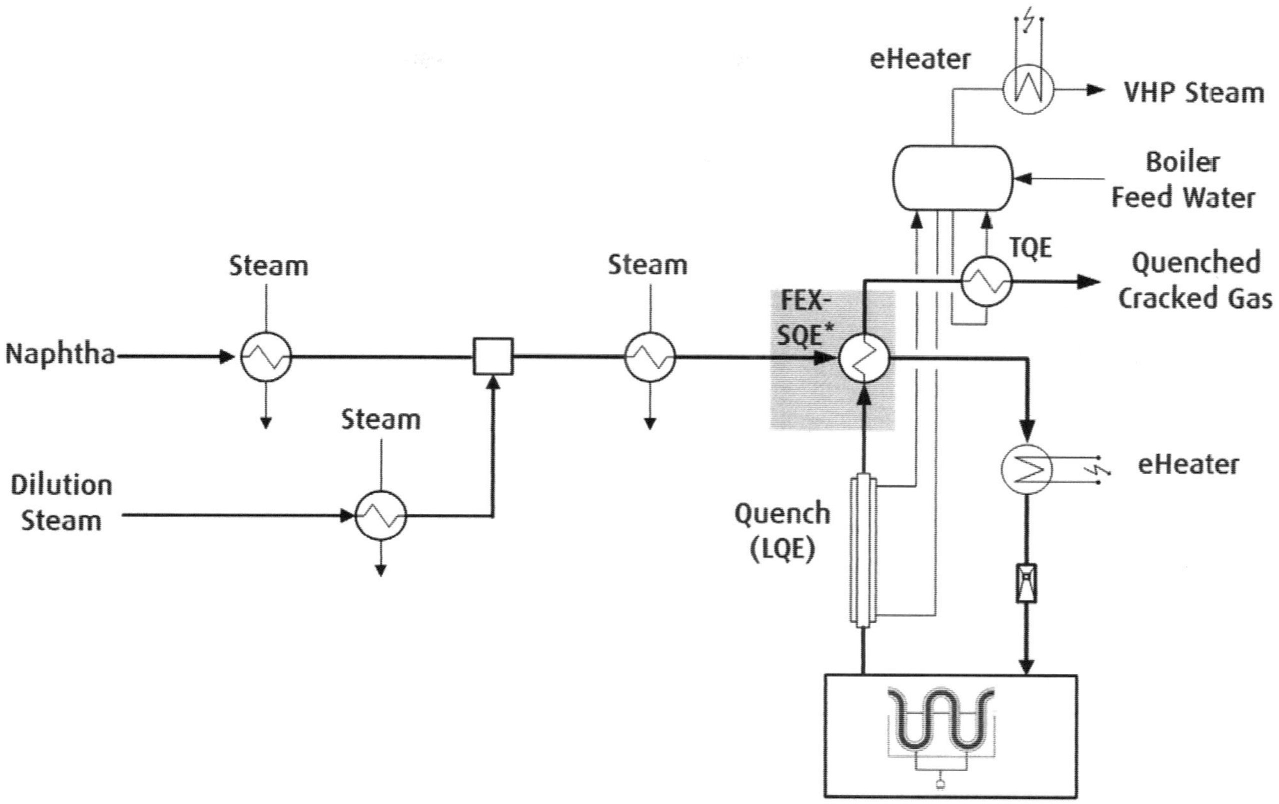

Figure 10: Cracked gas quench for advanced heat integration and lower steam production

A more advanced heat integration scheme is shown in Figure 10. The quench section of the cracked gas is splitted in three stages. Hot reactor effluent leaves the e-furnace coil box and enters a first quench exchanger, that is conventionally designed as transfer line exchanger (TLE) or linear quench exchanger (LQE) raising steam. The purpose of this first exchanger is to rapidly quench the cracked gas just below a critical temperature to stop secondary cracking reactions. Those secondary reactions are responsible for the loss of product yield and for the increased formation of coke that deposits on the walls of the TLE tubes leading to higher heat transfer resistance and higher pressure drops. In the shown example of naphtha, this corresponds to a primary quench outlet temperature of approx. 600°C. Accordingly, the length of the primary quench exchanger is far shorter than in the example above.

Now having quenched the cracked gas below the critical temperature, further cooling of cracked gas can subsequently be realized with less stringent requirements for cooling rate and residence time in subsequent quench stages. This allows for an application of a counter-current (gas phase) feed/effluent exchanger in the second position of the quench train. The temperature window in which the exchanger operates on feed shell side and cracked gas tube side, prevents effectively two unwanted phenomena, that have to be considered in general when transferring heat from reactor effluent to feed:

(1) The first phenomenon refers to the risk of pre-cracking hydrocarbon molecules at the hot end of the exchanger due to the contact of feed with a high temperature surface on the outside of the exchanger tubes. As the cracked gas is already quenched below 600°C when entering the apparatus, this risk is effectively avoided.

(2) The second phenomenon refers to the risk of "coking" at the cold end of the exchanger due to condensation of high boiling components present in the cracked gas, that may condense on the tube inside surface due to tube metal temperatures below dew point of said components. However, in the depicted scheme, preheated naphtha & dilution steam enters the feed/effluent exchanger at elevated temperatures of approx. ~300°C, thereby keeping the minimum tube metal temperature inside the tube above the critical condensation temperature.

Poor heat transfer coefficients are associated with heat exchange between two gaseous fluids leading to large heat exchange surface or high pressure drop. To counter this, various ideas have been implemented to achieve a realistic heat exchanger design with optimum performance.

First and most important, the differential temperature is increased by accepting a cracked gas outlet temperature around 450°C. This limits the amount of transferred heat and requires a third quench stage downstream of the feed/effluent exchanger – another unconventional approach for liquid feedstock cracked gas cooling.

Second, the feed/effluent quench (FEX) is based on a TEMA type straight tube heat exchanger design. The feed is routed in counter-current flow configuration on the shell side in a so-called no-tube-in-window design. This design arranges the tubes in the baffles in such a way that they always experience only a pure crossflow of the cooling medium, thereby preventing heat exchange in stagnation zones and swirling areas. Consequently, this design helps to avoid hotspots that increase the chances of fouling. The cracked gas enters and exits the tubes of the exchanger via inlet and outlet cones to prevent blockages of tube passes due to larger coke particles present in the gas. Tube design parameters are similarly as in conventional TLEs.

This exchanger enables the direct transfer of high temperature heat into the feed stream up to temperatures of ~approx. 450°C without compromising on product selectivity as the pressure drop of the device is not or hardly increased. Feed-side preheating can be implemented in such a way that pressure drop is not higher than in a conventional convection zone bundle.

The third quench stage is a conventional quench exchanger (TQE) that raises steam to the same steam drum as the primary quench exchanger and recovers the residual high-temperature heat, reducing the cracked gas to the temperature level of a single-stage quench outlet.

As the FEX outlet temperature of the feed is still below the desired coil inlet temperature, a final heating step in an electrical heater is necessary – similarly as in the scheme above but at reduced duty. Moreover, this electrical heater allows to compensate for operational fluctuations and to guarantee constant inlet temperature to the cracking coil.

In fractions of integrated cracked gas heat, this scheme allows to integrate 75-80% of the available high temperature cracked gas heat. Thereof, 25-35% are transferred via the newly implemented feed/effluent quench stage. This translates into a specific net steam surplus of 0.2 tons of VHP steam per ton of naphtha, which corresponds to approx. 25% of the cracked gas heat. The specific power consumption is reduced to 1.06 MWh/ton of naphtha due to more advanced heat integration.

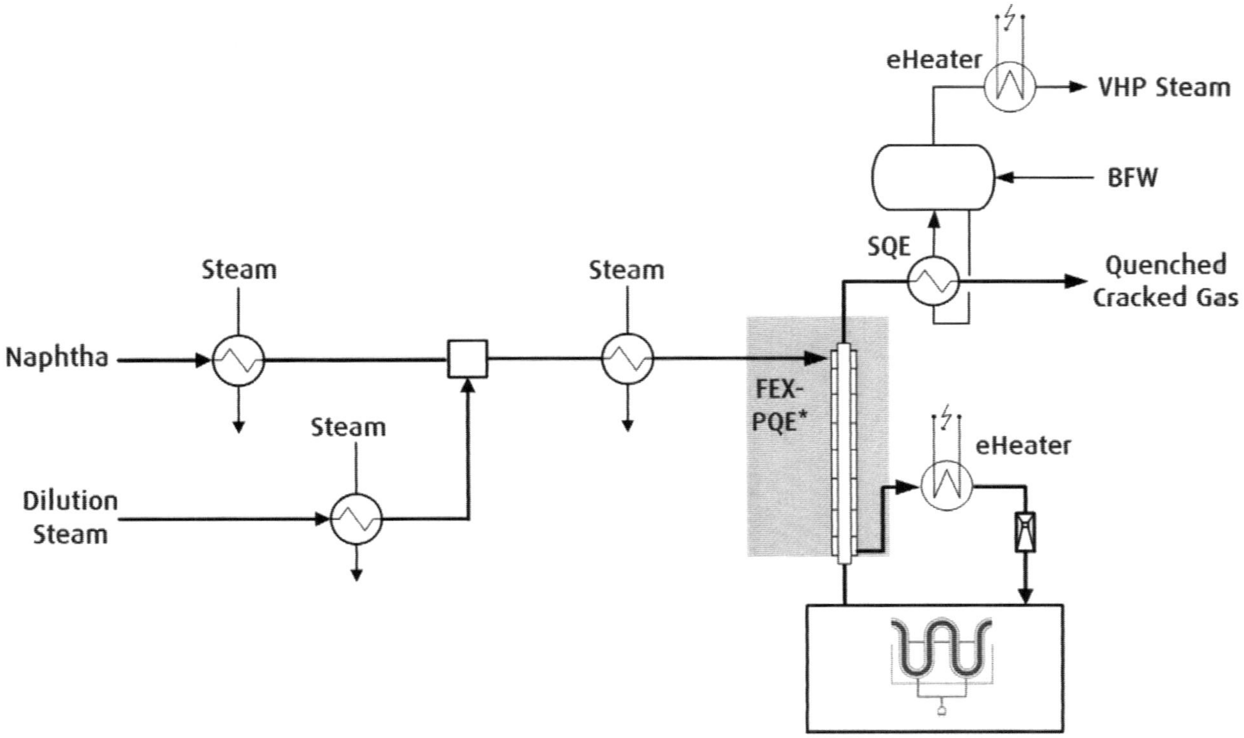

Figure 11: Cracked gas quench with intense heat integration and lowest steam production

A higher degree of heat integration >80%, in particular with respect to the very high temperature cracked gas heat, can only be achieved by means of special gas/gas quench exchanger designs. The quench in Figure 11 comprises two cooling stages, a feed/effluent quench exchanger in primary position (FEX-PQE) followed by a conventional quench in secondary position (SQE).

The feed/effluent exchanger (FEX) faces more stringent requirements in primary position than in secondary position: The reactor effluent coming from the cracking coils is at those high temperatures (~800-850°C) at which secondary side reactions continue to take place. To not compromise on product yields and operational performance due to rapid coking, residence time, cooling rate and pressure drop on cracked gas side must at least be comparable to a conventional TLE design.

Achieving such performance with maximized heat transfer rates in a gas/gas heat exchanger requires sophisticated measures, that go far beyond simple exchanger design tactics such as increasing differential temperature. One key element is heat transfer enhancement structures, that promote turbulence or increase the heat exchanger's surface area. Application of turbulence-promoting structures on the gaseous feed side allow to increase the rate of heat transfer from outside surface of the cracked gas tube to the flowing feed. Methods like fluid impingement or high shear-inducing geometries, have shown to be highly effective for optimal cracked gas quenching [7]. A collaborative effort between Linde, SABIC, ARVOS, and BASF resulted in the development of promising FEX designs utilizing said technologies. However, just like every turbulence-promoting feature, the benefits can only be drawn by expense of additional pressure drop on the feed side. The pressure-drop must be taken into account and aligned with project-specific boundary constraints.

Apart from the challenges related to heat transfer, the risk of pre-cracking feed hydrocarbons in the vicinity of an excessively hot exchanger tube surface (hot end) as well as the risk of condensation coking of cracked gas components on a too cold exchanger tube surface (cold end) are other pressing issues. These issues are of utmost importance and need to be taken into consideration when designing a feed/effluent quench exchanger.

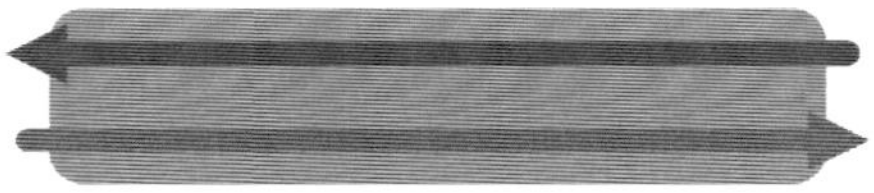

Figure 12: Countercurrent Flow

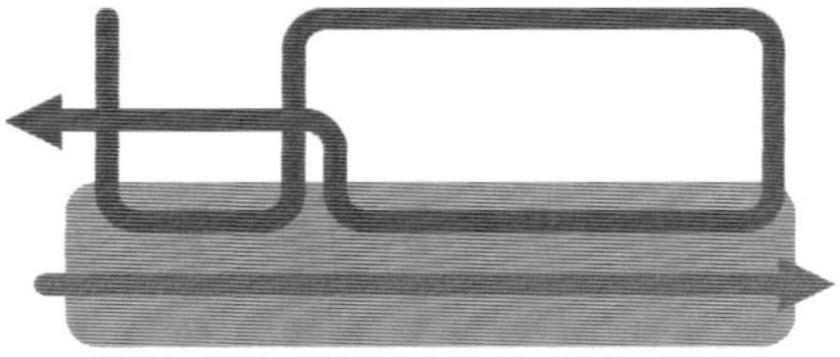

Figure 13: Co- and Countercurrent Flow (a)

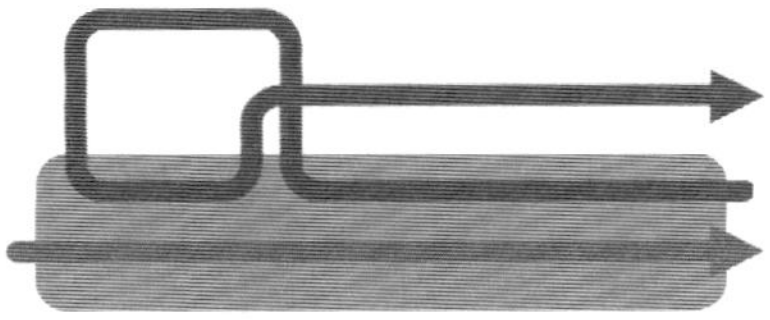

Figure 14: Co- and Countercurrent Flow (b)

By dividing the device from a single section unit with pure countercurrent flow (Figure 12), into multiple sections that include a co-current and a countercurrent stage in different configurations (Figure 13 & Figure 14), the issues mentioned above can be resolved. However, this again limits the degree of integrated high temperature heat, thus limiting similarly the maximum achievable feed preheating temperature.

To achieve the desired coil inlet temperature, a final electrical heating step in the electrical heater is required in an analogous way as in the schemes above but at further reduced duty. Decoupling of furnace operation from feed preheating is similarly possible as in the above schemes.

In fractions of integrated cracked gas heat, this scheme allows to integrate >90 % of the available high temperature CG heat. Thereof, approx. 45% are transferred via the innovative feed/effluent quench in primary position. This translates into a specific net steam surplus of 0.01 tons of VHP steam per ton of naphtha and corresponds to max. 5% of the cracked gas heat export. The specific power consumption is further reduced due to the intense heat integration to 0.85 MWh/ton of naphtha.

Conclusion & outlook

The electrification of cracking furnaces is set to play a key role in the transformation of the steam cracking industry towards net-zero GHG emissions. The large-scale demonstration of two applicable electrical heating technologies in an industrial plant setting is a significant milestone on this path. A successful demonstration can prove, that renewably produced electricity can be used in the future as main energy source in steam crackers while still ensuring reliable long-term furnace operation for base chemical production.

The EPC phase of the demonstration plant has enabled the development of technology-specific expertise in the fields of engineering, procurement and construction and facilitates now an efficient scaling of these exciting new technologies.

The demonstration plant is testing the DIR and IND e-Furnace technologies in a real plant setting throughout 2024 and beyond. Insights and learnings gained from this operation will be utilized to optimize the first commercial e-furnace generation.

A heat integration technology portfolio, featuring novel feed/effluent exchanger designs, has been developed for plant integration, providing a high flexibility to meet site-specific needs. This broad range of heat integration technologies is suitable for all forms of feeds, offering customized energy balance flexibility as well as operational flexibility.

In terms of commercialization, Linde Engineering is committed to provide commercial offers for the industrial scale deployment of the STARBRIDGE™ technology by the end of 2024. In the meantime, feasibility studies for integration of the e-furnace and periphery, as well as the balance-of-plant engineering, are in execution for various clients. These studies aim to develop the most effective approaches to scaling and implementation.

References

1. European Commission, JRC Science for Policy Report, "Energy efficiency and GHG emissions: Prospective scenarios for the Chemical and Petrochemical Industry", available under https://publications.jrc.ec.europa.eu/repository/bitstream/JRC105767/kj-na-28471-enn.pdf, p. 14, Table 3, 2017.

2. Lange, J. P., "Towards circular carbo-chemicals – the metamorphosis of petrochemicals", *Energy & Environmental Science*, 2021.

3. Amghizar, I., Dedeyne, J. N., Brown, D. J., Marin, G. B., Van Geem, K. M., "Sustainable innovations in steam cracking: CO_2 neutral olefin production", *Reaction Chemistry and Engineering*, 2020.

4. Middleton, J., "Decarbonisation of steam crackers", *Decarbonisation Technology*, 2021.

5. Zellhuber, M., Hofstaetter, M., Kemper, R., Kracker, G. "Electrification of steam cracker furnaces – Technology demonstration and the route to commercialization", AIChE Paper Number 27f, *Ethylene Producers Conference, 2023 Spring Meeting and 19th Global Congress on Process Safety*, 2023.

6. H. Cai, A.Krzywicki, M. C. Oballa, "Coke formation in steam crackers for ethylene production", *Chemical Engineering and Processing 41 (2002) 199–214*

7. Broekhuis, R. "Electrification of cracking furnaces", AIChE Keynote 29a, *3rd Topical Conference on Heat Exchangers, 2024 Spring Meeting and 20th Global Congress on Process Safety*, 2024.

SPRING24 +20TH GCPS

A Joint AIChE and CCPS Meeting

Paper Number: 120b

Ethylene Production through Electrochemical CO_2 Reduction Technology

Christine Gabardo
CTO, CERT Systems Inc
501 Alliance Ave, Suite 406, Toronto, ON, Canada

Abdollah Hajizadeh
Research Scientist II, CERT Systems Inc
501 Alliance Ave, Suite 406, Toronto, ON, Canada

Colin O'Brien
Principal Research Scientist, CERT Systems Inc
501 Alliance Ave, Suite 406, Toronto, ON, Canada

Daniel Young
Research Scientist II, CERT Systems Inc
501 Alliance Ave, Suite 406, Toronto, ON, Canada

Alexander Ip
CEO, CERT Systems Inc
501 Alliance Ave, Suite 406, Toronto, ON, Canada

Prepared for Presentation at
American Institute of Chemical Engineers
2024 Spring Meeting and 20th Global Congress on Process Safety
New Orleans, LA
March 24-28, 2024

Ethylene Production through Electrochemical CO_2 Reduction Technology

Christine Gabardo
CTO, CERT Systems Inc
501 Alliance Ave, Toronto, ON, Canada
christine@co2cert.com

Abdollah Hajizadeh
Research Scientist II, CERT Systems Inc
abdollah@co2cert.com

Colin O'Brien
Principal Research Scientist, CERT Systems Inc
colin@co2cert.com

Daniel Young
Research Scientist II, CERT Systems Inc
daniel@co2cert.com

Alexander Ip
CEO, CERT Systems Inc
alex@co2cert.com

Keywords: C_2H_4, CO_2 electrolysis, CO2RR, Pilot plant, Carbon capture and utilization

Abstract

Ethylene (C_2H_4) is a crucial feedstock in the petrochemical industry, essential for producing polyethylene, ethylene oxide, ethylene dichloride, and vinyl acetate. The conventional method of C_2H_4 production involves the conversion of ethane or naphtha into olefins through a steam cracking process. This process uses a significant amount of heat energy, which releases a substantial amount of carbon dioxide (CO_2) emissions, estimated at approximately 1.3 metric tons of CO_2-equivalent per metric ton of C_2H_4. This issue highlights the importance of finding a balance between C_2H_4's vital industrial function and the need to address its environmental impact by reducing CO_2 emissions. C_2H_4 production via electrochemical CO_2 reduction reaction (CO2RR) technology offers a pathway to lower carbon intensity C_2H_4 using renewable electricity, water, and CO_2 as process inputs, minimizing reliance on fossil fuels. Electrochemical CO_2 conversion provides advantages like modularity, scalability, and a single-step process. CERT Systems ("CERT") has advanced the CO2RR to ethylene process technological readiness level (TRL) through membrane electrode assembly (MEA) optimization and scaling. CERT, in collaboration with the University of Toronto, developed a pilot-scale electrolysis system for converting CO_2 into C_2H_4, demonstrated during the finals of the NRG COSIA Carbon XPRIZE competition. The pilot demonstration was the largest CO_2 to ethylene electrolyzer to date with a capacity to process 100 kg of CO_2 per day, and the first to convert industrial CO_2 emissions into ethylene.

1 Introduction

Ethylene (C_2H_4) is the most widely used and produced organic chemical in the petrochemical industry, serving as a key intermediate to produce various products like polyethylene, ethylene oxide, ethylene dichloride, and vinyl acetate. The worldwide ethylene production increased from 214.3 million tons per year in 2021 to 227.6 million tons per year in 2023, and is expected to reach to 287 million tons per year in 2030 [1]. Ethylene is produced primarily from steam cracking of ethane or naphtha, where the feedstock is thermally converted to ethylene, propylene, and hydrogen co-product at very high temperatures (> 750 °C) [2]. The conventional steam cracking process is highly endothermic and faces many challenges, such as coke formation, energy loss, and high carbon dioxide (CO_2) emissions. On average, 1.3 tonnes of CO_2 equivalent is emitted per tonne of C_2H_4 produced using the steam cracking process [3]. Hence, the development of alternative production technologies, such as electrochemical and biochemical processes, is a crucial step toward decreasing the carbon intensity of C_2H_4 [4].

Unlike steam cracking, where fossil fuels are used as both the chemical feedstock and energy feedstock, in CO2RR, the energy feedstock is renewable electricity, and the chemical feedstock is CO_2. The electrochemical conversion process involves a series of proton-coupled electron-transfer steps to produce various hydrocarbon products, such as C_2H_4, hydrogen (H_2), methane (CH_4), carbon monoxide (CO), formic acid (HCOOH), ethanol (C_2H_5OH), and acetic acid (CH_3COOH). The product selectivity can be tuned based on the electrolyzer operational parameters and the catalyst coated at the cathode. A CO_2 electrolysis product obtained by Gabardo et al. (2019) [5] consisted of 19.2% C_2H_4, 0.9% CH_4, 9.6% CO, 23.6% H_2, 45.6% CO_2, and 1.1% H_2O on a mole basis, but this ratio of products can be tuned based on the operational parameters and catalyst performance. Of the possible CO2RR products, ethylene is a promising commercial target due to its large market size and high potential for emissions reduction [6]. For every tonne of ethylene produced, over three tonnes of CO_2 are utilized on a stoichiometric basis.

CO_2 electrolysis involves a complex interaction of factors crucial to overall process performance, with selectivity, or Faradaic efficiency (FE), and reaction rate, or current density, being key metrics. The cathode catalyst and environment must be tuned to enable selective electroreduction to ethylene. In recent years, significant advancements have been made in selectivity and current density through materials design and cathode engineering [7–9]. Improvements in single pass utilization (SPU) of CO_2 in the reactor have likewise been achieved through the use of electrode designs that suppress CO_2 lost as carbonate [10,11]. These reports have demonstrated the technical viability of the process, but more work is needed to demonstrate commercial potential.

In order to advance the CO_2 electrolysis technology to industrial implementation, scalable systems and pilot demonstrations are needed. Among the common electrolyzer configurations, the membrane electrode assembly (MEA) stands out, utilizing an ion exchange membrane sandwiched between a cathode and anode in a zero-gap configuration [5]. This configuration minimizes ohmic resistances, alleviates issues with cathode flooding/ electrolyzer leaking, and can maintain high C_2H_4 selectivity over extended periods [5]. In addition, developing models to predict performance beyond the laboratory scale can be useful. In this regard, a semi-empirical electrolyzer model was developed in Aspen Custom Modeler (ACM) and validated against experimental results from single-cell lab-scale (5 cm^2), single-cell pilot-scale (800 cm^2), and stack (10x800 cm^2) pilot-scale electrolyzers [12]. These studies provide valuable insights into the factors influencing the commercialization potential of CO_2 electrolysis technology and highlight areas for further research and development to overcome barriers to scale. In summary, recent literature on CO_2 electrolysis scale-up attempts underscores the multifaceted nature of this field and the need for comprehensive approaches integrating catalyst development, reactor design, and system integration.

This paper highlights recent advancements in electrochemical CO_2 conversion to C_2H_4 and the efforts made towards scaling up this process. In Section 2, we describe the fundamentals of CO_2 electrolysis and the key metrics. In Section 3, we discuss CERT's journey scaling electrochemical CO_2 reduction technology, focusing on the 40,000 cm^2 electrolyzer stack pilot plant and key insights gained from this endeavor. Next, we discuss recent advances aimed at enhancing CO_2 electrolysis performance, including failure mode diagnosis, high CO_2 utilization, and durability improvements. Finally, we offer insights into the next developments for this technology.

2 Fundamentals of CO_2 Electroreduction

To conduct CO_2 electroreduction, the cathodic process of CO_2 reduction must be coupled with an anodic half-reaction. This overall process is not spontaneous, necessitating a voltage difference between the cathode and anode to drive the reaction. The electrochemical CO2RR can produce many different carbon containing compounds. The general form of the CO_2 reduction reactions in basic and acidic conditions, respectively, are as follows:

$$xCO_2 + yH_2O + ze^- \rightarrow C_xH_{2y-z}O_{2x+y-z} + zOH^- \tag{1}$$

$$xCO_2 + zH^+ + ze^- \rightarrow C_xH_{2y-z}O_{2x+y-z} + (z-y)H_2O \tag{2}$$

The stoichiometric coefficients x, y, and z for various common CO_2 electroreduction products and H_2 are illustrated in Table 1. These equations reveal that the reaction will elevate the local cathode pH, either by generating OH^- or by consuming H^+ ions, surpassing the pH of the bulk electrolyte solution. Furthermore, these equations underscore the three fundamental requirements for CO_2 reduction to transpire at a catalyst site: the presence of ions, electrons, and CO_2.

Table 1. Stoichiometric and thermodynamic properties for several common CO_2 electroreduction products and H_2.

Product	State at STP*	x (CO_2)	y (H_2O)	z (e^-)	Gibb's energy of formation (kJ/mol) [13]	E^0 (V vs. SHE**)
H_2	Gas	0	2	2	0.00	0.00
CO	Gas	1	1	2	-137.17	-0.10
HCOOH	Liquid	1	2	2	-351.00	-0.17
CH_4	Gas	1	6	8	-50.46	0.17
CH_3OH	Liquid	1	7	8	-166.27	0.01
C_2H_4	Gas	2	8	12	68.46	0.08
C_2H_5OH	Liquid	2	9	12	-174.78	0.08

* STP = Standard temperature and pressure

** SHE = Standard hydrogen electrode

The minimum voltage (E_i^0) required to drive the reaction at standard conditions can be calculated using thermodynamics:

$$E_i^0 = \frac{-\Delta G_{rxn,i}}{z_i F} \tag{3}$$

where $\Delta G_{rxn,i}$ is the Gibb's free energy change of component i at standard conditions; z_i is the number of electrons transferred per product molecule i; and F stands for Faraday's constant (equal to 96,485.3 C/mol).

The minimum voltage necessary for the electrochemical conversion of CO_2 into C_2H_4 is 1.15 V. This specific voltage level corresponds to a thermodynamic power consumption requirement of 13.21 kWh per kilogram of C_2H_4 produced [5]. It is important to emphasize that this figure represents the absolute minimum power consumption dictated by the underlying thermodynamics of the process.

The cathodic CO2RR necessitates an equivalent current of an anodic reaction. The oxygen evolution reaction (OER) is frequently employed in CO2RR experiments owing to the accessibility of its feedstock (water) and the presence of established catalysts. The chemical equations for OER in acidic and basic environments are as follows:

$$2H_2O \rightarrow 4H^+ + 4e^- + O_2 \tag{4}$$

$$4OH^- \rightarrow 2H_2O + 4e^- + O_2 \tag{5}$$

For the electroreduction of CO_2 to C_2H_4, the primary reactions are depicted in Figure 1. The first and second rows of reactions in this figure depict the CO_2 electroreduction reactions under basic and acidic conditions, respectively.

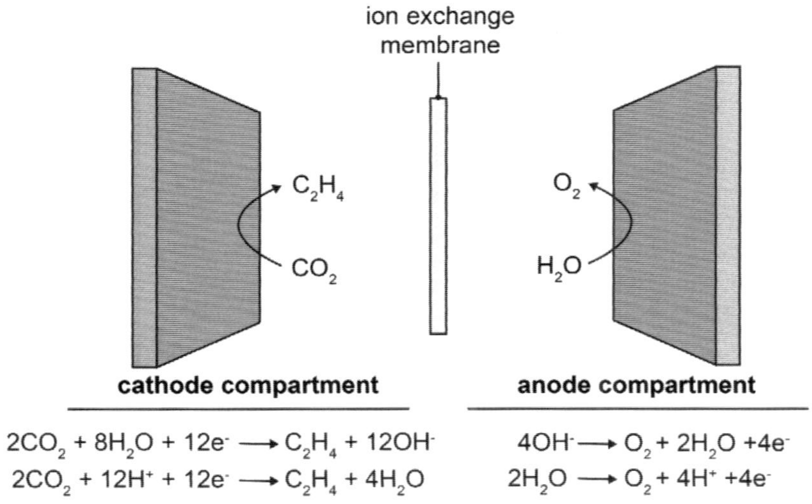

Figure 1. General reactions in a CO_2 electrolyzer.

There are several key metrics for CO_2 electrolysis systems as defined below:

Faradaic efficiency (FE). FE describes the percentage of total electrons that is used to make a certain product (e.g. C_2H_4). We also refer to this as selectivity of the system. High FE towards one product (e.g. 80% for C_2H_4) is desired compared to low selectivity towards multiple products (e.g. 20% H_2 + 20% CO + 20% HCOOH + 20% C_2H_5OH + 20% C_2H_4).

Voltage. Voltage describes how much energy is required for the system to run. There is a theoretical minimum value (E^0_{cell}) and the actual cell voltage ($E_{actual\ cell}$) is higher than minimum value due to various reasons. The desired case is $E_{actual\ cell} = E^0_{cell}$.

Energy efficiency (EE). EE is the product of the FE and voltage efficiency ($E^0_{cell}/E_{actual\ cell}$). High EE is always desired, and it means a high FE and/or a low voltage.

Current density. Current density refers to the number of electrons passing through a unit area within a unit amount of time. We can simplify this term into reaction rate. Ideally, high current density (~high reaction rate) operation is preferred.

Stability. Stability or durability mainly refers to how stable FE and voltage for the system are. A stable system exhibits consistent FE and voltage over time.

Single pass utilization (SPU). SPU represents the proportion of CO_2 that is converted within the electrolyzer. A high SPU is desired because a low SPU indicates that a significant portion of the CO_2 introduced into the electrolyzer remains unconverted and exits through the electrolyzer outlet.

3 CERT's XPRIZE Pilot Plant

The NRG COSIA (Canada's Oil Sands Innovation Alliance) Carbon XPRIZE [14], with a prize pool of $20 million, was an international competition aimed at fostering the development and expansion of innovative technologies capable of transforming carbon dioxide (CO_2) emissions into practical products. Its primary objective was to identify and scale the most promising CO_2 utilization technologies, assessing based on both the volume of CO_2 converted and the overall value of the resulting products.

Team CERT, composed of researchers from the University of Toronto, was established in 2016 to participate in the competition. Team CERT was among nearly 40 teams from around the world competing in the XPRIZE. Teams were judged solely on predetermined criteria focusing on CO_2 conversion and product value, along with meeting specific performance benchmarks for land and water applications, as outlined in the competition guidelines.

In Round 2, Team CERT developed a prototype unit that upscaled a laboratory reactor to a 400 cm^2 stack comprising four 100 cm^2 cells. This prototype featured an electrolyzer stack utilizing a flow cell configuration to electrochemically convert CO_2 into carbon monoxide (CO). Team CERT designed and scaled the cathodes for CO_2 to CO conversion, as well as engineered the electrolyzer stack and supporting infrastructure. The prototype's efficacy was demonstrated in late 2017 using simulated captured flue gas and verification was conducted by a third-party engineering firm.

In the spring of 2018, Team CERT was selected as one of ten finalists to advance to Round 3 of the competition, which comprised two distinct competition streams: one held in Wyoming at a coal-fired power plant and another in Calgary, Alberta, at a natural gas-fired power plant. Team CERT participated in the Calgary finals at the Alberta Carbon Conversion Technology Centre (ACCTC) alongside four other teams. Figure 2 presents the progress of CERT's electrochemical technology during the NRG COSIA Carbon XPRIZE competition.

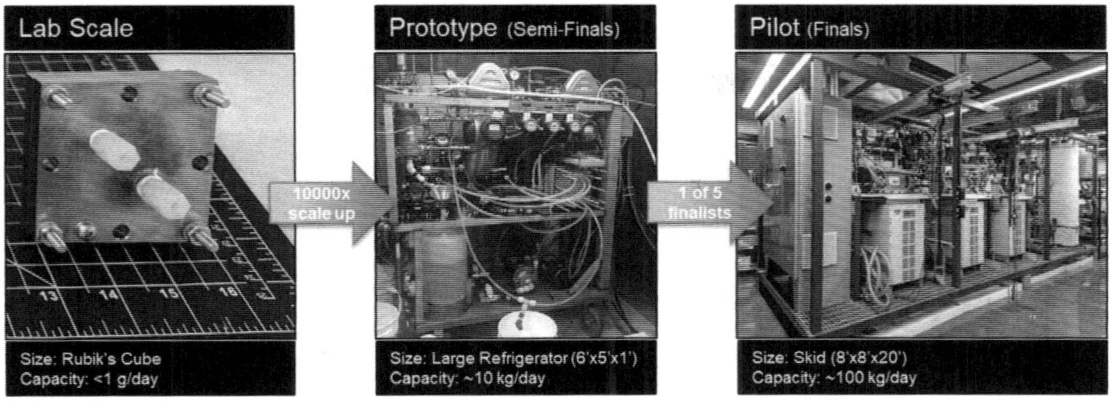

Figure 2. Team CERT's scaling through the XPRIZE competition Round 1 (lab scale), Round 2 (prototype), and Round 3 (pilot plant).

The design phase for the pilot plant, with a capacity of converting 100 kg of CO_2 per day, commenced in May 2019. For the finals, Team CERT altered the reactor design to employ a MEA architecture and shifted the product focus to syngas and C_2H_4 due to advancements that had been made at the laboratory scale to improve the efficiency and scalability of the CO_2 to C_2H_4 process. Throughout 2019, Team CERT focused on refining and scaling the membrane electrode assembly and electrolyzer stack designs. Team CERT increased the cell size to 800 cm^2 and developed five electrolyzer stacks, consisting of ten cells each (Figure 3). In parallel, by collaborating with a company specializing in modular plant design and construction, Team CERT developed the balance of plant.

By the end of 2019, the pilot plant was fully assembled and was transported to the demonstration site at the ACCTC in Calgary, Alberta. Located at the Shepard Energy Centre, the ACCTC provided each team with a designated gravel pad and utility connections, including power, natural gas, water, wastewater facilities, and either flue gas (4-6% CO_2) or captured CO_2 (98% CO_2).

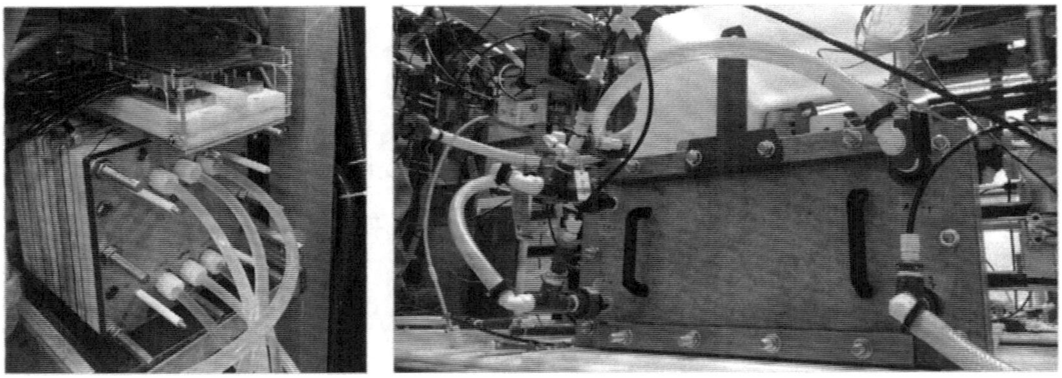

Figure 3. Photos of Team CERT's electrolyzer stacks in Round 2 (left) and Round 3 (right) of the NRG COSIA Carbon XPRIZE competition.

Team CERT relocated to the demonstration site in January 2020 and commenced commissioning the pilot plant in March 2020. However, due to the COVID-19 pandemic, the demonstration site was temporarily closed in mid-March. Operations resumed in July 2020, with the pilot plant being supplied with concentrated CO_2 captured by the ACCTC's amine-based capture unit (Figure 4). In accordance with competition guidelines, the teams were required to demonstrate 50 days of operation (24 hours/day), including 30 consecutive days. Team CERT commenced nearly continuous operation by mid-September and concluded the pilot demonstration in the first week of December 2020, with operation verified by the third-party engineering firm, 350 Solutions.

Through the XPRIZE competition, Team CERT logged nearly 2500 operating hours on the pilot system. For the first time, the team was able to demonstrate the electrochemical conversion of captured industrial emissions into ethylene.

Following the competition, CERT Systems Inc. established its own independent laboratory to further develop, scale, and commercialize CO_2 electrolysis devices.

Figure 4. Team CERT's pilot plant set up at the ACCTC, side view of the pilot plant with 3 bays showing, close up of one of the electrolyzer stacks, team photo of the operator team.

Key Insights & Future Directions. The journey through Round 3 of the NRG COSIA Carbon XPRIZE competition provided invaluable insights and paved the way for future developments. Here are the key learnings and directions moving forward:

- Scalability of Electrolyzer Design:
    - Our electrolyzer design was shown to be scalable, with the MEA architecture scaling up to a total electrolyzer area ~approximately 10,000 times larger than the system at the laboratory scale while maintaining similar selectivity and stability.
- Importance of Real-scale Testing:

  - Real-scale testing revealed phenomena not observable at laboratory scales or under ideal conditions.
  - Facing operational challenges, such as feedstock intermittency, during the competition spurred innovation, leading to the development of more robust CO_2 electrolysis systems.
  - Integration with upstream and downstream systems is crucial for optimizing the entire CO_2-to-product process efficiently.
- Optimization of the System:
  - The entire system, including CO_2 capture, conversion, and product separation, must be considered and optimized.
  - Energy consumption across all systems needs to be minimized. Tuning electrolyzer parameters, enhancing Faradaic efficiency, and reducing voltage can lower operating costs.
  - Improving single pass utilization and stability can decrease separation energy requirements and shift MEA materials from operational to capital costs, reducing downtime.

Informed by this pilot demonstration, CERT Systems has been focused on enhancing stability, single pass utilization, and faradaic efficiency in CO_2 electroreduction. With these improvements, we are planning our next scale pilot, bringing us closer to sustainable CO_2 utilization at scale.

4 CERT's Technology Advancements

4.1 Diagnosis of Failure Modes

While current CO_2 electrolyzers have made large improvements in selectivity and efficiency, stability remains a challenge. This limitation is partly due to difficulties in real-time monitoring of these electrolyzer systems and in determining the root causes of failure. Failures may arise from various factors such as initial conditions (e.g., over- or under-compression of the electrolyzer), gradual degradation of components (e.g., cathode or anode catalysts), accumulation of products or by-products, or sudden issues (e.g., membrane perforation or short circuits). Addressing these assembly-related, gradual, and sudden failure modes is crucial for enhancing both the lifespan and economic feasibility of CO2RR electrolyzers.

CERT Systems carried out an extensive study to diagnose the electrolyzer failure modes. In this project, in collaboration with Pulsenics [15], we utilized rapid in-line electrochemical impedance spectroscopy (EIS) to continuously monitor the performance of a CO2RR electrolyzer within a MEA throughout its operational lifespan. We employed simulations to replicate and analyze various failure modes that are commonly encountered in such systems, including assembly defects, gradual degradation, and immediate malfunctions. By fitting an equivalent electrical circuit (EEC) to the EIS data, we established correlations between alterations in Nyquist and Bode plots and specific electrochemical parameters, enabling the identification of distinctive signatures associated with each failure mode. The development of these failure mode signatures presents opportunities for pre-emptive measures to prevent failures, pinpoint malfunctioning cells, or redesign the system to mitigate underlying causes. Extending the stable operational duration of CO2RR electrolyzer cells is pivotal in advancing the commercial viability of this technology. We published our findings from this study in an article by Warkentin et al., (2023) [16], which provides detailed information on this project. Table 2 presents a summary of the impact of different failure modes on the cell operation.

Table 2. Failure modes and their key identifiers (reproduced with permission from publisher [16]).

Failure mode	Key identifiers (*: compared to baseline, ^{T}: with respect to time)								Detection time before visible in FE and voltage
	Electrochemical parameters				Voltage	FE			
	R0	R1	C_{eff}	W1	V	H_2	CO	C_2H_4	
Over-compressed	$\downarrow^{*}$	$\downarrow^{*}$	$\uparrow^{*}$			$\uparrow^{*}$	$\downarrow^{*}$		Immediate detection
Under-compressed	$\uparrow^{*}$	$\downarrow^{*}$	$\downarrow^{*}$			$\downarrow^{*}$	$\uparrow^{*}$		Immediate detection
Salt formation		$\downarrow^{T}$		$\uparrow^{T}$		$\uparrow^{T}$	$\downarrow^{T}$	$\uparrow^{T}$	~1 Hour
GDE flooding	$\uparrow^{T}$	$\uparrow^{T}$	$\downarrow^{T}$						Immediate detection
Cathode catalyst degradation	$\downarrow^{*}$		$\downarrow^{*}$ $\uparrow^{T}$			$\uparrow^{T}$	$\downarrow^{*T}$	$\uparrow^{*T}$	Immediate detection
Anode degradation		$\uparrow^{T}$	$\downarrow^{T}$		$\uparrow^{T}$	$\uparrow^{T}$			~11 Hours
Short circuit (prior)	$\uparrow^{T}$	$\uparrow^{T}$	$\uparrow^{T}$		$\downarrow^{T}$				~10 Hours
Short circuit (after)	$\downarrow^{T}$	$\downarrow^{T}$							

4.2 Increasing CO_2 Utilization and Durability

High CO_2 utilization and electrolyzer stability are key factors for the future scale-up of CO_2 electrolysis systems. One of the most significant benefits of higher CO_2 utilization in electrolysis systems is reduced need for downstream separation of unreacted CO_2. Optimizing this parameter not only simplifies the process but also reduces energy consumption and associated costs. Moreover, higher CO_2 utilization leads to increased production of valuable products, further enhancing the economic viability of the process. Additionally, electrolyzer stability, or durability, is crucial for the long-term operation of these systems. Ensuring that electrolyzers can withstand long usage without significant performance degradation or frequent maintenance is essential for cost-effectiveness and reduced Opex. We carried out several projects to address the technical barriers obstructing the commercialization of CO_2 conversion to C_2H_4. Specifically, we targeted improvement in single pass utilization (SPU) of CO_2, electrolyzer durability, and tolerance to low purity CO_2 feedstocks.

Through engineering and rigorous experimentation, CERT has been able to significantly improve the process. The main achievements of CERT Systems toward increasing CO_2 utilization and durability are highlighted below:

- Achieving over 50% utilization of CO_2 in a single pass leading to reducing the need for capture and recirculation of CO_2.

- Achieving over 1,400 h of stable operation. This is a necessary step on the path to commercial deployment, with water electrolyzers lasting >10,000 hours.
- Demonstrating high selectivity to C_2H_4 using feedstock with only 50% CO_2. This can reduce the energy costs associated with carbon capture.

5 CO_2 Electrolysis Future Scale Up

As the chemical manufacturing sector confronts the urgent need to meet net-zero targets, CO_2 electrolysis offers a promising solution. This technology not only helps reduce greenhouse gas emissions but also produces valuable fuels and chemicals. As research and development efforts intensify, the scalability of CO_2 electrolysis systems becomes a pivotal focus. Scaling up these technologies is essential to realize their full potential in sustainable chemical manufacturing and carbon management.

The CO_2 electrolysis process serves as the core of the CO_2-to-C_2H_4 pathway, yet it necessitates integration with both upstream and downstream processes. Figure 5 illustrates the overall block flow diagram (BFD) of a complete commercial plant designed to convert CO_2 to C_2H_4. The initial step involves capturing CO_2 from flue gas or the atmosphere. While this process is widely utilized, there remains room for efficiency improvement. Our findings indicate that achieving high selectivity does not always require highly purified CO_2, which opens opportunities for developing energy-efficient CO_2 capture systems.

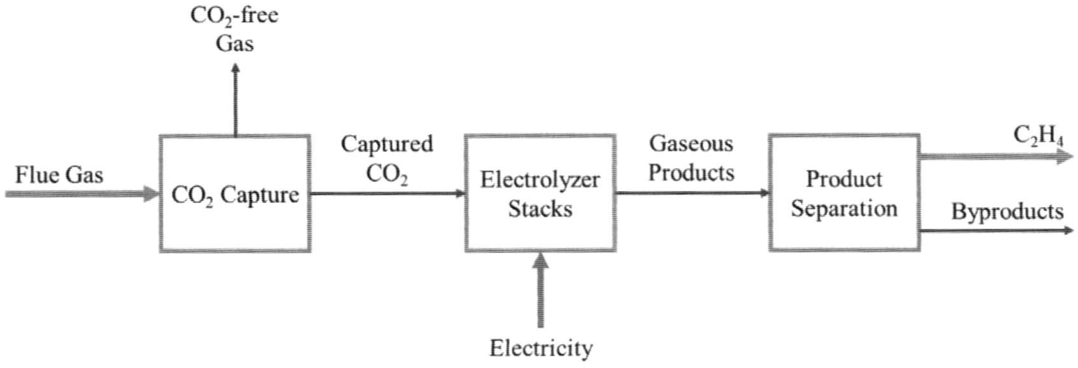

Figure 5. Overall future process to produce C_2H_4 from CO_2.

Moreover, reactive capture systems offer another avenue for implementation and advancement. Integrating reactive capture systems with carbonate electrolysis represents a promising approach to C_2H_4 production. In this integrated setup, the capture solution becomes the feedstock for the electrolyzer, eliminating the need for pure gaseous CO_2. Instead, the CO_2-rich capture solution is utilized directly in the carbonate electrolysis process to produce C_2H_4.

In addition to C_2H_4, the output from CO_2 electrolysis also includes other gases such as CO, H_2, and other impurities. To obtain purified C_2H_4 from this mixture, the development of efficient product separation and purification methods is crucial. The product separation unit not only focuses on purifying C_2H_4 but also facilitates the refining of byproducts, thereby enhancing the overall economy of the system. Notably, CO_2 electrolysis typically does not produce ethane as a byproduct, eliminating the need for costly C_2 separation.

To outline the necessary steps towards scaling up and reaching the pre-commercial stage for CO_2 electrolysis systems, several key actions must be undertaken:

1- Developing effective catalyst and membrane materials: Research and development efforts should focus on creating catalysts and membranes that enhance the efficiency and selectivity of CO_2 electrolysis, ensuring optimal conversion rates and product yields.
2- Designing efficient electrolyzer architecture: The design of the electrolyzer system plays a vital role in its performance and scalability. Developing innovative architectures that maximize surface area, facilitate efficient mass transfer, and ensure robust operation is essential for successful scale-up.
3- Increasing CO_2 utilization and durability: Enhancing CO_2 utilization and electrolyzer durability are critical factors for the commercial viability of CO_2 electrolysis systems. Strategies to improve CO_2 utilization rates and enhance the stability and longevity of electrolyzer components are essential for achieving cost-effective and sustainable operation.
4- Increasing scale for electrolyzers: Scaling up electrolyzer technology is imperative to meet the demand for large-scale CO_2 electrolysis systems. Research efforts should focus on developing scalable manufacturing processes and materials to enable the production of electrolyzers at a commercial scale.
5- Optimizing operation modes and electrolyzer performance: Fine-tuning operation modes and maximizing electrolyzer performance are essential for achieving high efficiency and productivity. Optimizing operating conditions, such as temperature, pressure, and flow rates, is essential to enhance the performance of electrolyzer systems.
6- Developing efficient product separation technologies: Efficient product separation technologies are necessary to purify ethylene and separate it from other gases produced during electrolysis. Developing advanced separation methods that minimize energy consumption and maximize product purity will be essential for optimizing the overall efficiency and economics of CO_2 electrolysis processes.
7- Integration of electrolysis with upstream production units: Integrating CO_2 electrolysis with upstream production units can streamline processes and improve overall efficiency. This includes exploring opportunities for integrating electrolysis with existing industrial processes to utilize waste streams and enhance resource efficiency.

In this work, we introduce a novel pathway towards sustainable ethylene production. CO_2 electrolysis enables electrification of chemical manufacturing through the use of CO_2 emissions, electricity and water as feedstocks for high value chemical production. We review work performed by CERT in the development and scale up of CO_2 electrolysis and provide pathways towards commercialization of this technology. By eliminating the use of highly-emitting fossil fuel feedstocks, CO_2 electrolysis holds potential for a net-zero chemical manufacturing industry.

Acknowledgments

We would like to extend our thanks to XPRIZE foundation, University of Toronto, TotalEnergies and Ontario Centre of Innovation (OCI) for supporting us and providing resources and facilities for the XPRIZE project. We would also like to acknowledge the great contributions of members of Team CERT throughout the competition.

References

[1] R. Chauhan, R. Sartape, N. Minocha, I. Goyal, M.R. Singh, Advancements in Environmentally Sustainable Technologies for Ethylene Production, Energy Fuels 37 (2023) 12589–12622. https://doi.org/10.1021/acs.energyfuels.3c01777.

[2] T. Ren, M. Patel, K. Blok, Olefins from conventional and heavy feedstocks: Energy use in steam cracking and alternative processes, Energy 31 (2006) 425–451. https://doi.org/10.1016/j.energy.2005.04.001.

[3] B. Young, T.R. Hawkins, C. Chiquelin, P. Sun, U.R. Gracida-Alvarez, A. Elgowainy, Environmental life cycle assessment of olefins and by-product hydrogen from steam cracking of natural gas liquids, naphtha, and gas oil, Journal of Cleaner Production 359 (2022) 131884. https://doi.org/10.1016/j.jclepro.2022.131884.

[4] T. Alerte, A. Gaona, J.P. Edwards, C.M. Gabardo, C.P. O'Brien, J. Wicks, L. Bonnenfant, A.S. Rasouli, D. Young, J. Abed, L. Kershaw, Y.C. Xiao, A. Sarkar, S.A. Jaffer, M.W. Schreiber, D. Sinton, H.L. MacLean, E.H. Sargent, Scale-Dependent Techno-Economic Analysis of CO2 Capture and Electroreduction to Ethylene, ACS Sustainable Chem. Eng. 11 (2023) 15651–15662. https://doi.org/10.1021/acssuschemeng.3c04373.

[5] C.M. Gabardo, C.P. O'Brien, J.P. Edwards, C. McCallum, Y. Xu, C.-T. Dinh, J. Li, E.H. Sargent, D. Sinton, Continuous carbon dioxide electroreduction to concentrated multi-carbon products using a membrane electrode assembly, Joule 3 (2019) 2777–2791.

[6] M.G. Kibria, J.P. Edwards, C.M. Gabardo, C.-T. Dinh, A. Seifitokaldani, D. Sinton, E.H. Sargent, Electrochemical CO2 Reduction into Chemical Feedstocks: From Mechanistic Electrocatalysis Models to System Design, Advanced Materials 31 (2019) 1807166. https://doi.org/10.1002/adma.201807166.

[7] C.-T. Dinh, T. Burdyny, M.G. Kibria, A. Seifitokaldani, C.M. Gabardo, F.P. García de Arquer, A. Kiani, J.P. Edwards, P. De Luna, O.S. Bushuyev, C. Zou, R. Quintero-Bermudez, Y. Pang, D. Sinton, E.H. Sargent, CO2 electroreduction to ethylene via hydroxide-mediated copper catalysis at an abrupt interface, Science 360 (2018) 783–787. https://doi.org/10.1126/science.aas9100.

[8] Y. Wang, Z. Wang, C.-T. Dinh, J. Li, A. Ozden, M. Golam Kibria, A. Seifitokaldani, C.-S. Tan, C.M. Gabardo, M. Luo, H. Zhou, F. Li, Y. Lum, C. McCallum, Y. Xu, M. Liu, A. Proppe, A. Johnston, P. Todorovic, T.-T. Zhuang, D. Sinton, S.O. Kelley, E.H. Sargent, Catalyst synthesis under CO2 electroreduction favours faceting and promotes renewable fuels electrosynthesis, Nat Catal 3 (2020) 98–106. https://doi.org/10.1038/s41929-019-0397-1.

[9] F.P. García de Arquer, C.-T. Dinh, A. Ozden, J. Wicks, C. McCallum, A.R. Kirmani, D.-H. Nam, C. Gabardo, A. Seifitokaldani, X. Wang, Y.C. Li, F. Li, J. Edwards, L.J. Richter, S.J. Thorpe, D. Sinton, E.H. Sargent, CO2 electrolysis to multicarbon products at activities greater than 1 A cm−2, Science 367 (2020) 661–666. https://doi.org/10.1126/science.aay4217.

[10] C.P. O'Brien, R.K. Miao, S. Liu, Y. Xu, G. Lee, A. Robb, J.E. Huang, K. Xie, K. Bertens, C.M. Gabardo, J.P. Edwards, C.-T. Dinh, E.H. Sargent, D. Sinton, Single Pass CO2 Conversion Exceeding 85% in the Electrosynthesis of Multicarbon Products via Local CO2 Regeneration, ACS Energy Letters (n.d.). https://pubs.acs.org/doi/full/10.1021/acsenergylett.1c01122 (accessed February 15, 2024).

[11] Y. Xu, R.K. Miao, J.P. Edwards, S. Liu, C.P. O'Brien, C.M. Gabardo, M. Fan, J.E. Huang, A. Robb, E.H. Sargent, D. Sinton, A microchanneled solid electrolyte for carbon-efficient CO2 electrolysis, Joule 6 (2022) 1333–1343. https://doi.org/10.1016/j.joule.2022.04.023.

[12] J.P. Edwards, T. Alerte, C.P. O'Brien, C.M. Gabardo, S. Liu, J. Wicks, A. Gaona, J. Abed, Y.C. Xiao, D. Young, A. Sedighian Rasouli, A. Sarkar, S.A. Jaffer, H.L. MacLean, E.H. Sargent, D. Sinton, Pilot-Scale CO2 Electrolysis Enables a Semi-empirical Electrolyzer Model, ACS Energy Lett. 8 (2023) 2576–2584. https://doi.org/10.1021/acsenergylett.3c00620.

[13] Introduction to Chemical Engineering Thermodynamics, 2021. https://www.mheducation.com/highered/product/introduction-chemical-engineering-thermodynamics-van-ness-abbott/M9781260721478.html (accessed February 26, 2024).

[14] Carbon XPRIZE | XPRIZE Foundation, XPRIZE (n.d.). https://www.xprize.org/prizes/carbon (accessed February 20, 2024).

[15] Pulsenics, (n.d.). https://www.pulsenics.com/ (accessed February 20, 2024).

[16] H. Warkentin, C.P. O'Brien, S. Holowka, B. Maxwell, M. Awara, M. Bouman, A.S. Zeraati, R. Nicholas, A.H. Ip, E.S. Elsahwi, C.M. Gabardo, D. Sinton, Early Warning for the Electrolyzer: Monitoring CO2 Reduction via In-Line Electrochemical Impedance Spectroscopy, ChemSusChem 16 (2023) e202300657. https://doi.org/10.1002/cssc.202300657.

Crossing the Chasm: Industrial Decarbonization and Advanced Nuclear

Paper 113c EPC Fundamentals technology

2024 Spring Meeting and 20th Global Congress on Process Safety

March 24, 2024 to March 28, 2024

New Orleans Ernest N. Morial Convention Center, New Orleans, LA

Scott J Bury, Ph.D. and Michael A. Curtis, Ph.D.

The Dow Chemical Company

Introduction

Every product begins with a Joule. The simple statement highlights that energy is the starting point for all manufacturing. The chemical process industry is a major energy consumer and hence a major emitter of CO2. Ethylene production stands out as a significant contributor to CO2 emissions in the chemical industry, particularly in the U.S. where it primarily comes from the process of steam cracking ethane gas. This process is instrumental in manufacturing vital chemical building blocks, such as ethylene, propylene, butadiene, benzene, toluene, and xylene (BTX), which are primarily used in the production of plastics. The CO2 emissions attributed to steam cracking, which is about 41MT and 90% of these emissions in the U.S. can be traced back to the production of ethylene alone. This underscores the substantial role that ethylene production plays in the carbon footprint of the chemical precursors generated through steam cracking. Steam cracking is one of the top 5 processes that contribute to 80% of the chemical and refining industries scope 1 & 2 emissions.[1]

Although the US DOE September 2023 Liftoff report, "Pathways to Commercial Liftoff: Decarbonizing Chemicals & Refining" highlights the critical economic role of the US Chemicals and Refining industry, it also strongly calls out the lack of progress towards the administration's decarbonization goals; "*The chemicals & refining sectors are not on pace to meet national decarbonization goals".[1]*

This comprehensive DOE report covers many options for decarbonization for the process industries from proven to low TRL technologies. The report views advanced nuclear technologies (ANT) as still nascent and requiring further development and demonstration. However, the DOE views ANT as important for meeting 2050 decarbonization goals and having a role for the process industries.

"Modular nuclear for combined heat and power (CHP) could replace fossil fuel sources, accounting for around 70% of chemicals & refining emissions. CHP from nuclear could substitute the need for electrification of heat and the CCS needed for dilute high-heat sources."[1]

Readers are encouraged to review the DOE Liftoff reports, "Pathways to Commercial Liftoff: Advanced Nuclear"[2] and "Pathways to Commercial Liftoff: Decarbonizing

Chemicals & Refining"[1]. Both are rich sources of information for thinking about decarbonization for the process industries and the role of nuclear power to meet that objective. They can be accessed through the main DOE Liftoff webpage:

Taking a bold step

The chemical manufacturing sector stands at a crossroads of increasing energy demands and pressing environmental sustainability targets especially for hard to decarbonize processes like steam cracking. Traditional fossil fuel-based energy sources are becoming less viable due to their environmental impact and the volatility of fuel prices. As such, there is a critical need for alternative energy solutions that are both reliable and have minimal carbon footprints. Small Modular Nuclear Reactors (SMRs), with their capability for combined heat and power (CHP) generation, present an innovative pathway to meet these energy requirements efficiently and sustainably. The integration of SMR technology for Combined Heat and Power (CHP) applications not only aligns with the global energy transition towards decarbonization but also meets the specific energy demands of chemical manufacturing processes.

To realize the potential of SMRs for CHP by putting steel and concrete on the ground, "*The Dow Chemical Company and X-Energy Reactor Company jointly announced the selection of Dow's UCC1 Seadrift Operations manufacturing site in Texas for their proposed advanced Small Modular Reactor (SMR) nuclear project on May 11, 2023. The project aims to provide the Seadrift site with safe, reliable, and zero carbon emissions power and steam as existing energy and steam assets near their end-of-life. This initiative is expected to reduce the Seadrift site's emissions by approximately 440,000 metric tons of CO2 equivalent per year. Dow and X-energy will now prepare and submit a Construction Permit application to the U.S. Nuclear Regulatory Commission (NRC), marking an important milestone in bringing the project to fruition. Construction on the four-reactor project is anticipated to begin in 2026 and be completed by the end of this decade.*" [3]

The remainder of this report discusses the criteria for selecting the appropriate SMR nuclear technology to facilitate CHP in chemical manufacturing facilities, ensuring an efficient, sustainable, and economically viable energy supply. Examples of the application of the criteria to the selection of the X-energy high temperature gas reactor technology(HTGR) will be highlighted.

Overview of SMRs

SMRs are a class of nuclear reactors with a power capacity of up to 300 MW per unit, significantly smaller than traditional nuclear power plants. Their modular nature allows for fabrication in a factory setting and on-site assembly, reducing construction times and costs. SMRs are designed with advanced safety features, making them a secure choice for proximity to industrial applications, that is, at the fence line. SMRs are one of the advanced nuclear technologies that the DOE considers as a key part of the Department's goal to develop safe, clean, and affordable nuclear power options. There is extensive information and background at the DOE website: https://www.energy.gov/ne/advanced-small-modular-reactors-smrs.

SMR technology is varied from scaled down versions of operating, larger pressurized water reactors like the Westinghouse AP-300 using low enriched uranium fuel to novel designs with novel coolants and fuels like HALEU, high assay, low enriched uranium used by X-Energy's high temperature gas reactor, or recycled waste nuclear fuel like Moltex Energy's Stable Salt Reactor.

The DOE defines Advanced nuclear in their LiftReport [2] as follows: *"…includes two generations of reactors: Gen III+ and Gen IV. Gen III+ reactors are similar to the reactors currently operating in the U.S.—they use water as a coolant and low-enriched uranium (LEU) as fuel. Because of these shared characteristics with the current operating nuclear fleet in the U.S., deployment of most Gen III+ reactors is likely to be nearer-term than other, more innovative, reactor types. Gen IV reactors offer additional use cases and supplemental safety features. They will use novel coolants, and many will use HALEU as fuel, which has not been used by the existing U.S. commercial nuclear fleet. Advanced nuclear reactors are categorized into three size groups: large, small modular reactors (SMRs), or microreactors. For the purposes of this report, large nuclear plants are defined as having ~1,000 MW capacity, SMRs as having ~50-300 MW capacity, and microreactors as having less than ~50 MW. For reference, 1 MW of nuclear capacity can power about 800 homes per year."[2]*

The following table is a partial list of active reactor designers showing a cross section of the proposed technologies. It is important to note that although many of these technologies have been piloted in the past – usually decades ago, the only recently built full-scale reactors classified as SMRs have been in Russia and China. Regardless, these companies and their innovative reactor designs represent significant advancements in the development of small modular reactors, aiming to provide safer, more economical, and more adaptable nuclear power solutions across the globe. The World Nuclear Association has a comprehensive listing, technology description and development history of SMRs and advanced nuclear technologies that can be found at https://www.world-nuclear.org/information-library/nuclear-fuel-cycle/nuclear-power-reactors/small-nuclear-power-reactors.aspx [4]

Table 1: Examples of SMR Technology

Company	**Technology/Reactor Type**	**Description**	**Reference**
NuScale Power	NuScale Power Module (PWR)	Pressurized water reactor, scalable system deployable in units of up to 12 modules, each with a capacity of 60 MW.	https://www.nuscalepower.com/en/products
Holtec International	SMR-160 (PWR)	Pressurized water reactor design with a 160 MW capacity, deployable in units of up to six modules.	https://holtecinternational.com/2020/07/20/smr-160-at-age-10/

GE Hitachi Nuclear Energy	BWRX-300 (BWR)	Boiling water reactor design with a 300 MW capacity, emphasizing simplified design for safety and efficiency.	https://www.gevernova.com/nuclear/carbon-free-power/bwrx-300-small-modular-reactor
Rosatom	KLT-40S (PWR)	Pressurized water reactor primarily used for marine propulsion but also suitable for electricity generation, with a 35 MW capacity.	https://en.wikipedia.org/wiki/KLT-40_reactor
Westinghouse Electric Company	eVinci Micro Reactor (HP) & AP300 (PWR)	High-temperature heat pipe reactor capable of generating 5 MW of electricity. AP300 plans for a 300 Megawatt capacity modular reactor.	https://www.westinghousenuclear.com/energy-systems/evinci-microreactor/ https://www.westinghousenuclear.com/energy-systems/ap300-smr/
TerraPower	Natrium Reactor (Sodium Fast Reactor with Molten Storage)	Next-generation reactor design focusing on safety, affordability, and scalability, selected for a demonstration plant in Wyoming.	https://www.terrapower.com/our-work/natriumpower/
ARC (Canada)	Sodium Cooled Fast Reactor	A simple, inherently safe design, operating at atmospheric pressure, yields lower plant cost. Uses metallic UR alloy that is simpler to fabricate	https://www.arc-cleantech.com/technology
Rolls-Royce SMR	UK SMR (PWR design)	Close-coupled three-loop PWR design with an output initially planned to be 440 MWe, later increased to 470 MWe.	https://www.rolls-royce-smr.com/
Kairos Power	Fluoride Salt-Cooled High Temperature Reactor (KP-FHR)	Betting on high-temperature molten salt reactors to provide competitive, reliable, and responsible nuclear energy with a near-zero carbon footprint.	https://kairospower.com/technology/

Moltex Energy	Stable Salt Reactor-Wasteburner (SSR-WB)	Molten salt reactor design and waste nuclear fuel recycling	https://www.moltexenergy.com/technology-suite/
X-energy	Xe-100 (HTGR)	Advanced He cooled reactor using TRISO pebbles.	https://x-energy.com/reactors/xe-100
Ultra Safe Nuclear	Micro Modular Reactor (MMR®) Fully Ceramic Micro-encapsulated (FCM®) nuclear fuel	Fission Battery concept using TRISO fuel	https://www.usnc.com/

The Case for SMRs in Chemical Manufacturing

Chemical manufacturing processes require substantial amounts of heat and electricity. Traditionally, these energy needs are met through classic systems such as fossil fuel co-generation plants. The first deployment option for SMR's is as drop-in CHP replacements, efficiently providing high-temperature heat and electricity simultaneously, but without the CO2 footprint. The second deployment option is to be dedicated, clean firm-power for a more highly electrified site that has eliminated tradition steam boilers in favor of heat pumps for example or implemented e-crackers or developed large scale electrochemistry processes.[1] Another potential use is hydrogen production.[5] The high-temperature heat generated by SMRs can be harnessed to produce hydrogen through thermochemical processes, such as high-temperature steam electrolysis (HTSE) or sulfur-iodine (SI) cycles, both of which produce no greenhouse gas emissions[5]. A dedicated electrical generation source decouples the site from the larger grid and guards against disruptions and the need to over subscribe to intermittent renewables.

The integration of SMRs into chemical manufacturing for CHP offers several advantages:[6]

- Energy Efficiency: SMRs can achieve overall energy efficiencies of up to 85% by utilizing the waste heat from electricity generation for industrial processes.
- Reduced Carbon Footprint: By replacing fossil fuel-based energy sources, SMRs significantly lower greenhouse gas emissions, aligning with global sustainability targets.
- Reliability and Stability: SMRs provide a constant energy output, essential for the continuous operation of chemical manufacturing facilities, and are less susceptible to the fluctuations of fuel prices.
- Scalability and Flexibility: The modular nature of SMRs allows for scalable and flexible deployment, tailored to the specific energy demands of a facility.

Heat and Power Demand

The primary criterion for selecting an SMR for CHP applications in chemical manufacturing is the facility's specific energy requirements. Chemical processes often require high-grade heat at various temperatures and significant amounts of electricity. An appropriate SMR technology must, therefore, offer a thermal output that matches the process heat requirements and be capable of efficient electricity generation. The compatibility of the SMR's thermal output with the facility's temperature requirements for different processes (e.g., distillation, drying, or chemical reactions) is crucial. Gen III+ SMR's using water-cooling are limited to 300°C unless followed by compression and heating. Non-water-cooled reactors can reach higher steam temperatures directly without additional energy input and are likely to appeal to sites that currently operate with high temperature steam.

Reliability and Availability

Chemical manufacturing processes can have variable energy demands. The chosen SMR technology should have a good load-following capability, meaning it can adjust its output in response to the changing energy demands of the facility without compromising operational efficiency or safety. The use of multiple small reactors can provide the needed redundancy and availability to provide uninterrupted heat and power. The expected refueling cycle effects the reliability and availability of the SMRs with an apparent advantage to long fuel cycles (10-20 year) and on-line fueling like proposed by X-Energy and Karios.

Safety and Regulatory Compliance

Given the proximity of SMR installations to industrial and sometimes populated areas, the selected technology must incorporate advanced safety features. The leading SMR designs all use passive safety systems that rely on natural physical principles to ensure the reactor's safe shutdown in emergencies, reducing the risk of accidents and minimizing radiation release. Conversely, the ability to locate the SMR near a site may also be a function of the potential negative events from the chemical manufacturing site.

Licensing and Regulatory Approval

The SMR technology must be in a mature stage of development, with a clear path to regulatory approval. The approval process is country dependent and is governed by a complex set of legal regulations and government agencies. This is perhaps the most challenging aspect of the implementation of SMRs for CHP in the process industries.

Environmental Impact

One of the motivations for adopting SMR technology in chemical manufacturing is to reduce the carbon footprint. The selected SMR must demonstrate a significant reduction in greenhouse gas emissions compared to fossil fuel-based systems, supporting the industry's environmental sustainability goals.

The management of nuclear waste is a critical consideration. Different SMR technologies have different fuel efficiently and their spent fuel will have different potential for recycling. Spent fuel storage and final disposal is determined by regulation and part of the licensing process.

Economic Viability

The economic feasibility of implementing an SMR-based CHP system in a chemical manufacturing facility is pivotal. The selected technology should offer competitive capital costs, including construction and commissioning, and operational costs. Financial models must account for the lifecycle costs of the plant, including decommissioning and waste management expenses.

The competitive nature of the petrochemical business has led to the capability of process companies to successfully engineer, procure, construct and start-up complicated plants. The utility model (electric grid) for nuclear has faced challenges for traditional grids as shown by the NuScale challenges and it will not work for SMR for process CHP. The process industries will have to work to make SMRs for process CHP viable. The expertise of the SMR vendors is nuclear reactors, not plant construction.

Scalability and Flexibility

The modular nature of SMRs offers scalability and flexibility, allowing for phased construction and expansion to match the growth of the chemical facility. The selected technology should facilitate modular deployment, enabling a tailored energy solution that can evolve with the facility's needs. Again, the experience of the process industries can be brought bear here to improve the design and layout of the system to ensure the lowest construction and operating cost. This is complementary to the world-class nuclear engineering provided by the reactor designers. It recognizes that there is an opportunity to reduce the cost to put steel and concrete in the ground and put a SMR into operation.

Reduction to Practice

This section will illustrate how the proceeding criteria were used to select the Xe-100 technology for Project Long Mott, the first of kind SMR for combined heat and power at a chemical facility.

The Seadrift site is a long-established production facility that utilizes a traditional fossil fuel cogeneration system to provide electrical power and high-pressure steam (565°C) to the site. The cogen asset is approaching end of life and needs to be replaced. One decarbonization alternative is to go with a gas-fired plant with carbon capture but this also requires readily accessible infrastructure for the captured CO2.

Of potential SMR technologies, high-temperature gas reactors and their coated fuels (TRISO) have been studied since the 1960's. One of the obvious appeals is that they have the capability to provide high temperatures directly without compression and heating. Extensive studies and evaluation of HTGR technology at Idaho National Labs [7] describes modular HTGR as:

- Inherently safe core cannot melt
- High outlet temperature for more efficient electricity production and process heat

- Minimal radiological or dynamic coupling between the reactor and the collocated process heat application
- Environmentally benign, reliable, mature (for a non-LWR)

This positions HTGR as a suitable technology option for Seadrift, and X-energy is the most advanced vendor for this technology (Figures 1 & 2).

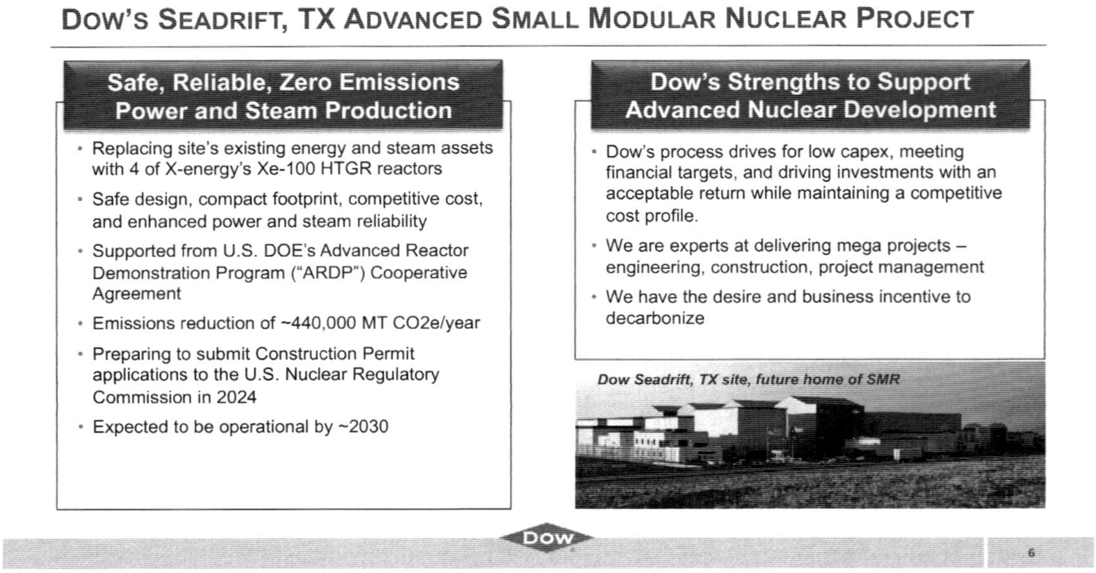

Figure 1 The Long Mott Project: The FOAK SMR for CHP at a Chemical Production Site

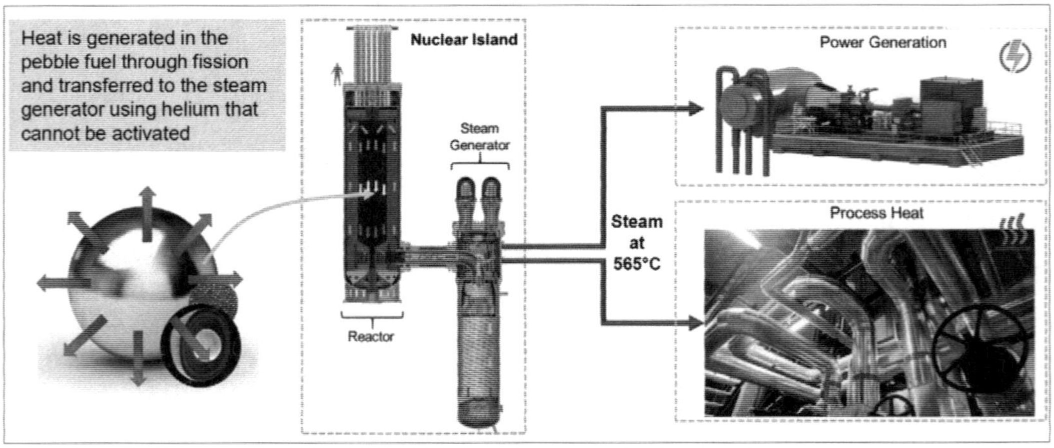

Source: https://www.energy.gov/ne/articles/x-energy-developing-pebble-bed-reactor-they-say-cant-melt-down

Figure 2: The X-energy HTGR technology meets the CHP specifications.

As mentioned earlier, the modular nature of SMR's can be important to building in redundancy for CHP applications that require 24/7 availability and no extended outages. X-energy's base design, the Xe-100, is readily scalable and a four-reactor configuration meets the Dow requirements for redundancy and reliability for both the steam and electrical power.

The Xe-100 reactor meets the technical requirements as a source for CHP but it would not be viable if it could not be co-located with the site with minimum safety risk, and have minimum negative environmental impact. The Xe-100 HTGR technology does this with aplomb. The following is the basis for the safety design that allows for a minimum exclusion zone (~400m).[7]

- Utilize inherent material properties as basis for safety
  - Helium coolant – neutronically transparent, chemically inert, low heat capacity, single phase
  - Ceramic coated (TRISO) particle fuel – high temperature capability, high radionuclide retention
  - Graphite moderator – high temperature stability, large heat capacity, long thermal response times
- Simple reactor design with inherent and passive safety features
  - Retain most radionuclides at source (i.e., within fuel)
  - Shape and size reactor to allow passive heat removal from reactor core using uninsulated reactor vessel
  - Heat is still removed if system is depressurized due to breach in reactor helium pressure boundary (HPB)
  - Heat is radiated from reactor vessel to Reactor Cavity Cooling System panels – also passive
  - Large negative temperature coefficient supports intrinsic reactor shutdown
  - No reliance on AC-power to perform required safety functions
  - No reliance on operator intervention; insensitive to incorrect operator actions or inactions

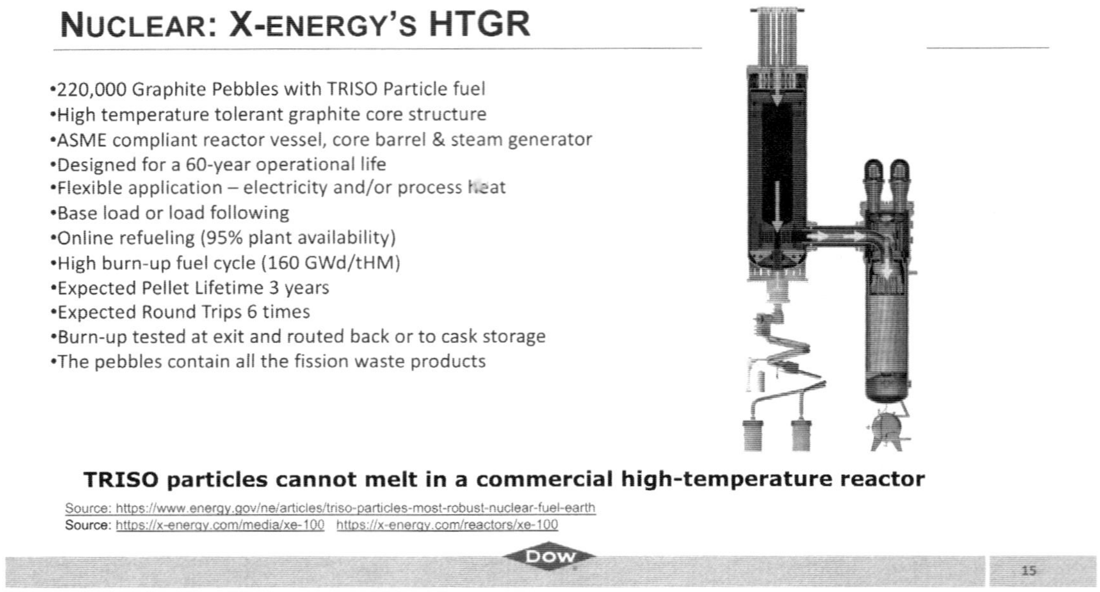

Figure 3: The Xe-100 robust and reliable design basis

At the heart of the system is the TRISO fuel. This is the foundation of the reactor safety as it can't melt and retains all the fission by-products. [8]

- TRISO stands for TRi-structural ISOtropic particle fuel.
- The kernel is encapsulated by three layers of carbon- and ceramic-based materials that prevent the release of radioactive fission products.
- About the size of a poppy seed
- They can be fabricated into cylindrical pellets or billiard ball-sized spheres called "pebbles" for use in either high temperature gas or molten salt-cooled reactors.
- TRISO fuels are structurally more resistant to neutron irradiation, corrosion, oxidation and high temperatures than traditional reactor fuels.
- Each particle acts as its own containment system thanks to its triple-coated layers.
- Retains fission products under all reactor conditions.

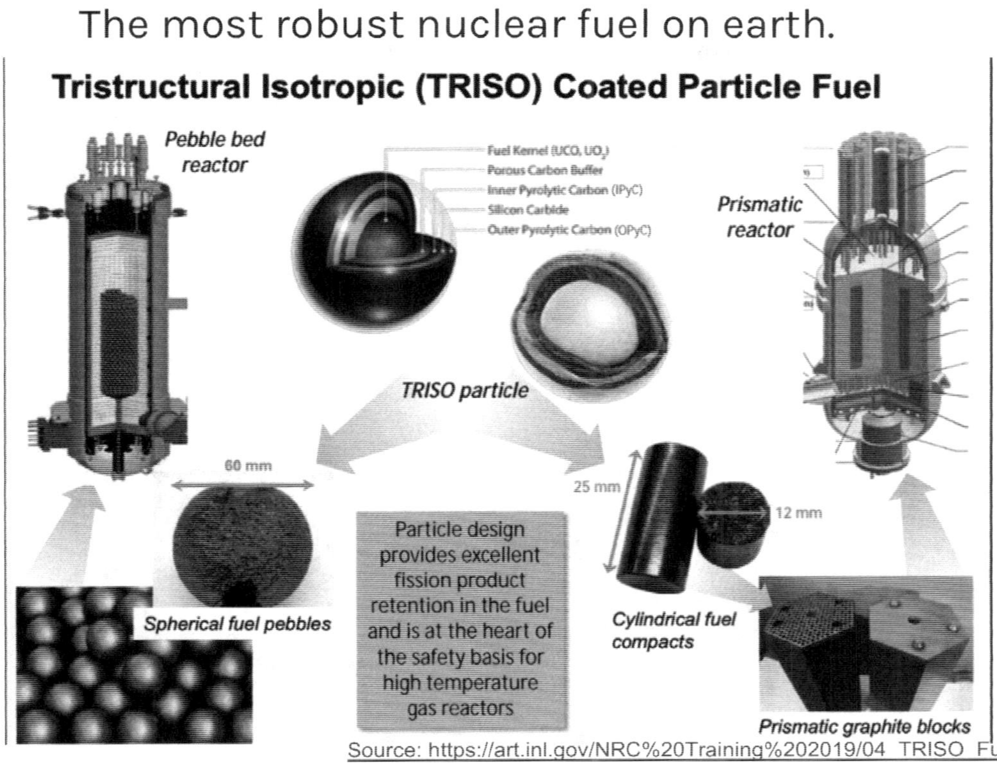

Source: https://art.inl.gov/NRC%20Training%202019/04_TRISO_Fuel.pdf

Figure 4: TRISO fuel

Conclusion

Ethylene producers and the rest of the chemical and refining sectors face significant challenges to economically reach net-zero. Emerging SMR technologies can play a role in meeting theses targets. Selecting the appropriate SMR technology for CHP applications in chemical manufacturing facilities requires a comprehensive evaluation of energy requirements, safety and regulatory compliance, environmental impact, economic viability, and scalability. By meticulously assessing these criteria, stakeholders can identify an SMR solution that not only meets the energy needs of chemical manufacturing processes but also advances the goals of sustainability, safety, and economic efficiency. As the ethylene and chemical industry moves towards a more sustainable future, the adoption of SMR technology for CHP presents a promising pathway to achieving these objectives.

References

[1] https://liftoff.energy.gov/wp-content/uploads/2023/12/20230921-Pathways-to-Commercial-Liftoff-Chemicals-Refining.pdf

[2] Pathways to Commercial Liftoff - Advanced Nuclear - Mar 20_UPDATED (energy.gov)

[3] https://investors.dow.com/en/news/news-details/2023/Dows-Seadrift-Texas-location-selected-for-X-energy-advanced-SMR-nuclear-project-to-deliver-safe-reliable-zero-carbon-emissions-power-and-steam-production/default.aspx

[4] https://www.world-nuclear.org/information-library/nuclear-fuel-cycle/nuclear-power-reactors/small-nuclear-power-reactors.aspx

[5] https://www.iaea.org/topics/non-electric-applications/nuclear-hydrogen-production

[6] https://www.energy.gov/ne/advanced-small-modular-reactors-smrs

[7] https://art.inl.gov/NRC Training 2019/05_HTGR_Safety_Design.pdf

[8] https://art.inl.gov/NRC Training 2019/04_TRISO_Fuel.pdf

AICHE PAPER NUMBER: 113D

LUMMUS' HYBRID CRACKING HEATER

Baozhong Zhao
Director of Engineering of Lummus Heat Transfer
Lummus Technology LLC, Bloomfield NJ, USA

Alejandro S. Carrillo
Vice President of Lummus Heat Transfer
Lummus Technology LLC, Bloomfield NJ, USA

Peter D. Kuzma
Senior Director Technology for Ethylene
Lummus Technology LLC, Bloomfield NJ, USA

Richard Jibb
Technology Director
Lummus Technology LLC, Bloomfield NJ, USA

Prepared for Presentation at the 2024 Spring National Meeting
New Orleans, LA, March 24-28, 2024

LUMMUS' HYBRID CRACKING HEATER

Baozhong Zhao
Director of Engineering of Lummus Heat Transfer
Lummus Technology LLC, Bloomfield NJ, USA

Alejandro S. Carrillo
Vice President of Lummus Heat Transfer
Lummus Technology LLC, Bloomfield NJ, USA

Peter D. Kuzma
Senior Director Technology for Ethylene
Lummus Technology LLC, Bloomfield NJ, USA

Richard Jibb
Technology Director
Lummus Technology LLC, Bloomfield NJ, USA

Abstract:

Reducing carbon dioxide emissions is a major effort in the Petrochemical Industry. The cracking heaters in an ethylene plant are the major sources of CO_2 emission. However, other than low carbon-bearing fuel firing, any firing reduction in a fired cracking heater, such as air preheater application, will have an impact not only on the operation of the heaters, but also on the plant recovery section. This paper will introduce Lummus' Hybrid Cracking Heater technology (SRT-h™) for revamp of existing heaters with minimum modifications or for new heater designs. Lummus' Hybrid Cracking Heater technology combines fuel firing and electrical heating to effectively reduce the hydrocarbon fuel firing for both gas and liquid cracking heaters, which will reduce the CO_2 emission whilst minimizing any consequential impacts on the recovery section. The hybrid cracking heater can reduce CO_2 emission by 15-40% or higher without negative impact on the plant capacity or the steam balance. Heater performance improvements that can be realized by the hybrid heater design will also be demonstrated. The SRT-h Hybrid Cracking Heater technology can be applied in stages-initially for partial CO_2 reduction and can later be extended to zero CO_2 emission when combined with non-carbon fuel firing technologies.

Introduction

Ethylene production is an energy intensive chemical process. Fuel for the cracking heater is typically from plant off-gas, supplemented with fuel import. Depending upon the feed stock, the fuel composition may be different. For ethane cracking and without any hydrogen export, the fuel may contain about 80% hydrogen. For liquid cracking, the fuel will be mainly methane offgas. The CO_2 emission will range from 0.35-0.9kg CO_2/kg of olefins (ethylene and propylene) with the fuel of about 80% hydrogen to natural gas.

Clearly, low carbon containing fuel is preferred for low CO2 emission, such as firing a fuel with high hydrogen content. However, hydrogen is a valuable chemical for petrochemical process and will cost more as a fuel compared with firing natural gas. On the other hand, hydrogen may not be readily available.

For a modern cracking heater in an ethylene plant, the thermal efficiency can be 94% or higher. There will be about 65-75% of the total heat input for the feed preheat and cracking, about 20-30% of the remaining heat for boiler feed water (BFW) preheat and high or super high pressure steam production. To lower the CO_2 emission, i.e. reduce the fuel firing, either (i) the heat in the flue gas through the convection section needs to be recycled back to the radiant box or (ii) the heat carried over from the radiant section to the convection section needs to be reduced – in either case, without negative impact on the process crossover temperature.

For the heat recycle from the convection section to the radiant section, combustion air preheat is one of the main options. With the combustion air preheat option, part of the heat in the flue gas will be extracted for preheating the combustion air and the remaining heat will be used for process feed and BFW/steam. Since the heat to the process stream needs to be maintained, the remaining heat to BFW/steam will be reduced which will impact the superheated steam production and, therefore, impact the plant recovery section arrangement especially if the plant is trying to revamp the heater with the air preheat option. Hence, a combustion air preheater will reduce the fuel firing, but result in lower steam production. The plant will require more electrical power consumption to drive the compressors in the recovery section.

Description of Hybrid Cracking Heater Concept

Instead of the combustion air preheat option, a hybrid heater design will reduce the fuel firing by providing the energy requirement via partial electrical heating. The electrical heating and fuel firing will co-exist in the radiant section of the cracking heater. The electrical heating will cover 15-25% of the total heat requirements and the fuel firing will cover 75-85% of the total heat requirement.

When hydrocarbon fuel is fired in the radiant box, the combustion exhaust will contain H_2O and CO_2, which will shield the direct radiation from the electrical heating elements or the refractory behind the heating element and reduce the radiation efficiency. This will lead to a higher heating element temperature which could cause more electrical power consumption and shorter heating element service life. In addition, the presence of moisture or hydrocarbon radicals prior to completing the combustion process may lead to chemical attack to the heating elements. To ensure efficient radiant heat transfer by the electrical heating elements and avoid chemical attacks to the heating elements, it is desirable to have the electrical heating and fuel firing in two separate zones such that the working environment for the electrical heating can be adjusted to minimize the H_2O and CO_2 impact on the electrical heating elements.

Lummus' Hybrid Cracking Heater (SRT-h) technology has been developed in parallel with other Lummus low CO2 technologies. The electrical heat input will use the components and methods developed for the fully electric SRT-e™ Electric Cracking Heater ([1] Jibb et al. 2023). Both the SRT-e and SRT-h technologies use external resistance heating elements to convert electrical energy to heat. In the SRT-h hybrid cracking heater, the radiant section is divided into two zones. The bottom section will be for electrical heating and the upper section will be for fuel firing as shown in Figure 1.

Air purge connections will be provided in the electrical heating section to protect it from H_2O and CO_2 migration from the fuel combustion zone into the electrical heating zone. The purge can also be used for heater startup prior to the burner light off.

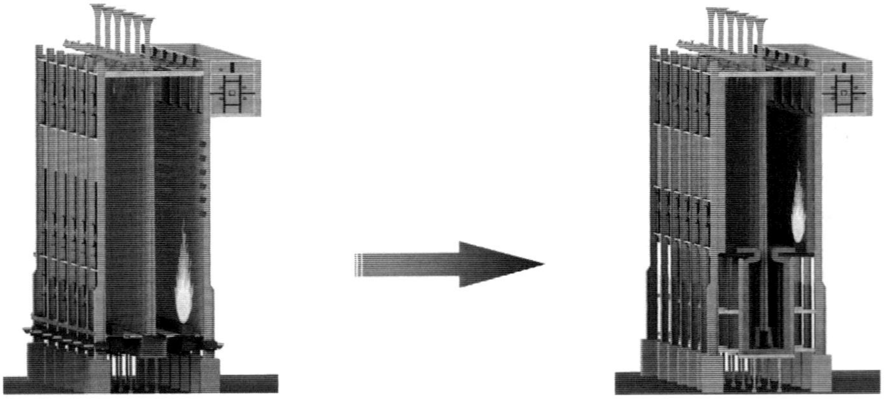

Figure 1: Conventional cracking heater (left) and hybrid cracking heater (right)

Lummus SRT-h™ can be applied to new heaters and for heater revamps for CO_2 reduction with both gas cracking and liquid cracking, whilst Lummus SRT-e is a complete electrical cracking heater which eliminates all CO_2 emissions. SRT-h electrical heating builds on the design approach of the SRT-e, i.e., it uses external resistance heating elements to heat part of the same radiant coil as the conventional SRT heater and uses proven and available electrical components. For a given capacity, the electrical power requirement of the SRT-e will clearly be higher than for the SRT-h™. If the electricity is from a renewable green source, and the power is readily available, SRT-e™ will be a good option for CO_2 reduction. However, if the available power supply is not sufficient for full power supply to the ethylene plant, SRT-h may provide an excellent choice that (i) meets the production requirement with minimum changes to the plant operation (ii) provides significant CO_2 reduction and (iii) can be effectively applied to both gas and liquid cracking.

Performance of Hybrid Cracking Heater

The Lummus' Hybrid Cracking Heater uses the same short residence time SRT® radiant coil as used in a conventional cracking heater. The convection section design and piping layout, including the crossover piping, will also be the same as a conventional cracking heater. The only change from a conventional cracking heater is in the lower radiant section that is used for the electrical heating. Compared to the heat flux profile with burner firing in a conventional cracking heater, the electrical heating elements provide a more even heat flux profile which maximizes the heat input and minimizes the risk of overheating the radiant tube and process streams.
To compare the hybrid cracking heater with the conventional cracking heater, a commercial ethane cracker was selected. Then, the same heater is considered following conversion to a proposed hybrid cracking heater. The commercial heater is a Lummus SRT III ethane cracker with both hearth and side wall firing. The total heat input is about 89.7MW. The fuel split between the hearth and wall is about 90:10. The proposed hybrid heater will be based on the same commercial heater with modifications to the lower part of the radiant box only. With the electrical energy input, the total fuel fired will be reduced from

89.7MW to 73.6MW. The hearth burners will be relocated to a higher elevation. The electrical heating elements will be installed at the lower radiant section. The electrical power input will be about 16.1MW, which represents about 17.9% of the total heat input. The fuel flow through the wall burners will no longer be necessary to meet the flame pattern and heat flux profile requirement. The heat liberation per hearth burner remains close enough to the hearth burner capacity in the conventional heater. If an existing conventional heater is converted to hybrid cracking heater, the hearth burners may be reused. To avoid any impact to the recovery section and maintain the same steam production, the hybrid heater will be operated at higher excess air at about 42%. This ensures that the convection section receives about the same heat input as the conventional heater, resulting in the same steam production

The performance comparison between the conventional and hybrid cracking heaters was based on a Computational Fluid Dynamics (CFD) model. Lummus has validated CFD models and applied these CFD models for all projects. The CFD results match the plant operation very well and provide valuable information for design optimization and troubleshooting as part of the engineering design for all cracking heaters, including heater revamps.

The CFD simulation selected a section of the heater, including 4 SRT III coils, 12 hearth burners and 12 WST for the conventional cracking heater. For the hybrid heater, it will include 4 SRT III coils, 12 hearth burners and 5.37MW electrical power input. The fluid domain is shown as in Figure 2 for both the conventional and Hybrid cracking heaters.

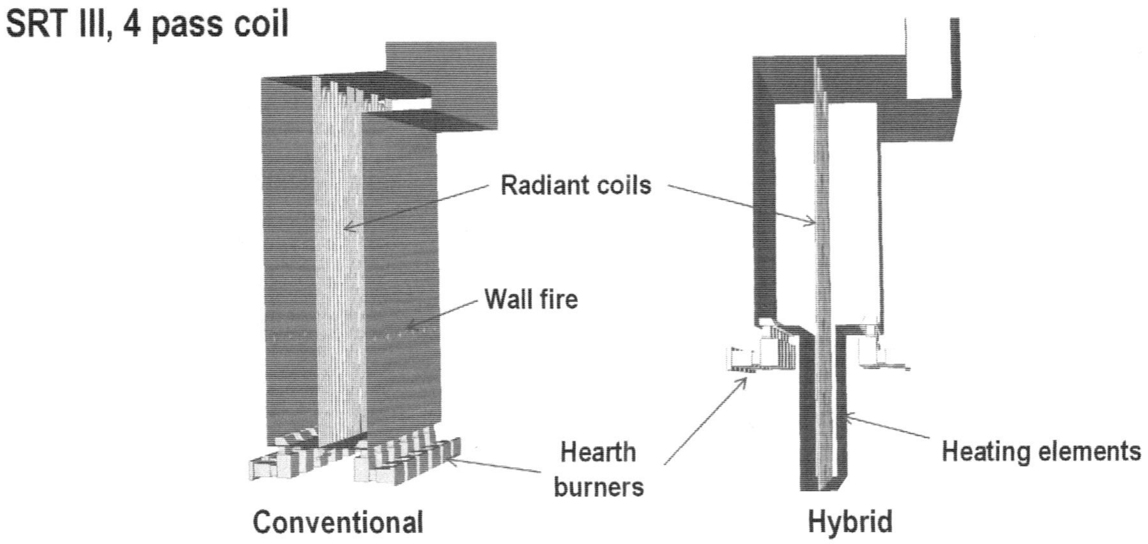

Figure 2: CFD domain for conventional cracking heater and Hybrid cracking heater

When the heater is changed to hybrid heat input, the radiant box geometry will be different than the geometry of the conventional heater, which will lead to different aerodynamics in the radiant box. In addition, the room for fuel combustion will be reduced. The CFD model was used to verify that the modified radiant section for the hybrid cracking heater will be adequate for complete fuel combustion (i.e. to avoid combustion in the convection section)

Figure 3 and Figure 4 show the flame pattern and the selected CO envelope at 1000ppm. The flame moves up along the heater side wall without any sign of flame rollover or mal distribution on the far side and near sides of the convection section. The CO envelope shows the combustion is completed before the combustion exhaust flows into the convection section.

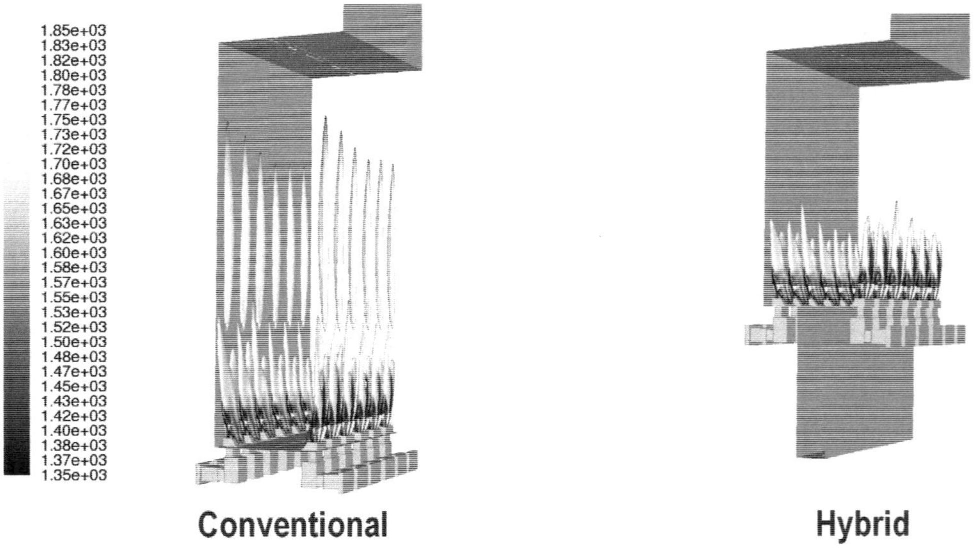

Figure 3: Flame pattern in conventional and hybrid cracking heaters

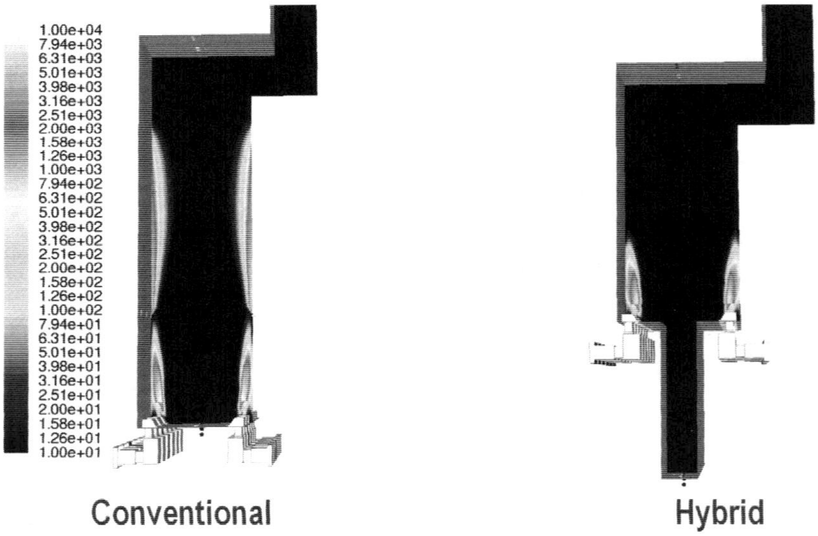

Figure 4: CO profile in conventional and hybrid cracking heaters

Figure 5 shows the flue gas temperature profile. About one third of the radiant heating surface is heated by electrical power and two thirds of the surface is heated by combustion. The total amount of combustion exhaust is about the same as that in the conventional cracking heater due to higher excess air applied to the hybrid cracking heater. The relatively large amount of combustion exhaust reduces the temperature gradient and provides more uniform temperature distribution in the combustion zone.

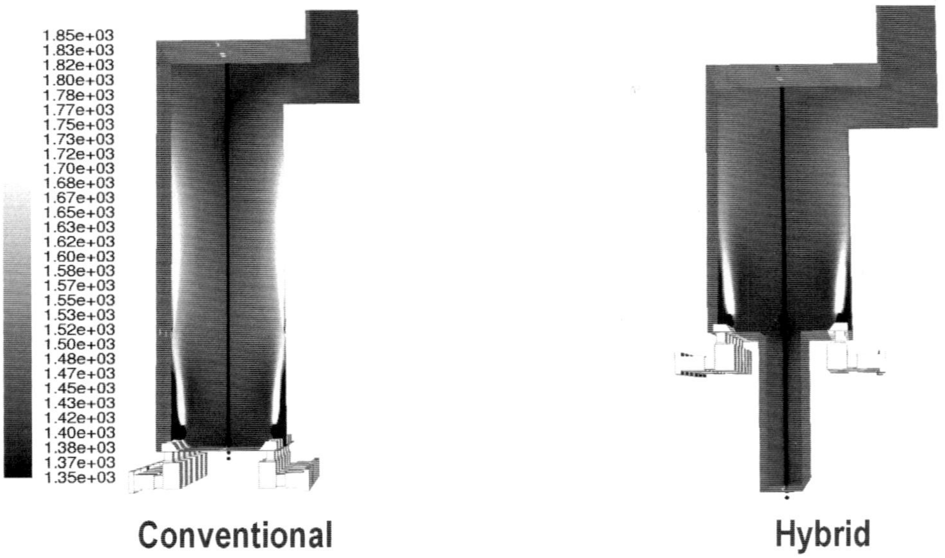

Figure 5: Flue gas temperature profile in conventional and hybrid cracking heaters

Figure 6 shows the refractory wall temperature profile. Consistent with the flue gas temperature profile shown in Figure 5 as well as the inherent uniformity of the electrical heating, the refractory hot face temperature profile is more uniform when compared with the conventional fuel fired cracking heater.

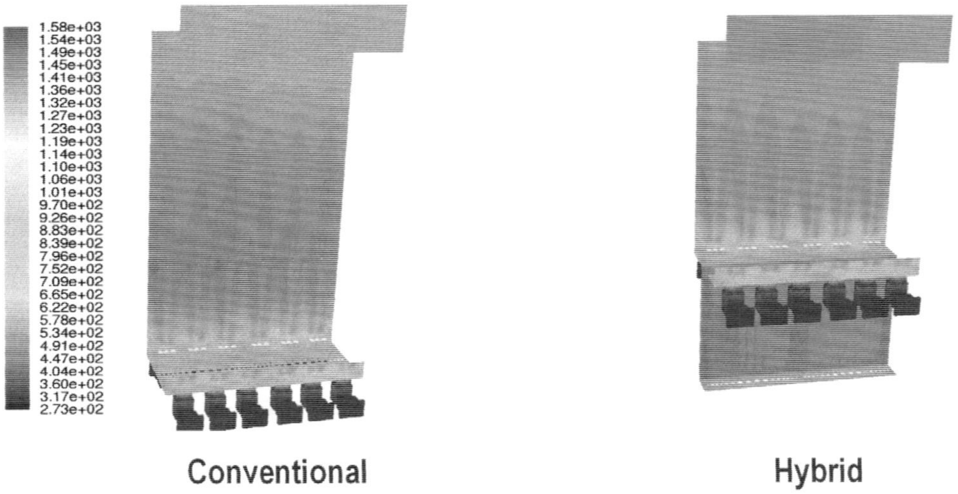

Figure 6: Refractory temperature profile in conventional and hybrid cracking heaters

Figure 7 shows the radiant coil temperature profile. The radiation heat transfer to the radiant coil is determined by the combustion exhaust temperature, flue gas composition, and refractory surface temperature. Since the hybrid heater has a more uniform combustion exhaust temperature distribution and more uniform refractory hot face temperature profile, the radiant coil Tube Metal Temperature (TMT) will have less variation from the minimum to maximum. Figure 8 shows the radiant tube circumferential temperature band of each pass of the coil. Under the same process flow conditions, the minimum TMT will be higher, and the maximum TMT will be lower for the hybrid heater compared with the TMT range in the conventional heater due to improved heat flux profile in the hybrid cracking heater.

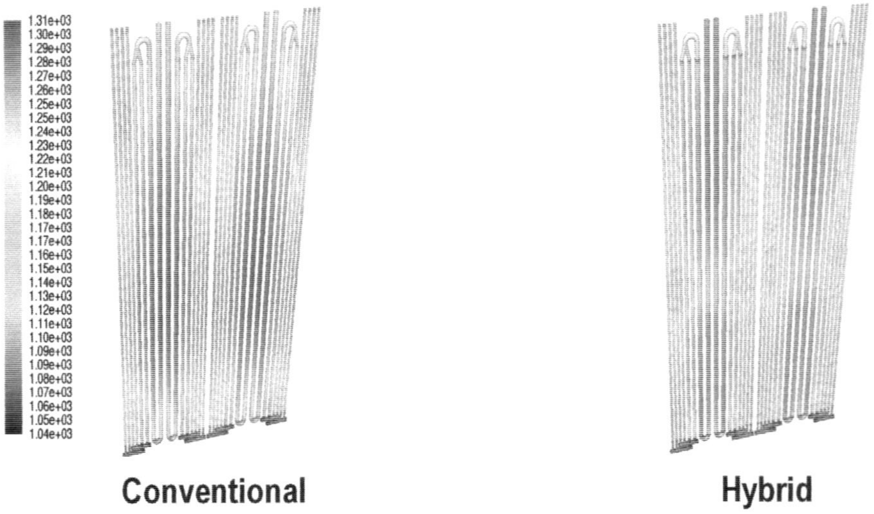

Figure 7: Radiant coil temperature profile in conventional and hybrid cracking heaters

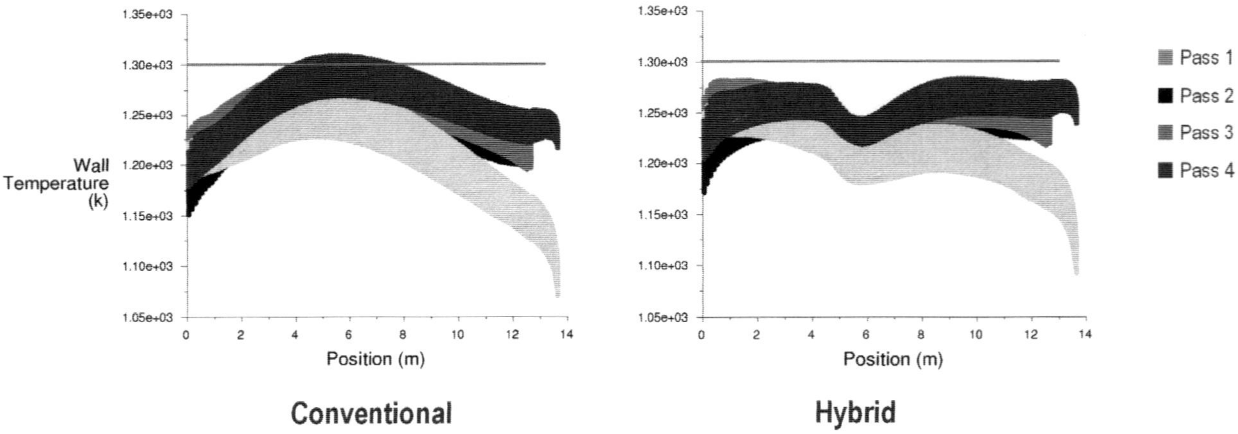

Figure 8: Radiant tube temperature band in conventional and hybrid cracking heaters

Table 1 shows the performance comparison between the selected conventional cracking heater and the proposed hybrid cracking heater. The hybrid heater will have a higher firebox temperature and slightly lower radiant efficiency. The maximum TMT will be lower than the TMT in the conventional heater. The radiant tube TMT is one of the critical factors that affects the heater run length. With a lower TMT, the heater will achieve a longer run length and the coil will have a longer service life.

Case	Conventional Design	Hybrid Design
Radiant coil inlet temperature, °C	681	681
Radiant coil outlet temperature,°C	836	836
Radiant efficiency, %	45.5	44.8
Maximum tube metal temperature (SOR), °C	1036	1011
Bridgewall temperature., °C	1125	1178

Table 1: Performance comparison between the conventional and hybrid cracking heaters

Case Study of Different Low CO_2 Emission Options

Lummus offers various technical options to lower CO_2 emissions, including very high efficiency designs , Lummus External Air Preheater (LEAP™), Lummus' Hybrid Cracking Heater (SRT-h) and complete electrical heating (Lummus SRT-e Electric Cracking Heater). LEAP is an innovative technology that was developed to provide an air preheat option for an existing cracking heater revamp or for grassroots designs. LEAP is based on heating surface optimization in the convection section and will not change the existing structure and foundation loadings. For a heater revamp, part of the existing heating surface will be re-arranged for heating a thermal fluid and, then, the thermal fluid will be heating up the combustion air through an external air preheater. LEAP can achieve up to 20% firing reduction.

The case study selected the same conventional commercial ethane cracker used in the earlier CFD analysis as the base case, in order to compare two revamp options - (i) the same commercial heater revamped with LEAP and (ii) the same heater revamped with SRT-h (see Figure 9). For SRT-h, two cases are considered. The first is to limit the changes to the radiant section only and, at the same time, maintain the steam production to minimize any impact to the recovery section. The second case is to illustrate the possibility of further reductions in CO_2 emissions with slight changes to the heating surface in the convection section. The results are summarized in Table 2.

Figure 9: Conventional heater, LEAP and SRT-h

Options	Conventional design	Revamp convection with LEAP	Revamp radiant section only	Revamp both radiant and convection
	Base	LEAP	Hybrid 1	Hybrid 2
Total heat input, MW	89.7	75.4	89.7	83.7
Electricity, MW			16.1	16.1
Excess air, %	10	10	42	33
Thermal efficiency, %	93.8	94.7	94.0	94.1
Steam production, ton/h	61.8	45.1	62.0	55.1
Firing reduction, %	0	15.9	17.9	24.7
CO_2 reduction w/ PSA, %	0	37	41	56
CO_2 reduction. w/o PSA, %	0	25	27	34

Table 2: CO_2 emission reduction with different Lummus low CO_2 technologies

Compared with the base case of the conventional ethane cracker, LEAP can reduce the firing by 16%. If there is no significant export of hydrogen from the plant this reduction in fuel requirement will make the plant self-sufficient. The methane content in the fuel will be lower, therefore, the total CO_2 emissions will be reduced by about 25%. If a PSA unit is installed, the H_2 content in the fuel will increase, the methane and other carbon bearing components in the fuel will reduce. The net maximum CO2 reduction can be as high as 37%. Since less fuel is being fired, less flue gas will flow through the convection section, there will be less heat for steam production. Compared with the base case, the steam production reduced from 61.8t/h to 45.1t/h. If the compressors in the recovery section are driven by steam turbines, with reduced steam flow, an electrical motor driver for one compressor may be necessary.

For SRT-h option 1, with radiant section modification only and maintaining the steam production, the fuel firing reduction is about 17.9%, resulting in CO_2 reduction similar to the LEAP option (27% for SRT-h and 25% for LEAP). With PSA unit, the CO_2 emission reduction can be as high as 41%. Compared with the base case, the stream production is maintained at 62.0t/h. Other than the radiant section modification and addition of the electrical power system, there is no need to make any additional changes to the plant, including the recovery section. The CO_2 reduction is significant with minimal changes to the heater and no changes to any other part of the plant.

For the SRT-h option 2, in addition to the changes required for the radiant box to accommodate the electrical heating, the heater convection section is also modified. The excess air for the fuel combustion is set at 33%. The total fuel firing is reduced by 24.7%, and the CO_2 emission is reduced by 34%. With a PSA unit, the CO_2 reduction can be as high as 56%. The steam production is reduced from 61.8t/h in the base case to 55.1t/h with SRT-h™ and additional changes to the convection section. Compared with the base case and SRT-h option 1, the total energy input is reduced from about 89.7MW to 83.7MW. With further optimization of the convective heating surface as well as changes to the Transfer Line Exchangers (TLE), the CO_2 reduction can be much more than 56%.

For ethane cracking, the H_2 in the off gas is capable of supplying about 65% of the heat required. With SRT-h technology and small addition of zero carbon fuel or feed preheat optimization, the heater can realize zero CO_2 emission. Lummus' Hybrid Cracking Heater can be a helpful step from partial CO_2 reduction to complete decarburization. The methane in the offgas can be a feed stock for olefin production using Lummus technology of Oxidative Coupling of Methane (OCM).

The above discussion and comparison were mainly on the CO_2 reduction. LEAP offers benefits more than just air preheat. It also allows the heater to be more flexible for changes in the operating conditions and different feed stocks. Compared with LEAP, SRT-h can reduce the CO_2 emission with the least changes to the plant and, at the same time, further improve the reliability and safety for the heater operation. Each of the new Lummus low CO_2 emission solutions has its own technical advantages to meet different client needs.

Test Facility

Lummus low CO_2 technologies are based on existing technologies that have been proven in over a thousand of Lummus cracking heaters installed worldwide. Although electrical heating has not been implemented in the past cracking heaters, electrical heating is a proven technology and has been applied in similar high temperature applications in other industries. Incorporating electrical heating to cracking heaters can improve the heat distribution and increase the heater operation stability.

Lummus electric cracking heater options include the fully electric SRT-e and the hybrid SRT-h technology options. The resistance heating elements and associated electrical components are currently being rigorously tested in a dedicated Electrified Component Demonstration Facility (ECDF):

- The electrical heating elements for both technologies are being tested under the same condition and with the same power input and control system as in the commercial cracking heater.
- The test program includes startup, normal operation, and shutdown conditions.
- The testing also includes a monitoring of the temperature, pressure, electrical current, voltage and the radiant chamber environment and the chamber purging system.

With the firsthand experience of the electrical heating element performance provided by the testing facility combined with the successful experience of SRT technology, both SRT-h and SRT-e are ready for commercialization.

Summary

The SRT-h Hybrid Cracking Heater will combine fuel firing and electrical power input. All safety measures developed over the years for the conventional cracking heaters can be applied to the SRT-h heater and will provide the same or improved reliability. If there is any power supply issue to the heating elements, the energy fluctuation in the electrical power supply system can be compensated by the fuel firing. In case of loss the fuel, the electrical power input to the heater will be capable of maintaining the radiant box in hot condition to minimize the thermal shock to the high alloy radiant coil, which is one of the major factors of coil rupture or coke spalling. The hybrid heating improves the heater safety and reliability and support more stable heater operation.

For an existing ethylene plant, the heater revamp with Lummus' Hybrid Cracking technology can be implemented step by step without a complete plant shutdown, as the SRT-h can maintain the same heater performance, including the steam production, to meet the requirements of the recovery section. It will not require any modifications to the other existing equipment in the plant

In summary, SRT-h provides the following advantages:

- Reduced CO_2 emissions
- More desirable heat flux profile
- Improved reliability and flexibility of heater operation through hybrid operation

- Equal or higher thermal efficiency than conventional cracking heater
- Lower total NOx emission with less fuel firing
- More simplified electrical power control in the case of base load electrical heating
- Can be applied for revamps to existing heaters or to new heaters
- No impact on the recovery section of the plant
- Minimum modifications to the heaters
- Revamp can be implemented without a complete plant shutdown
- Zero CO_2 emission is possible when combined with carbon free fuel sources or optimization of feed preheat.

Reference

[1] AIChE technical paper reference number: 27e "ELECTRIFICATION OF STEAM CRACKING – SRT-eTM ELECTRIC CRACKING HEATER" Jibb R., Zhao B. Carrillo A. Thakurdin K. Kawalski C. 2023.

Paper # 120e

Development, Installation, and Operation of a Chemical Looping Pilot Plant for Super-Dry Reforming of Methane

Lukas C. Buelens
Postdoctoral Researcher
Laboratory for Chemical Technology, Ghent University

Michiel W.F. Van Cauwelaert
Doctoral Researcher
Laboratory for Chemical Technology, Ghent University

Hilde Poelman
Doctoral Assistant
Laboratory for Chemical Technology, Ghent University

Vladimir V. Galvita
Associate Professor
Laboratory for Chemical Technology, Ghent University

Kevin M. Van Geem
Full Professor
Laboratory for Chemical Technology, Ghent University

Prepared for Presentation at the 2024 Spring National Meeting
New Orleans, Louisiana, March, 26, 2024

Development, Installation, and Operation of a Chemical Looping Pilot Plant for Super-Dry Reforming of Methane

Lukas C. Buelens
Postdoctoral Researcher
Laboratory for Chemical Technology, Ghent University

Michiel W.F. Van Cauwelaert
Doctoral Researcher
Laboratory for Chemical Technology, Ghent University

Hilde Poelman
Doctoral Assistant
Laboratory for Chemical Technology, Ghent University

Vladimir V. Galvita
Associate Professor
Laboratory for Chemical Technology, Ghent University

Kevin M. Van Geem
Full Professor
Laboratory for Chemical Technology, Ghent University

Abstract:

Super-dry reforming is a promising CO_2-to-CO conversion process which could become a key enabler of the CO_2-to-ethylene value chain, the adoption of which could decrease the carbon footprint of ethylene production by up to 67% compared to conventional routes. At Ghent University's Laboratory for Chemical Technology, a chemical looping pilot plant was built to investigate the upscaling of super-dry reforming from gram to kilogram scale. In this work, the specifications and initial (non-)reactive operation of this pilot plant is investigated, highlighting the requirements and design characteristics of this pilot plant. The importance of transient behavior in chemical looping fixed bed reactors is reflected in this work's focus on dynamic pilot plant operation.

Introduction

Super-dry reforming of methane [1] has been developed at the Laboratory for Chemical Technology (LCT) at Ghent University as a sustainable chemical process complimentary to the well-established steam reforming (equations 1 and 2) and dry reforming (equation 3) process. Indeed, whereas steam reforming focuses on the production of pure H_2 or syngas with high H_2:CO ratio (3:1), and dry reforming on the production of syngas with low H_2:CO ratio (1:1), super-dry reforming (equation 4) maximizes the CO yield at the cost of H_2.

$$CH_4 + 2H_2O \longrightarrow 4H_2 + CO_2 \qquad \Delta H^\circ{}_{298K} = 165\ kJ\ mol^{-1}_{CH_4} \qquad (1)$$

$$CH_4 + H_2O \longrightarrow 3H_2 + CO \qquad \Delta H^\circ{}_{298K} = 206\ kJ\ mol^{-1}_{CH_4} \qquad (2)$$

$$CH_4 + CO_2 \longrightarrow 2H_2 + 2CO \qquad \Delta H^\circ{}_{298K} = 247\ kJ\ mol^{-1}_{CH_4} \qquad (3)$$

$$CH_4 + 3CO_2 \longrightarrow 4CO + 2H_2O \qquad \Delta H^\circ{}_{298K} = 330\ kJ\ mol^{-1}_{CH_4} \qquad (4)$$

$$2CO + 4H_2 \longrightarrow C_2H_4 + 2H_2O \qquad \Delta H^\circ{}_{298K} = -210\ kJ\ mol^{-1}_{C_2H_4} \qquad (5)$$

The development of an effective CO_2-to-CO conversion technique provides a path towards the creation of new chemical value chains based on CO as a platform molecule, such as the value chain towards chemicals such as ethanol and ethylene through gas fermentation and dehydration [2], as schematically illustrated in Figure 1 and global reaction equation 5. This value chain to ethylene has been estimated to hold the potential to reduce the carbon footprint of ethylene production from 2.4 to 0.8-1.0 $kg_{CO2,eq}/kg_{ethylene}$ when compared with its conventional counterpart [2].

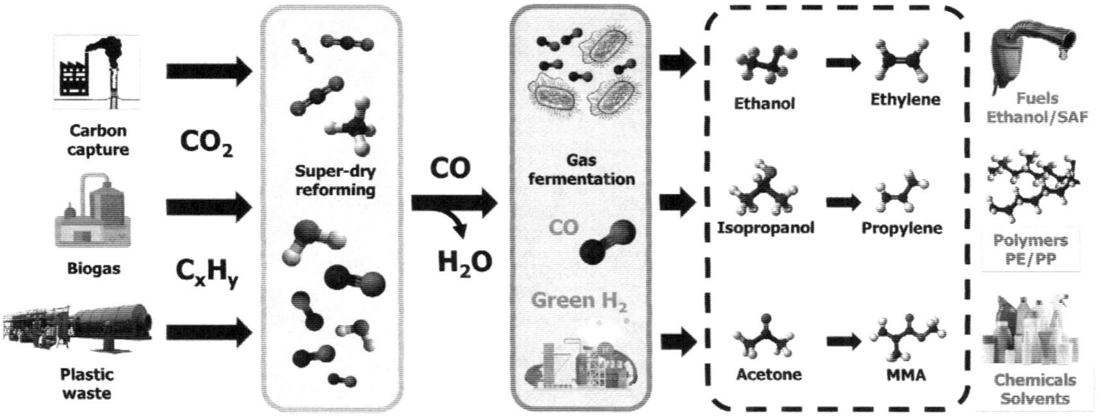

Figure 1. *Value chain for the production of carbon-neutral fuels and chemicals. Adapted with permission from [2].*

To achieve its selectivity towards CO, the super-dry reforming process makes use of three processes: (1) catalytic dry reforming of methane, (2) chemical looping CO_2 sorption, and (3) chemical looping redox reactions. Catalytic dry reforming of methane, equation 3, is well known as the conversion of CH_4 and CO_2 into syngas with a H_2:CO ratio close to 1, typically realized over a Ni or noble metal based catalyst bed. The concept of chemical looping CO_2 sorption using the reversible carbonation of CaO is illustrated in Figure 2A and equation 6. The concept of chemical looping redox reactions using the reversible reduction of metal oxides such as Fe_3O_4 and Mn_3O_4 for executing the reverse water-gas shift reaction, equation 7, and methane combustion, equation 8, is illustrated by Figure 2, B and C. The main advantages of the chemical looping approach, in which global reactions are split into gas-solid sub reactions, are the possibility to improve product selectivity and inherent separation of products owing to the separation of these sub reactions in space or time.

$$CaO + CO_2 \rightleftarrows CaCO_3 \qquad \Delta H^{\circ}_{298K} = -178\ kJ\ mol^{-1}_{CO_2} \tag{6}$$

$$CO_2 + H_2 \rightleftarrows CO + H_2O \qquad \Delta H^{\circ}_{298K} = 41\ kJ\ mol^{-1}_{CO_2} \tag{7}$$

$$CH_4 + 2O_2 \longrightarrow CO_2 + 2H_2O \qquad \Delta H^{\circ}_{298K} = -802\ kJ\ mol^{-1}_{CH_4} \tag{8}$$

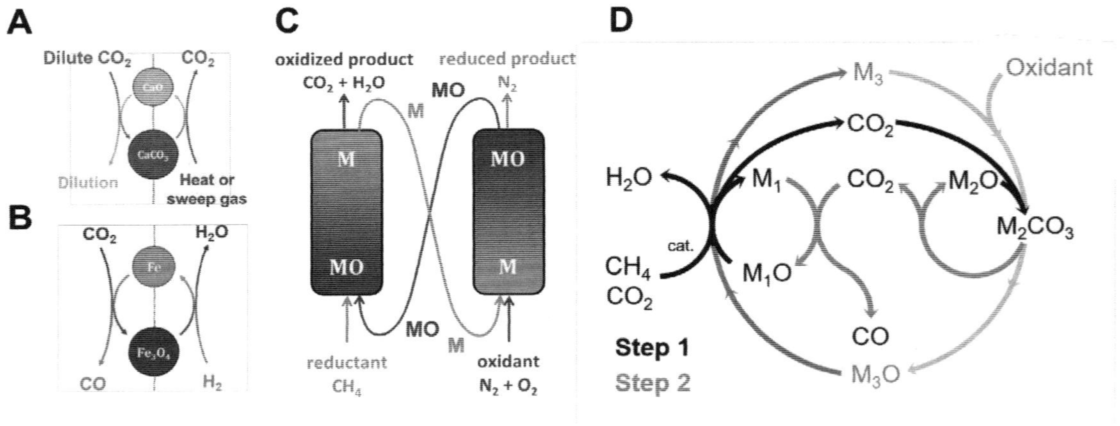

Figure 2. *Schematic representation of: (A) chemical looping CO_2 separation using the reversible carbonation of CaO into $CaCO_3$ (schematic adapted from [3]); (B) chemical looping redox reaction (reverse water-gas shift) using the reversible reduction of Fe_3O_4 into Fe (schematic adapted from [3]) with H_2 as reductant and CO_2 as oxidant; (C) chemical looping redox reaction (combustion) using the reversible reduction of generic metal oxide (MO) into metal (M) with CH_4 as reductant and air as oxidant; (D) the (autothermal) super-dry reforming process adapted with permission from [4]. Typical metal oxides contain Fe_3O_4 (M_1O) as oxygen carrier for CO_2 conversion in accordance with schematic (B), CaO (M_2O) as CO_2 sorbent in accordance with schematic (A), and Mn_3O_4 (M_3O) as oxygen carrier for heat generation in autothermal super-dry reforming (outer loop), in accordance with schematic (C). Typical oxidants are oxygen and air.*

Super-dry reforming leverages the chemical looping approach, using a two-step cycle to avoid the water-gas shift reaction, as illustrated by Figure 2D, to maximize the selectivity of CH_4 and CO_2 conversion into CO as per equation 4. In the first step of the cycle, CH_4 is reformed with CO_2 over a reforming catalyst to produce syngas, which reduces a metal oxide M_1O such as Fe_3O_4 with the formation of H_2O and CO_2 (equations 9-12), the latter of which carbonates a second metal oxide M_2O such as CaO (equation 6). The removal of CO_2 enhances the reducibility of M_1O. In comparison with conventional dry-reforming (equation 3), an expensive noble metal catalyst is not needed to mitigate carbon deposition. Since the typical operating conditions of temperature (750°C), pressure (1 bar), and feed composition (3:1 CO_2/CH_4 ratio) correspond with conditions under which carbon deposition is thermodynamically unfavorable, a Ni-based catalyst can be applied. The global reaction occurring in this first step is given by equation 13. The second step of the cycle is commenced by introducing a driving force for the metal carbonate decomposition (reverse of equation 6), either by an inert sweep gas [1], temperature increase [5] or pressure decrease [6]. The accompanying release of CO_2 results in the oxidation of iron with formation of CO (reverse of equations 11-12). By performing this CO production step in absence of H_2O through the use of a CO_2 sorbent such as CaO and a redox active material such as Fe_3O_4 [1], the CO production step is separated from the H_2O production step (equation 14).

$$H_2 + Fe_3O_4 \rightleftarrows H_2O + 3FeO \qquad \Delta H^\circ{}_{1023K} = 62\ kJ\ mol_{H_2}^{-1} \qquad (9)$$

$$H_2 + \frac{1}{4}Fe_3O_4 \rightleftarrows H_2O + \frac{3}{4}Fe \qquad \Delta H^\circ{}_{1023K} = 25\ kJ\ mol_{H_2}^{-1} \qquad (10)$$

$$CO + Fe_3O_4 \rightleftarrows CO_2 + 3FeO \qquad \Delta H^\circ{}_{1023K} = 27\ kJ\ mol_{CO}^{-1} \qquad (11)$$

$$CO + \frac{1}{4}Fe_3O_4 \rightleftarrows CO_2 + \frac{3}{4}Fe \qquad \Delta H^\circ{}_{1023K} = -7\ kJ\ mol_{CO}^{-1} \qquad (12)$$

$$CH_4 + 3CO_2 + Fe_3O_4 + 4CaO \leftrightharpoons 3Fe + 4CaCO_3 + 2H_2O \qquad (13)$$

$$inert + 3Fe + 4CaCO_3 \leftrightharpoons Fe_3O_4 + 4CaO + 4CO + inert \qquad (14)$$

The super-dry reforming process has been shown to possess economic potential for the valorization of biogas and biomethane [7], but also for steel mill gases such as the blast furnace gas [4,5], demonstrating the flexibility of the chemical looping approach towards resource valorization [8]. Although super-dry reforming is more endothermic than dry reforming, the amount of reaction heat required per mole of CO_2 utilized is significantly lower (110 vs. 247 kJ per mole CO_2). Still, heat management will be an important technical challenge to address in upscaling the super-dry reforming process. By using a third metal oxide such as Mn_3O_4 or NiO, Singh et al. showed that the super-dry reforming process could be run in an autothermal way (equation 15) by introducing a chemical looping combustion (Figure 2C and equation 8) functionality within the two-step process [4], illustrated by the outer loop in Figure 2D. Applying the chemical looping combustion cycle *in situ* of the super-dry reforming process to provide the heat of reaction, particularly for the CO_2 release during decarbonation of $CaCO_3$, requires a relatively pure source of oxygen to avoid dilution of the product stream. An interesting way around this has been proposed in which chemical looping combustion, calcium looping and chemical looping air separation are combined in an energy-neutral chemical looping process for the capture and release of CO_2 from steel mill gases [9].

$$CH_4 + 1.9\,CO_2 + 0.55\,O_2 \longrightarrow 2.9\,CO + 2H_2O \qquad \Delta H^\circ{}_{298K} = 0\ kJ\ mol^{-1}_{CH_4} \qquad (15)$$

The transient nature of chemical looping processes and the use of functional materials pose specific requirements towards the infrastructure required for technology upscaling. Through funding by the Flemish Moonshot Program, the LCT had the opportunity to design and build a chemical looping pilot plant.

Design and Installation of a Fixed Bed Chemical Looping Pilot Plant

The chemical looping pilot plant for super-dry reforming at the LCT was designed and built according to the basic flow diagram as shown in Figure 3.

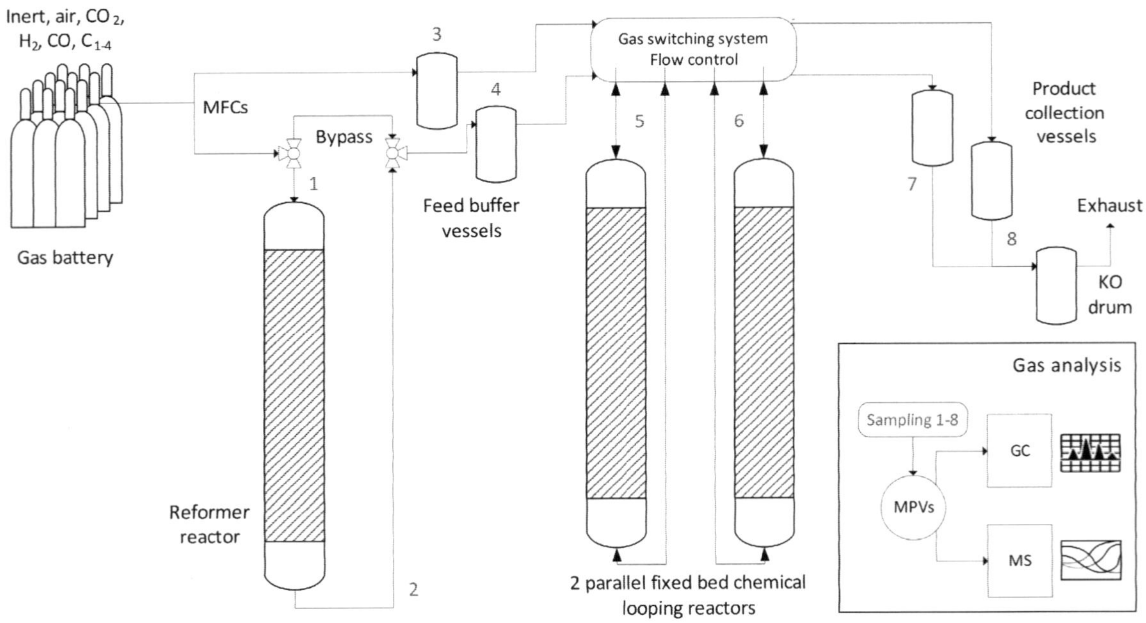

Figure 3. *Basic process flow diagram of the chemical looping pilot plant for super-dry reforming installed at the LCT.*

From the outside gas battery and storage area, various mixtures of inert (N_2, Ar, He), oxidizing (CO_2, air, O_2) and reducing gases (H_2, CO, C_1-C_4) can be dosed into the pilot plant using a total of 11 mass flow controllers (MFCs). In addition, the setup allows to feed liquids through a pump and evaporator combination (not shown in Figure 3). The setup houses two heated gas feed buffer vessels, each of which can be separately filled with the feed gas required for the 2 individual steps of the chemical looping cycle. Through the availability of a fixed bed reformer reactor upstream of one of the buffer vessels, the effect of separating the catalytic dry reforming part from the chemical looping part (Figure 2D) can be investigated. A gas switching and flow control system allows to control the flow rate and flow direction of feed gas into the two parallel fixed bed chemical looping reactors, as well as the direction of the product streams to one of either product collection vessels. All three reactors have an internal diameter of 45 mm, allow for a maximum bed height of 2000 mm, and are equipped with a 6-point thermowell at the center of the reactor tube. The split shell furnace houses 5 heating zones which allow to heat the reactor up to 900°C within 6-7 hours. The system is designed to operate the reformer reactor up to 20 barg and the parallel chemical looping reactors up to 12 barg. The nominal design capacity is 1 kg_{CO2}/h (equivalent to ca. 8 ton_{CO2}/yr for a continuously operated plant), but the total flow rate can readily go up to a maximum of 3 kg_{CO2}/h (equivalent to ca. 24 ton_{CO2}/yr).

The gas analysis section consists of a Hiden ExQ mass spectrometer (MS) and an Interscience CompactGC$^{4.0}$ gas chromatograph (GC), preceded by a combination of several flow-through multiport valves allowing to take samples from 8 different locations within the pilot unit (labeled 1 – 8 in Figure 3). The closing of the mass balances over each reactor is done by means of a coriolis mass flow meter at the outlet of each of the 3 reactors. Sample points 1 and 2 correspond with the reformer reactor inlet and outlet gas, samples 3 and 4 correspond with the feed buffer vessels outlet gas, samples 5 and 6 correspond with the parallel chemical looping reactor outlet gases, and samples 7 and 8 correspond with the product collection vessels outlet gas. In addition, the multiport valve system allows quick switching to feed 2 different gas calibration mixtures into the gas analysis devices for calibration. Figure 4 shows the as built pilot plant at the manufacturers premises and at the LCT premises.

Figure 4. *Chemical looping pilot plant for super-dry reforming installed at the LCT: as-built pilot plant at the manufacturers premises (left) and at the LCT premises (right).*

Operation of a Fixed Bed Chemical Looping Pilot Plant

The transient nature of the chemical looping cycle in general, and that of super-dry reforming in particular, requires a dynamic approach to the operation of the pilot plant. In this paragraph, the dynamic operation of a chemical looping pilot plant with two parallel fixed bed reactors (Figure 4) will be illustrated through a number of (non-)reactive experimental tests.

Dynamic flow control to parallel fixed bed reactors

In a first series of functional tests (Figure 5), the effect of periodically switching the feed to the parallel reactors between composition A and B through the feed buffer vessels is investigated. In a first experiment (Figure 5, left), each halfcycle is equal in length, and the parallel reactors are alternately fed with A and B so that A is fed to reactor 1 while B is fed to reactor 2, and vice versa. During these experiments, the feed buffer vessels pressure control is relatively stable under the applied control settings. In a second experiment (Figure 5, right), the duration of the A feeding step is three times longer than that of B. As seen from the initial flow deviations in the first 20-30 seconds of each halfcycle, the flow control system is not completely on point yet satisfactory for our purpose. By choosing different durations for the A and B feeding steps, it is clear that the pressure control on the feed buffer vessels becomes more complex, particularly for composition B, which does not flow to any of the two reactors half of the time. This indicates the importance of optimized control systems to deal with the dynamic operation of parallel fixed bed chemical looping reactors.

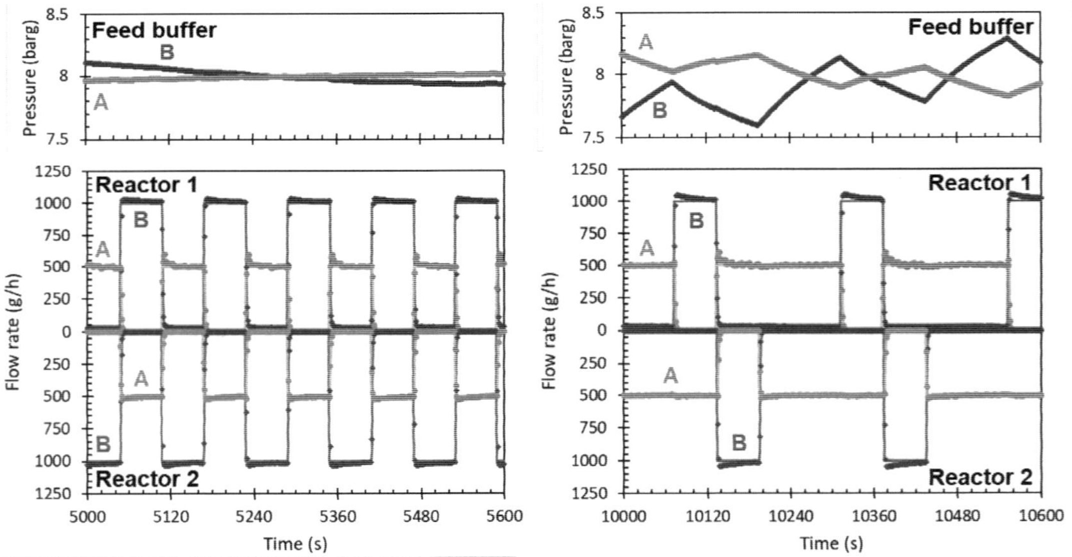

Figure 5. *Feed buffer tank pressure (top) and flow rate (bottom) to parallel reactors 1 and 2 during cycles with equal (left) or different (right) step durations. The reactor temperature and pressure is 20°C and 1.5 barg. The composition of gas feed A corresponds with 5%* H_2 *in Ar and that of gas feed B corresponds with 21%* O_2 *in* N_2*.*

Reactor temperature control during periodic switching of flow direction

In the next functional test, one of the parallel looping reactors was heated to 800°C. Figure 6 shows the temperatures of a 6-point thermowell at the center of the reactor as a function of time, indicating that a heating ramp of 150°C/hour can be achieved. Moreover, the gas flow direction was switched once per hour (see inset of Figure 6 for detail). It is clear that the temperature in the inlet zone of the reactor is about 25-30 °C lower than the central zones, and the outlet zone of the reactor is around 5-10 °C lower than the central zones. When switching the flow direction from bottom to top or vice versa, the temperatures in these outer zones quickly approach a new steady state, whereas the temperature in the central zones remains largely unaffected.

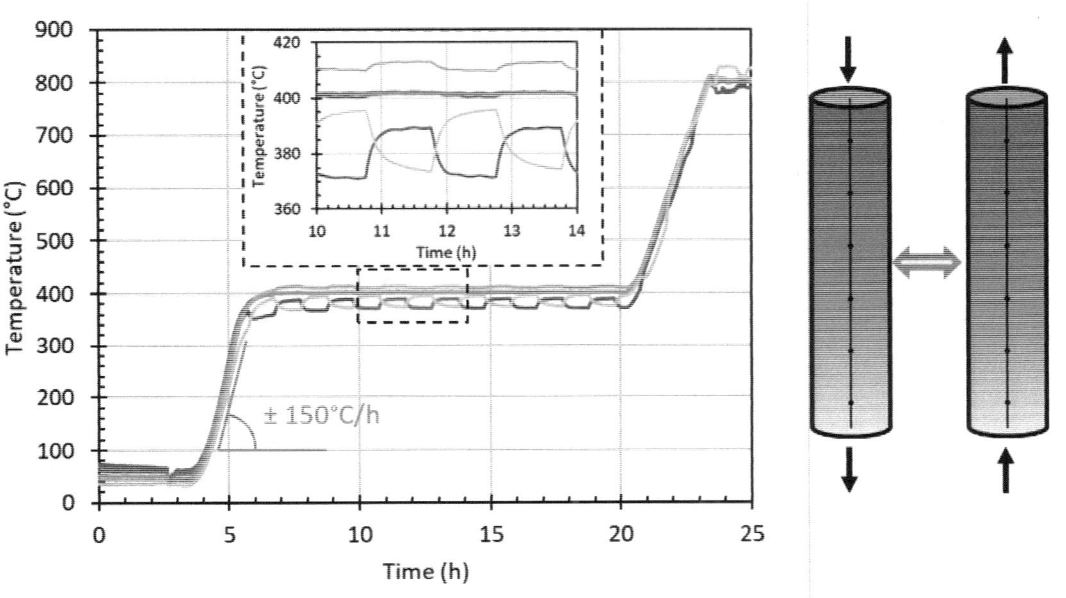

Figure 6. *Temperature as a function of time for chemical looping fixed bed reactor 2 with periodic switching of flow direction. The reactor bed consists of inert SiO_2 extrudates (2 mm diameter). Flow rate and pressure were maintained at 200 g/h synthetic air and 1.5 barg.*

Chemical looping reverse water-gas shift experiments

As model system for the first pilot plant experiments, the chemical looping reverse water-gas shift reaction was chosen. This chemical looping process corresponds with the M_1O-M_1 loop in Figure 2, where M_1O (e.g. Fe_3O_4) is reduced by H_2 in a first step, and the resulting M_1 (Fe) is reoxidized by CO_2 in a second step with formation of CO, completing the redox cycle. Reactor 2 was loaded with a bed of 3.5 kg of $Fe_2O_3/MgAl_2O_4$ extrudates (2 mm diameter), separated by quartz wool discs. Figure 7 shows the buffer vessel pressure control, reactor temperature, inlet and outlet flow rate of reactor during periodic switching between H_2 and CO_2 at 250°C (left) and at 700°C (right). Each cycle consists of 30 minutes reduction and 30 minutes oxidation. The gas flow direction was kept from top to bottom, except for the second reduction by H_2 at the 1 hour mark in Figure 7, before which the gas flow direction was switched and gas was fed from bottom to top.

The results of this experiment show a limited effect on the H_2 conversion into H_2O of the flow direction as witnessed by the consistent reactor outlet flow rate, whereas the impact on the CO_2 conversion into CO is very significant. Indeed, based on the outlet and inlet flow rate during the oxidation step, the overall CO_2 conversion the second cycle (two-directional flow) is twice as high as that of the first cycle (single directional flow). Given that the bed remains partially reduced after the first cycle as visible from the discrepancy between inlet and outlet flow rate at the end of the cycle, it is likely that such accumulated reduced species contribute to the CO_2 conversion during the second cycle. Moreover, it has been demonstrated in the past through simulation work [10,11] and lab-scale experimentation [12,13] that the driving force for this kind of chemical looping redox process can be drastically increased by operating the reactor bed in two-directional flow. To the best of our knowledge, our results presented in Figure 7 provide the first experimental pilot-scale demonstration of the merit of this operational strategy in chemical looping processes, all the while validating the functionality of our recently installed pilot plant using an excellent model reaction system.

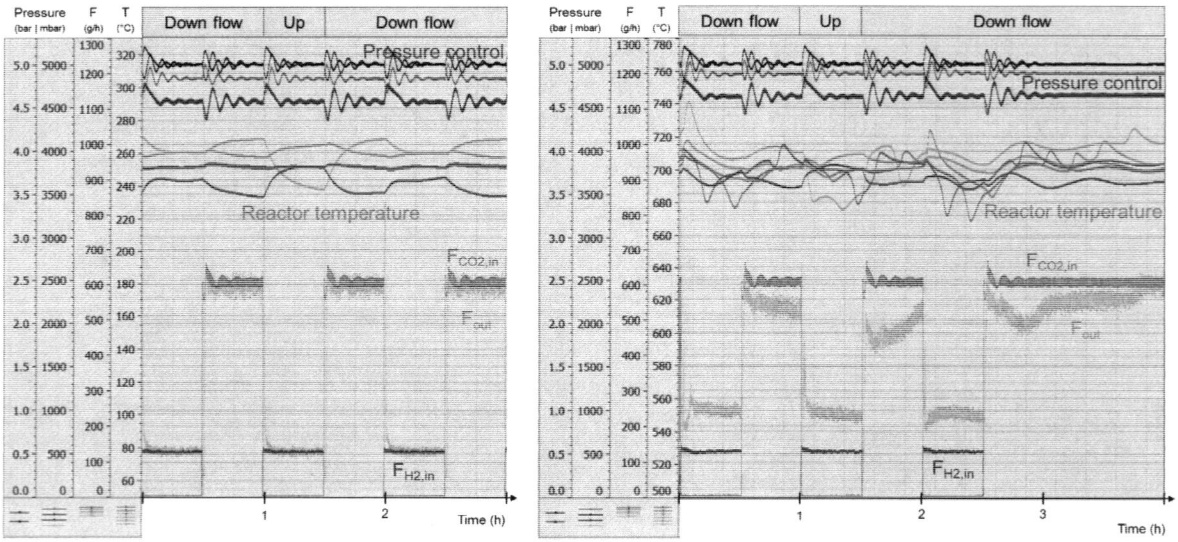

Figure 7. *Process flow, temperature and pressure control at 250°C (left) and 700°C (right) during cycles using H_2 (85% in N_2) and CO_2 (85% in N_2) over a 3.5 kg Fe_2O_3/$MgAl_2O_4$ bed at 1 barg. All gas mixtures are fed from top to bottom except during the H_2 feed step of the second cycle. The difference in reactor inlet (green, cyan) and outlet (light grey) mass flow rate corresponds with O exchange from the oxygen carrier with the gas phase to yield H_2O (during reduction with H_2) and CO (during oxidation with CO_2).*

Conclusion

The development of the super-dry reforming process has come a significant way since its conception and is on the verge of being experimentally tested at TRL 5. Initial pilot plant tests indicate that the plant has the right functionality to deal with the transient behavior of chemical looping processes when operated in multiple fixed bed reactors and to provide high quality experimental data. Using the chemical looping reverse water-gas shift reaction as model system, it was shown that the operation of chemical looping redox processes in two-directional flow results in significantly higher CO_2 conversion levels in comparison with single directional flow. Given the endothermicity of super-dry reforming and CO_2 utilization in general, the development of strategies for heat management and autothermal operation is a crucial piece of the technological puzzle for future studies. To this end, experimental pilot plant campaigns will provide directions for further upscaling.

Acknowledgements

Lukas Buelens and Michiel Van Cauwelaert acknowledge financial support from the Fund for Scientific Research Flanders (FWO Flanders) through (post-)doctoral fellowship grants 12E5623N and 1160723N. Vladimir Galvita acknowledges a personal grant from the Research Fund of Ghent University (BOF; 01N16319). The authors acknowledge funding from the European Research Council under the European Union's Horizon 2020 research and innovation program / ERC grant agreement n° 818607. We gratefully acknowledge the financial support of the Flemish Government and Flanders Innovation & Entrepreneurship (VLAIO) through the Moonshot project LSI SDR HBC.2019.0140. Finally, the authors are grateful to all contributors from Zeton and LCT's engineering team, in particular but not exclusively Bart Zaalberg (Zeton), Mark Wolters (Zeton), Bert Depuydt (LCT) and Tom Verspeelt (LCT), for their valuable input during the design, construction and commissioning of this pilot unit.

References

1. Buelens L.C., Galvita V.V., Poelman H., Detavernier C. and Marin G.B. (2016) "Super-dry reforming of methane intensifies CO_2 utilization via Le Chatelier's principle", *Science* 354(6311), pp. 449-452.

2. Van Cauwelaert M., Delikonstantis E., Buelens L., Poelman H., Van Geem K.M. and Galvita V. (2022) "From steel to ethylene : super-dry reforming of CH_4", *34th Ethylene Producers' Conference*, San Antonio, USA.

3. Buelens L. C., Poelman H., Marin G.B. and Galvita V.V. (2019) "110th Anniversary: Carbon Dioxide and Chemical Looping: Current Research Trends", *Industrial & Engineering Chemistry Research* 58(36), pp. 16235-16257.

4. Singh V., Buelens L.C., Poelman H., Saeys M., Marin G.B. and Galvita V.V. (2022) "Upcycling the carbon emissions from the steel industry into chemicals using three metal oxide loops", *Energy Advances* 1(6), pp. 367-384.

5. Singh V., Buelens L.C., Poelman H., Saeys M., Marin G.B. and Galvita V.V. (2022) "Carbon monoxide production using a steel mill gas in a combined chemical looping process", *Journal of Energy Chemistry* 68, pp. 811-825.

6. Flores-Granobles M. and Saeys M. (2022) "Dynamic pressure-swing chemical looping process for the recovery of CO from blast furnace gas", *Energy Conversion and Management* 258, pp. 115515.

7. Verbeeck K., Buelens L.C., Galvita V.V., Marin G.B., Van Geem K.M. and Rabaey K. (2018) "Upgrading the value of anaerobic digestion via chemical production from grid injected biomethane", *Energy & Environmental Science* 11(7), pp. 1788-1802.

8. Singh V., Buelens L.C., Poelman H., Marin G.B. and Galvita V.V. (2023) "Chemical looping: a technology platform for upcycling low-grade industrial resources", *Discover Chemical Engineering* 3(1), pp. 12.

9. Singh V., Buelens L.C., Poelman H., Saeys M., Marin G.B. and Galvita V.V. (2022) "Decarbonisation of steel mill gases in an energy-neutral chemical looping process", *Energy Conversion and Management* 254, pp. 115248.

10. Heidebrecht P., Hertel C. and Sundmacher K. (2008) "Conceptual analysis of a cyclic water gas shift reactor", *International Journal of Chemical Reactor Engineering* 6, pp. 1-17.

11. Wenzel M., Rihko-Struckmann L. and Sundmacher K. (2018) "Continuous production of CO from CO_2 by RWGS chemical looping in fixed and fluidized bed reactors", *Chemical Engineering Journal* 336, pp. 278-296.

12. Metcalfe I.S., Ray B., Dejoie C., Hu W., de Leeuwe C., Dueso C., García-García F.R., Mak C.-M., Papaioannou E.I., Thompson C.R. and Evans J.S.O. (2019) "Overcoming chemical equilibrium limitations using a thermodynamically reversible chemical reactor", *Nature Chemistry* 11(7), pp. 638-643.

13. Bulfin B., Zuber M., Gräub O. and Steinfeld A. (2023), "Intensification of the reverse water–gas shift process using a countercurrent chemical looping regenerative reactor", *Chemical Engineering Journal* 461, pp. 141896.

AIChE Paper Number: 120f

Integration of Coolbrook's RotoDynamic Reactor (RDR) into existing steam crackers

Dr. Tuomas Ouni,
Head of Process Development,
Coolbrook Oy

Dr. Gunther Kracker,
Head of Product Management and Cost Estimation
Sustainable Hydrocarbon Solutions, Linde Engineering

Tobias Sinn
Process Design Engineer
Sustainable Hydrocarbon Solutions, Linde Engineering

Prepared for Presentation at the
2024 Spring Meeting and 20th Global Congress on Process Safety
New Orleans, Louisiana, March 26, 2024

AIChE Paper Number: 120f

Integration of Coolbrook's RotoDynamic Reactor (RDR) into existing steam crackers

Dr. Tuomas Ouni,
Head of Process Development,
Coolbrook Oy

Dr. Gunther Kracker,
Head of Product Management and Cost Estimation
Sustainable Hydrocarbon Solutions, Linde Engineering

Tobias Sinn
Process Design Engineer
Sustainable Hydrocarbon Solutions, Linde Engineering

Abstract

Coolbrook's RotoDynamic ReactorTM (RDR) can reach 100% CO2 free olefin production. It is an electrically driven turbomachine where one or a cascade of stator-rotor-diffuser sequences heat gaseous steam-hydrocarbon mixture efficiently and with extremely low residence time into reaction temperature. The heat increase occurs intrinsically in the heated gas and results from conversion of rotor blade -induced kinetic energy of the gas into thermal energy by rapidly slowing the velocity of the gas. As the velocity of the gas slows down from supersonic to subsonic, a shockwave occurs that enhances the conversion of kinetic energy into heat. Using RDR with renewable electric power makes it one of the most promising technologies capable of cutting 100% of process emissions (CO2, NOx and particulate emissions).

In 2023 Coolbrook and Linde Engineering have entered into a strategic partnership to collaborate on the development and deployment of Coolbrook's RotoDynamic Reactor to jointly develop an offering for petrochemical companies.

RDR can be implemented step by step also at existing cracker sites into the existing infrastructure. The conference contribution will demonstrate different options for integrating Coolbrook RDR into existing steam crackers, ranging from concepts with steam generation as required for existing plant conditions to solutions with minimized specific electricity consumption per ton feedstock.

Coolbrook's RDR Pilot plant in Brightlands Campus in Geleen, Netherlands was built to test and develop RDR Technology for electrified steam cracking of hydrocarbon feedstock. The test program consists of a variety of test runs divided over several basic raw materials. The pilot will function as a step towards commercial demonstration of RDR technology. The project will deliver design tools and proof-of-concept for commercial implementation, which will have industrially relevant size and will be connected to an existing and operating ethylene cracker. The capabilities of the pilot plant and a summary of available results will be presented.

Introduction

I

Coolbrook is a Finnish-Dutch-British technology and engineering company on a mission to de-carbonize major industrial sectors such as petrochemicals and chemicals, iron and steel and cement. Coolbrook's proprietary, revolutionary RotoDynamic Technology combines principles of rocket science, turbomachinery, and chemical engineering to electrify and replace burning of fossil fuels in hard-to-abate industrial processes. RotoDynamic technology generates process heating for high temperature processes with CO_2-free electricity, and once implemented at scale, the technology has the potential to cut 2.4 billion tons of annual global CO_2 emissions in major industrial sectors, equal to 7% of all global CO_2 emissions.

The technology has two main applications: RotoDynamic Reactor™ (RDR™) to reach 100% CO_2 free olefin production, and RotoDynamic Heater™ (RDH™) to provide carbon-free process heating to e.g. iron and steel, cement and chemicals production. Backed by a growing number of governments, technology partners, industrial producers, and environmental regulators, Coolbrook technology is on path to become the new global standard in industrial electrification.

Instead of heating the feedstock mixture from outside the reaction zone, RotoDynamic Reactor's high-velocity rotor blades create thermal energy to heat the mixture inside the reaction zone – quickly and much more efficiently. RDR uses renewable electric power, making it the only technology capable of cutting 100% of direct process emissions (CO_2, NO_x and particulate emissions).

Linde Engineering is a leading technology partner for plant engineering and construction worldwide. Linde Engineering has supplied its Ethylene Technology for eighty years with proven success and a well-known reputation. Linde can offer all kinds of supplies from feasibility studies, Pre-FEED/FEED packages up to the complete scope of work for an ethylene complex including procurement, construction, and commissioning/start-up services. Linde has a tremendous knowledge not only on how to design ethylene plants but more importantly on how to build a reliable ethylene plant.

Global success of Linde Engineering is built on extensive process engineering expertise in the planning, project development and construction of turnkey industrial plants. Linde's attitude to take the lead in providing solutions for difficult challenges during execution phase to the complete satisfaction of the customers is a key to success.

Coolbrook and Linde Engineering are in a strategic partnership to collaborate on the development and deployment of Coolbrook's RotoDynamic Reactor technology to replace burning fossil fuels with clean electricity in ethylene plants world-wide. Coolbrook and Linde Engineering are jointly developing an offering for petrochemical companies to be able to achieve a significant reduction in emissions and to boost profitability through increased olefin yields and improved uptime through reduced coking.

In addition to greenfield installation, RDR can be implemented step by step also at existing cracker sites into the existing infrastructure. In this way both the environmental and commercial viability are maximized. Coolbrook/Linde integration concepts are flexible towards the available and required utilities of the existing plant set-up and allow steam crackers to minimize the integration efforts in transition towards electrified steam cracking.

CO2 reduction by RDR technology

Coolbrook's RotoDynamic Reactor is an electrically driven turbomachine where one or a cascade of stator-rotor-diffuser sequences heats gaseous steam-hydrocarbon mixture efficiently and with extremely low residence time into reaction temperature. The heat increase occurs intrinsically in the heated gas and results from conversion of rotor blade - induced kinetic energy of the gas into thermal energy by rapidly slowing the velocity of the gas. As the velocity of the gas slows down from supersonic to subsonic, a shockwave occurs that enhances the conversion of kinetic energy into heat.

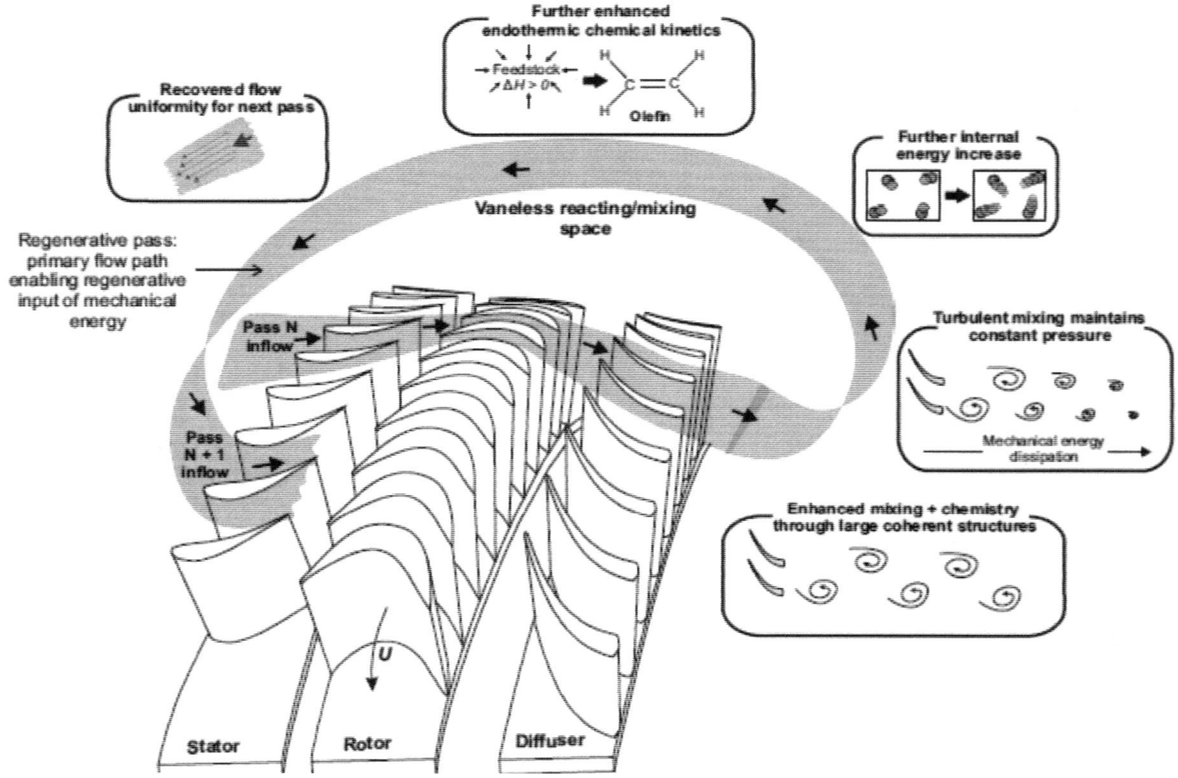

Figure 1: Close view on a single passage through a bladed section in a RotoDynamic Reactor

The intrinsic heat generation of RDR offers considerable benefits in terms of steam cracking chemistry and coke formation. Heat is generated in the bladed section of RDR through dissipation of kinetic energy into thermal energy, hence the distribution of heat is uniform across the flow path. As cracking reaction rates are dependent on the temperature, also the concentration profile remains uniform across the flow path, which allows improved control of the chemical reactions.

Furthermore, as heat is not conducted into the reaction mixture through hot metal surfaces in RDR, the wall temperatures remain at or slightly below reaction temperature. This means that the wall temperatures in RDR are 100-200 K lower than in conventional cracking tubes, decreasing the coking rate at the walls and subsequent carburization significantly. Furthermore, the extremely turbulent conditions in RDR reduce the thickness of the stagnant boundary layer next to metal surfaces, reducing the residence time of coke precursors at wall conditions.

As a result, the coke formation in RDR is expected to be low, and majority of the coke formed will be flushed away with the turbulent flow. Signs of such 'self-cleaning' mechanism were observed in the test runs.

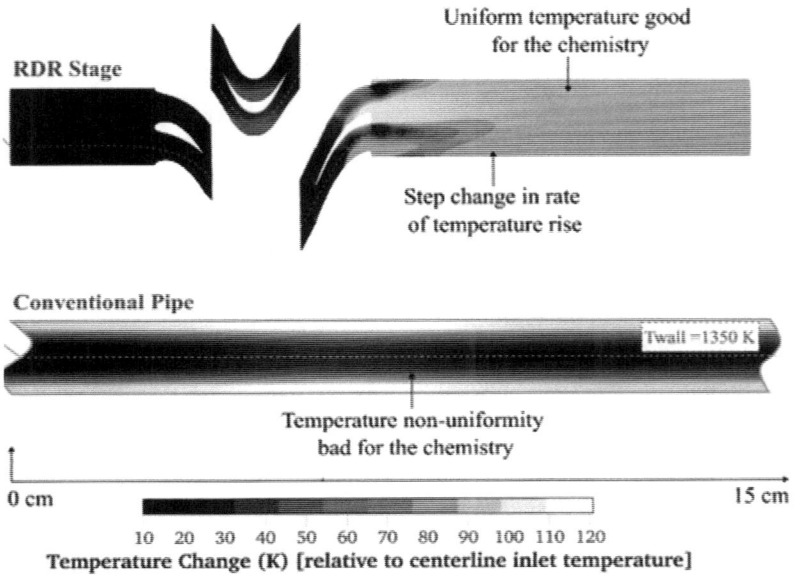

Figure 2: cross-sectional diagram of heat distribution in conventional cracking coil and RDR

RDR can be implemented step by step also at existing cracker sites into the existing infrastructure. In this way both the environmental and commercial viability are maximized. Currently, it is economically more viable than any other existing or announced furnace replacement technology to reduce CO2 emissions.

Strategic Partnership for commercialization and scale-up of RDR

In March 2023, Coolbrook and Linde Engineering have entered into a strategic partnership to collaborate on the development and deployment of Coolbrook's RotoDynamic Reactor technology to replace burning fossil fuels with clean electricity in ethylene plants world-wide.

It was agreed that Coolbrook and Linde Engineering will be jointly developing an offering for petrochemical companies to accelerate the development, industrialisation, and deployment of Coolbrook's RDR technology and its integration to existing ethylene plants as well as greenfield projects.

In this collaboration, Coolbrook will have the responsibility regarding the RDR package which consists of the RDR unit, its electric drive and the first stage of product cooling equipment. Linde Engineering will lead the integration of the technology to the wider process unit including the mass and energy integration with the downstream plant and any engineering or design activities related to that (Figure 3):

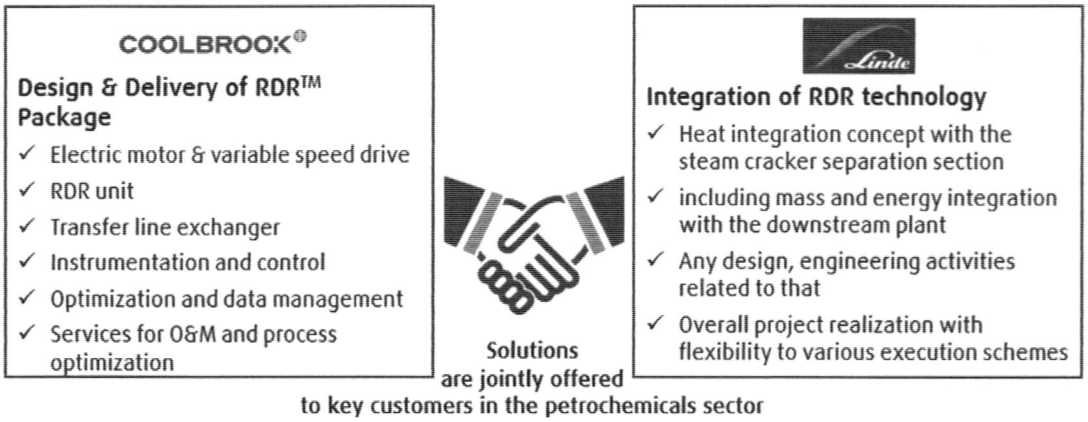

Figure 3: Roles and responsibilities within the strategic partnership between Coolbrook and Linde Engineering

Conceptual integration of a RotoDynamic Reactor into existing steam crackers

Conventional fired furnaces form the heart of a steam cracking plant, not only because they host the cracking reaction, but also because of their central role in the plant's energy system. Large quantities of heat are recovered from the furnace flue gas and cracked gas streams leaving the cracking coils. This heat is used to preheat furnace feed streams and to generate high pressure utility steam, which is later used in turbines and heat exchangers for mechanical work and heat supply in the downstream separation section.

In contrast to that, a RotoDynamic Reactor generates heat for the cracking reaction by converting mechanical shaft power into kinetic energy of gas, which is then deliberately dissipated as heat. The RDR mechanism results in an efficient energy transfer where at least 90% of the consumed electrical energy is converted to heat and transferred to the process fluid.

In a RDR, as there is no flue gas waste heat, alternative methods of heat supply for feed and dilution steam heating, boiler feed water heating, and high-pressure steam superheating are necessary. Such a system, which replaces the function of a conventional convection zone, utilizes process inherent heat sources that are now restricted to just cracked gas heat. However, additional heating through electrical power can be used if needed for balancing.

The elimination of the previously inherent coupling of the reaction and preheating systems in conventional fired cracking furnaces now provides new perspectives on steam cracking improvement. Feed inlet temperature to the RDR reaction section can be adjusted in the new preheating system according to the requirements of the RDR cracking process. This allows for adjustments to operational parameters and eventually for further optimization of product yields.

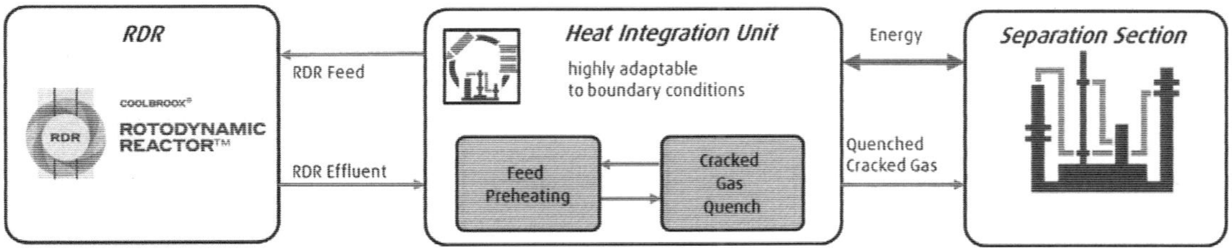

Figure 4: Sketch of an RDR with heat integration system and separation section for product purification

Figure 4 illustrates schematically the system components, comprising RDR, heat integration unit and product purification, as well as the exchanged feed, effluent and energy streams. The design of said heat integration unit is quite flexible and can be adjusted to application/site specific requirements. This includes aspects like the overall plant energy balance (e.g. steam demand), the RDR & furnace configuration and ultimately feed types to be processed.

The utilization of cracked gas heat essentially depends on the energy balance of the plant, and the purpose it serves, whether it is desired for preheating the feed or required as steam export to the separation section of the plant (e.g., in steam operated heaters or turbine drives). It is to be remembered that in such configuration, cracked gas heat and electrical power are the only heat sources. As such, the intensity of the heat integration of cracked gas is determined by the consideration "steam produced vs. power consumed". It must be noted that, when compared to a conventional fired furnace, steam surplus from an RDR & heat integration unit decreases even for low cracked gas heat utilization, as part of the produced steam may be required for e.g., feed vaporization purposes.

In case of grassroot projects, these central heat integration units are likely to be very different than in existing plants, with high degrees of electrification throughout the plant and reduced utility steam circulation. In case of plant revamps, the changes in energy balance for a single fired furnace replacement may still be within the plant's operational margins whereas multiple furnace replacements may necessitate the need for simultaneous modifications in the separation section of the plant. The lack of excessive heat from a convection section and the availability of electrical power from renewable sources increase the incentive to utilize energy recovered from cracked gas cooling only for process heating instead of powering pumps and compressors. The superior efficiency of electrically powered drives for rotating equipment, typically over 90%, compared to conventional steam turbines, typically only 30 to 40%, encourages the adoption of such systems to further improve the overall plant efficiency.

Exemplary integration scenarios

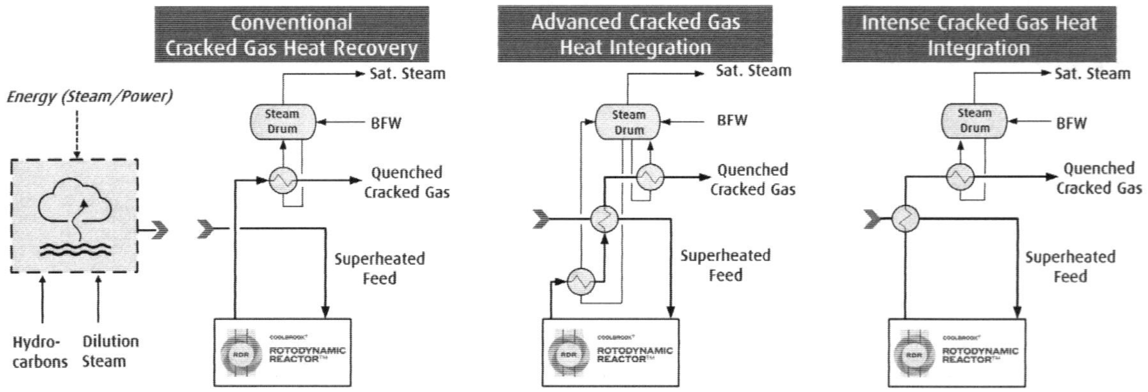

Figure 5: Three different levels of cracked gas heat integration, each relying on different quench exchanger technology

Figure 5 displays three exemplary options for integrating Coolbrook's RDR into existing steam crackers with different process topologies regarding feed pre-heating and cracked gas quenching:

- Concept 1: Full steam export – in this concept the production of very high-pressure steam is maximized for downstream use
- Concept 2: Medium steam export - in this concept the energy from transfer line exchanger (TLE) is directed in a flexible way for both feedstock pre-heating and VHP steam generation.
- Concept 3: No steam export – in this concept the energy of TLE is fully captured in feedstock and dilution steam pre-heating and the specific electricity consumption per ton feedstock is optimized.

Shortest process residence time and shockwave heating enable the RotoDynamic Reactor to process a mix of hydrocarbon feedstock and dilution steam at far lower inlet temperatures when compared to steam cracking in conventional coils (compare left scheme of Figure 8). The feed must be entirely in a vapor state while maintaining sufficient distance to the dew point. This is a big differentiator compared to steam cracking in coils, where inlet temperature of 560-690°C high above the dew point are required in any case.

Low RDR inlet temperatures are particularly advantageous for scenarios with considerable steam demand of the separation section, especially in case of conventional liquid cracker integration. This implies that additional shaft power is required to bridge the gap in heat until the cracking reaction initiates. However, a significant benefit of this approach is that high-temperature heat originates from a medium voltage power supply instead of a low voltage power supply including transformer losses, as required for conventional resistive e-heaters. Moreover, it results in a simple and compact heat integration that is reduced in equipment count and plot space.

Low RDR feed inlet temperatures are not mandatory and can be increased by means of external preheating up to a level close to conventional coil inlet temperature, leading to a saving in electrical power consumption of the RDR. As a result, a higher degree of cracked gas heat integration will be required (compare middle and right scheme of Figure 5) resulting in reduced steam surplus. As such, those schemes are suitable for scenarios with moderate, low or no steam demand of the separation section (e.g. Naphtha cracker with e-drives or gas cracker).

The three schemes of Figure 5 rely on distinct quench exchanger technologies. The option on the far left utilizes a conventional quench exchanger, which generates steam on the shell side. On the other hand, the middle and right scheme feature combinations of conventional quenching and novel gas/gas exchangers, specifically developed to maintain the same performance level as traditional cracked gas cooling. This allows for the direct integration of high temperature cracked gas heat.

In general, the design of the feed preheating is very flexible for the entire range of cracker feedstock – from gaseous to liquid and heavy feedstock (e.g., chemically recycled plastics). However, the fouling and coking behavior of particularly heavy feedstocks may impact the design of the cracked gas quench, potentially limiting the intensity of heat integration to some extent.

RDR Pilot Plant

Coolbrook's RotoDynamic Technology Pilot plant in Brightlands Campus in Geleen, Netherlands was built to test and develop the Technology for electrified steam cracking of hydrocarbon feedstock.

Figure 6: Coolbrook RDR pilot reactor in operation at Brightlands Chemelot Campus, Netherlands

The test program consists of a variety of test runs divided over several basic raw materials. Duration of test runs vary between a few hours needed for yield testing and a few weeks needed for testing coking tendency and reliability. Project will function as a step towards commercial demonstration of RDR technology. The project will deliver design tools and proof-of-concept for commercial implementation, which will have industrially relevant size and will be connected to an existing and operating ethylene cracker.

At the time of writing of this article in early 2024, the test campaign for RDR was in progress. In 2023, Coolbrook had successfully cracked naphtha in the RotoDynamic Reactor, demonstrating electric steam cracking. Preliminary results from naphtha cracking in RDR indicate a significant improvement in yields for high value chemicals, and reduced coking. Further test campaign will bring further proof for these observations and will allow Coolbrook to model the chemistry of cracking and coke formation in detail together with the aerodynamic performance of the RotoDynamic Reactor.

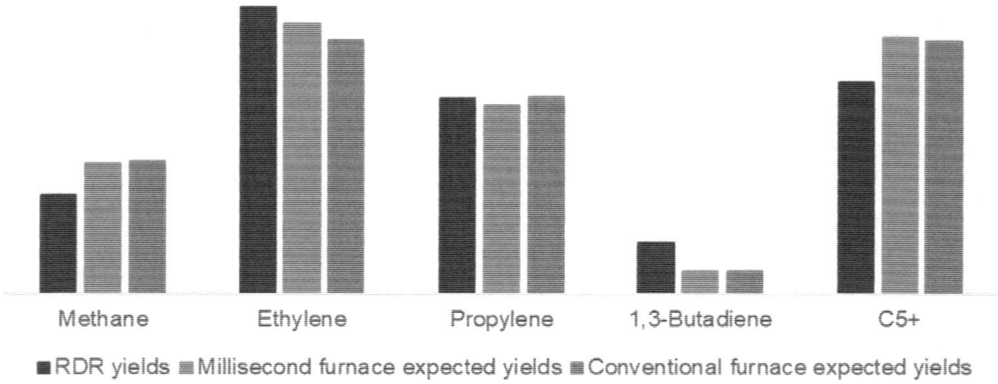

Figure 7: Experimental product yields at low severity naphtha cracking in the RotoDynamic Reactor™ compared to simulated yields from same feedstock in conventional and millisecond furnaces

Summary and next steps

Coolbrook is a Finnish-Dutch-British technology and engineering company developing Rotodynamic Reactor™ (RDR™), an electrically driven turbomachine technology for electrification of steam cracking. In addition to full decarbonization, benefits of the RDR technology include improved ethylene and propylene yield and significantly reduced coking

Coolbrook has performed first naphtha cracking tests at its pilot plant in the Netherlands, with results preliminarily validating the benefits of the technology

Linde and Coolbrook collaborate on development and piloting, the commercial deployment of the technology, and providing process integration technology for Coolbrook - Linde joint RDR offering. Coolbrook and Linde are working towards following steps in the development of RDR technology:

- 2024 finalization of pilot tests & design of large scale equipment

- 2024 - 2025 start large scale demonstration projects together with lead customers
- 2026 onwards: RDR commercially available to replace cracking furnaces in industrial scale

SPRING24
+20TH GCPS
A Joint AIChE and CCPS Meeting

Enhancing Asset Effectiveness through Hybrid Modeling

Presenter/Principal Author: Sameer Thorat
Director of Process Analytics and Engineering, North America
Ingenero Inc.
4615 Southwest Frwy., Suite 320, Houston, TX-77494
sthorat@ingenero.com

Other Author: Nishant Jaiswal
Manager Process Technology
Ingenero Technologies
njaiswal@ingenero.com

Prepared for Presentation at
American Institute of Chemical Engineers
2024 Spring Meeting and 20th Global Congress on Process Safety
New Orleans, LA
March 24-28, 2024

Enhancing Asset Effectiveness through Hybrid Modeling or Data-driven Modeling

Keywords: Ethylene, Digitalization, AI, ML, Digital Twin, Big data analytics, IIOT, Reliability, Multi-model techniques, OAE, Hybrid modeling

Abstract

Big data analytics and IIOT can enhance plant performance and reliability by using multi-model techniques and systematic analysis of plant operation. Multi-modeling is essential, but it also needs to be integrated with data collection, model-based information and knowledge generation, visualization and dashboards for interpretation and early detection, and implementation and tracking for benefit realization.

Successful application of a good overall asset effectiveness (OAE) solution considerably helps in improving the asset effectiveness resulting in,

- Longer mean time between machine overhauls/ replacement → extending T/A intervals
- Improving efficiency and overall asset effectiveness → increasing plant production capacity depending on the severity of operating constraints
- Operating plant to optimal best levels while ensuring reliability and safety

Using a combination of first principle, empirical models, and predictive analytics, smart solutions can detect equipment constraints and deterioration in real time. They also provide useful information for making timely and appropriate decisions to optimize asset effectiveness and ensure operational safety. Industry specific advanced analytics of historic and real-time data enable improvements that are challenging to achieve with traditional monitoring approaches.

First principle models are based on established scientific principles, while empirical models capture complex relationships that may not be easily described mathematically. By combining them, we can improve the accuracy and robustness of our predictions and simulations. This approach is called hybrid modeling, and it is successfully used in manufacturing process operations. However, it is important to understand the problem, the available data, and the physical processes when the first principle and empirical models are applied together.

Hybrid modeling enables continuous evaluation of measured and derived parameters against established engineering limits. This can reveal opportunities and potential reliability issues. The performance is tracked using suitable techniques, tools and equipment models, and the deviation from normal is quantified. Moreover, predictive and prescriptive analytics can help assess the impact of deviation, the need for corrective action and the assistance in finding challenging solutions.

Data analytics can enhance the plant operating strategy and lead to substantial economic gains in the ethylene process industry. This paper will discuss how an effective predictive analytics solution driven by different techniques, tools and models is effectively utilized to optimize production and asset performance of an ethylene plant.

1 Introduction

With the emergence of big data analytics and IIoT, the potential to attain operational excellence has significantly expanded, with particular importance within the ethylene processing industry. The advent of big data analytics and Industrial Internet of Things (IIOT) has ushered in a new era of opportunity, promising enhanced plant performance and reliability through sophisticated modeling techniques and systematic analysis of plant operations. Central to this paradigm shift is the concept of asset effectiveness, a multifaceted measure encompassing factors such as equipment reliability, operational efficiency, and overall production capacity.

The integration of hybrid modeling or data-driven modeling approaches, which combine the strengths of first principle models, empirical models, and predictive analytics is becoming a vital approach to improve asset effectiveness. By harnessing the power of these diverse methodologies, operations stakeholders are empowered to not only detect equipment constraints and deterioration in real-time but also to make informed, proactive decisions aimed at optimizing asset effectiveness and ensuring operational reliability. Such approaches are applicable in the ethylene production industry, where the complexities of process dynamics demand innovative solutions for maximizing efficiency while maintaining stringent reliability standards.

This paper explores the intricacies of enhancing asset effectiveness through harnessing data analytics in the form of hybrid modeling in ethylene process. Drawing upon industry-specific insights and real-world case studies, it delves into the methodologies, applications, challenges, and future directions of these transformative approaches. Through a comprehensive analysis, we aim to underscore the pivotal role of predictive analytics in driving tangible economic gains and operational excellence in the ethylene process industry. As we navigate the complexities of modern industrial operations, the integration of advanced modeling techniques emerges as a cornerstone in the pursuit of sustainable, high-performance ethylene manufacturing.

2 Asset Effectiveness

Let’s delve into the concept of Asset Effectiveness within the context of the ethylene processing industry. Asset effectiveness can be defined as a function/measure of productivity over time. It is a multifaceted measure that encompasses factors such as equipment reliability, operational efficiency, and overall production capacity.

“Asset Effectiveness is a function of productivity over time.”

- ❑ **Asset effectiveness →** Defines how well the efficiency and productivity of assets are utilized to achieve operational goals and deliver value to the business.

- ❑ **Productivity →** Tracking productivity over time provides insights into asset performance and effectiveness.

Asset effectiveness hinges on a series of critical factors that collectively shape its performance and contribution to operational goals.

Critical Factors influencing Asset effectiveness:

- Maintenance practices: strategies and procedures employed to ensure reliability.
- Operational processes: efficient utilization of assets within given constraints and operational challenges.
- Automation and Technology: leveraging advanced technology in terms of design, control, operation, and maintenance.
- Organization skills and expertise: developing knowledge work force well trained in implementing best practices, addressing issues, and maximizing asset effectiveness.
- **Data Analytics and Monitoring**:
  - → Leveraging data analytics and real-time monitoring systems can provide valuable insights into asset performance and condition.
  - → Predictive analytics, condition monitoring, and performance KPIs enable proactive maintenance strategies and optimization efforts to enhance asset effectiveness.

An essential aspect of understanding asset effectiveness is through evaluation. This can be quantified by utilizing the Overall Asset Effectiveness (OAE) metric as a reliable benchmark.

2.1 OAE Metric

The Overall Asset Effectiveness (OAE) metric serves as a valuable benchmark for evaluating efficiency and productivity at both plant and equipment levels. Calculated as the product of Throughput, Availability, and Yield, the OAE metric provides a thorough assessment of asset performance, pinpointing areas for process improvement. Especially when we compare actual performance to design, historical best, and ideal.

Overall Asset Effectiveness (OAE) = Throughput X Availability X Yield

Key Calculations:

- **Throughput**: (Actual Capacity/Ideal Capacity)
- **Availability**: (Operating Time/Planned Time)
- **Yield**: (Real Efficiency/Desired Efficiency)

<u>**Key Benchmark Comparisons:**</u>

- **Design**: Actual vs Design
- **Historical Best:** Actual vs Previous Best
- **Ideal:** Actual vs Ideal

By utilizing the OAE metric and its key calculations, organizations can identify areas for process enhancement, streamline operations, and reduce downtime. Organizations can also benchmark performance, identify improvement opportunities, and optimize asset utilization to achieve operational goals and deliver value to the business. The implementation of the OAE metric is facilitated through software solutions or manual data collection methods, enabling organizations to track productivity over time and gain insights into asset performance and effectiveness.

While effective multi-modeling is crucial, it must be complemented by a robust data collection process. This data is then transformed into actionable information and knowledge using models, visualization, and dashboards to aid interpretation and early detection. Subsequently, an implementation and tracking process ensures the realization of benefits. This holistic approach has proven highly effective in enhancing the health of critical equipment in ethylene units, resulting in:

- Extended mean time between machine overhauls/replacements, thereby prolonging T/A intervals.
- Increased efficiency and overall effectiveness, leading to augmented plant production capacity, tailored to the severity of operating constraints.

2.2 Key Components of an Effective Overall Asset Effectiveness (OAE) Solution

A. Data Collection and Integration:

Effective OAE solutions begin with robust data collection and integration from various sources, ensuring quality and reliability. Post-collection, tools reconcile, validate, and consolidate data, feeding it into machine learning models to derive useful analytics maximizing asset effectiveness.

B. Model-Based Information and Knowledge Generation:

Leveraging models and analytics, organizations develop insights and knowledge from collected data. Advanced algorithms identify patterns, predict issues, and highlight improvement opportunities, enhancing asset performance and efficiency.

C. Visualization and Dashboards for Interpretation:

Smart visualization tools translate complex data into easily understandable formats, avoiding information overload. Interactive dashboards provide insights tailored to critical areas, aiding decision-making, and operational optimization. Alerts highlight areas needing attention.

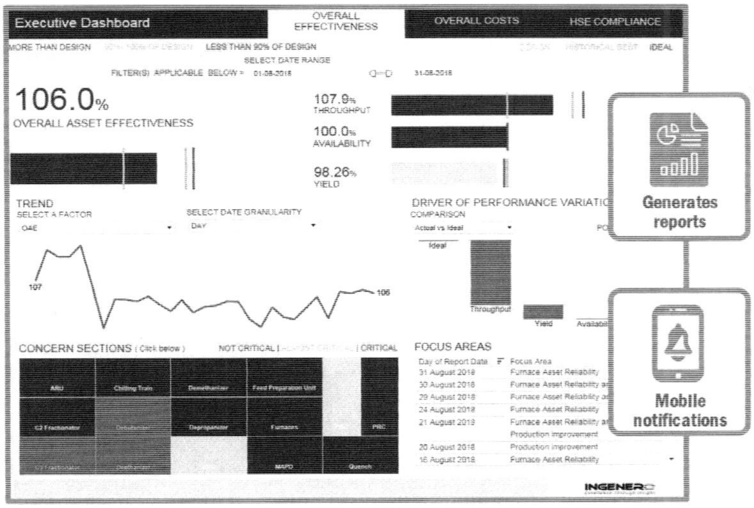

Enhancing Data Quality:

- Elimination of data errors from raw data

Leveraging Generative AI:

- Utilizing model-driven KPI benchmarks and insights

Interactive Smart Dashboards:

- Featuring interactive date range selection
- Color-coded indicators highlighting degree of criticality
- Drill-down capability for deeper analysis

Real-time Benefit Tracking:

- Conducting live gap analysis "Actual vs Ideal" with degree of change "Profit"

D. Implementation and Tracking for Benefit Realization:

OAE solutions aim for tangible benefits by implementing recommended strategies derived from data analysis. Establishing mechanisms for tracking and monitoring changes ensures desired benefits are realized and sustained, fostering performance enhancement and operational efficiency.

Integrating these components enables organizations to harness data and analytics effectively, optimizing asset performance and driving continuous improvement.

The importance of following a comprehensive approach to implementing an OAE solution is key and must look to, encompassing data collection, model-based insights, visualization tools, and robust tracking mechanisms. By incorporating these key elements, organizations can optimize asset performance, streamline operations, and drive continuous improvements in asset effectiveness within the ethylene processing industry.

3 Hybrid Modeling

Applying analytics to gain real-time insights is a formidable undertaking, especially within the intricate domain of ethylene processing. Considerations such as enhancing furnace yields and capacity for economic viability must be balanced against maintaining furnace availability, including optimal run length and coil life. The dynamic nature of yield benchmarks is crucial for effectiveness, as plants may not meet the same benchmarks under varying conditions, such as seasonal changes or differences in cleanliness levels. Additionally, ensuring controller effectiveness, capturing instrument drifts for stable operation, early identification of equipment problems, and predictive maintenance to ensure reliability and continuous optimal operation are paramount. Incorporating considerations of economics and profitability further underscores the complexity of process optimization, where opportunities must be identified within defined boundaries encompassing current economics, process capability, equipment health, and safety.

Relying solely on individual models often proves insufficient in consistently achieving success across diverse scenarios. However, integrating data science methodologies with first-principle models holds promise in optimizing ethylene processes to their fullest potential. By simultaneously considering multiple objectives such as recovery, capacity, reliability, and economics, this hybrid approach aims for comprehensive process optimization. Therefore, the integration of hybrid modeling techniques is essential to overcome challenges and drive improvements in asset effectiveness.

To ensure success, adopting the right approach and utilizing appropriate technology are critical. An intelligent analytical solution offers real-time decision insights to operations, driving actionable outcomes and fostering new levels of operational excellence. The key lies in integrating the right analytics tools and customizing a solution that delivers tangible value.

By combining Augmented Intelligence, Machine Learning Techniques, and first-principle models, organizations can tackle challenges and enhance operational efficiency and gain valuable, non-intuitive insights often concealed within large data volumes in various ethylene processing areas. This practical application sets the stage for exploring specific applications of Hybrid Modeling in ethylene production processes to drive operational excellence, maximize asset performance, facilitate troubleshooting, addresses complex operational issues, and provides augmented intelligence to optimize operations.

4 Ethylene Process Practical Applications

a. Improving Ethylene Furnace Performance

The performance of Ethylene furnaces is dictated by various parameters such as throughput, yield, run length, coil life, and energy efficiency. However, enhancing one aspect often comes at the expense of others. Leveraging Big Data analytical techniques has emerged as a promising approach to optimize furnace operations while balancing these criteria effectively. Big Data analytics offer insights into finding optimal settings to meet operational objectives. Yet, for Ethylene furnaces, successful implementation requires integration with the first principle reaction kinetic models. While empirical data models excel in predicting instrument failures and managing availability issues, they may face challenges in accurately replicating furnace performance across different feedstocks and conditions. First principles models, on the other hand, provide a deeper understanding of relationships and can closely replicate performance. The choice between empirical data models and first principles models depends on the complexity and time constraints involved in model development and maintenance. By combining these approaches, it becomes possible to enhance production and run length effectively.

Furnace Run Prediction Models: Optimizing Operations for Improved Performance

In industrial furnace operations, the rate of coking in furnace coils plays a pivotal role in determining various parameters such as Coil Inlet Pressure (CIP), Tube Metal Temperatures (TMTs), and Coil Outlet Pressure (COP). These parameters, in turn, significantly impact the furnace's run length and overall performance. Accurate prediction of these parameters over time is crucial for effectively managing furnace operations, optimizing performance, and maximizing profits.

Traditionally, operational decisions in furnace optimization have been made based on standard operating guidelines, often relying on the judgment of operational and technical experts. However, effectiveness of these methods can vary greatly and may not always consider all parameters affecting run length. To enhance the reliability of this process, predictive models can be employed to forecast key furnace parameters such as CIP/COP, TMT, and even factors like ethylene yield.

Predictive modeling serves as the foundation for identifying the rate of coking, differential coking, and expected run length of the furnace, aiding in scheduling decoking and planned maintenance activities.

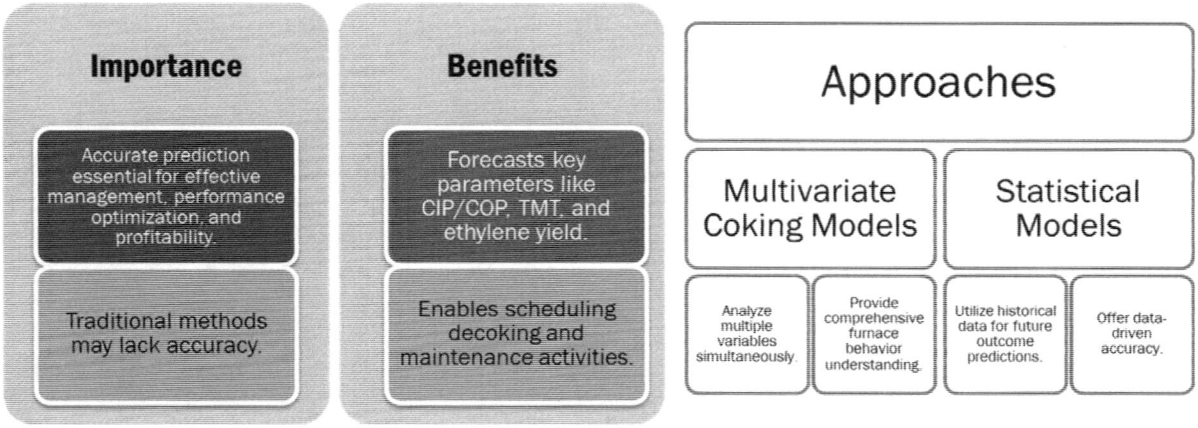

Two main approaches, either used individually or in combination, are commonly employed for this purpose: Multivariate coking models and Statistical models.

I. Multivariate Coking Model:

The multivariate coking model integrates furnace kinetics with advanced algorithms and statistical techniques to estimate coking rates accurately. The model predicts the actual coking rate, which depends on various factors such as gas phase reactions, feed composition and rate, conversion, dilution steam, and process conditions.

While traditional kinetics models have limitations in accurately predicting coke deposition, the multivariate coking model addresses these challenges by considering dynamic operating conditions and adjusting parameters based on actual furnace data. This model undergoes two main steps: data tuning for coking rate estimation and prediction based on coking, active sites, and thickness. Benefits of this approach include timely actions to address constraints indicated by the model, such as adjusting conversion or feed rates, burner adjustments, and fuel gas pressure, ultimately extending furnace run length and optimizing performance.

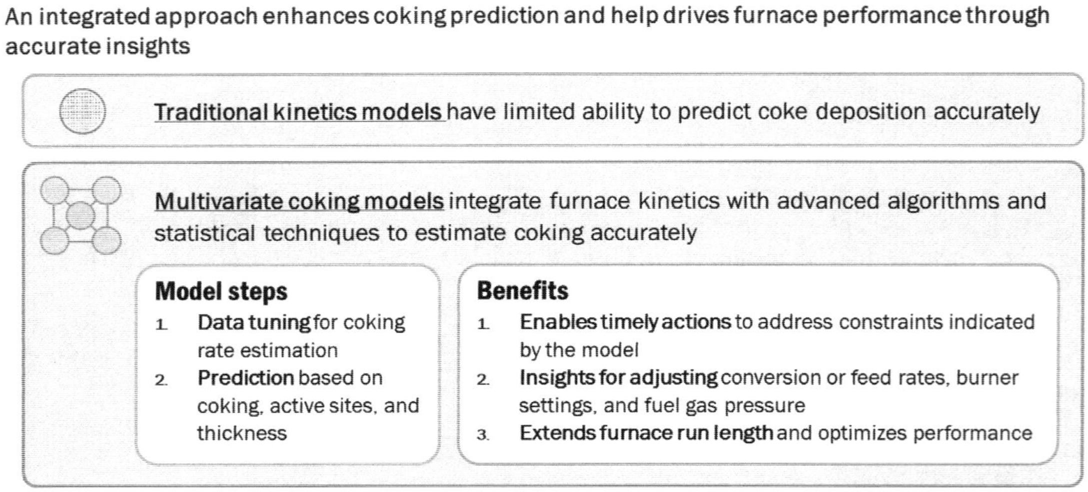

II. Statistical Model:

The statistical model for predicting CIP relies on developing correlations between CIP and other furnace parameters, such as feed flow, specific heat capacity (SHC), and conversion, based on current operating conditions. As the furnace run progresses, the model refines these correlations, while requiring a minimum initial dataset for reasonable accuracy, with additional data further improving results.

This model assumes stability in furnace parameters like feed rate, conversion, and SHC ratio, predicting CIP rise over the forecast period. However, it can adapt to changes in plant operations by incorporating updated data and redefining correlations, albeit requiring a minimal dataset of changed conditions for accurate predictions.

Applications and Value - Statistical optimization approaches using coking models offer immense value in various scenarios, including improving or achieving desired run lengths, maximizing production based on decoke schedules and economic factors, and optimizing furnace operations based on downstream unit constraints.

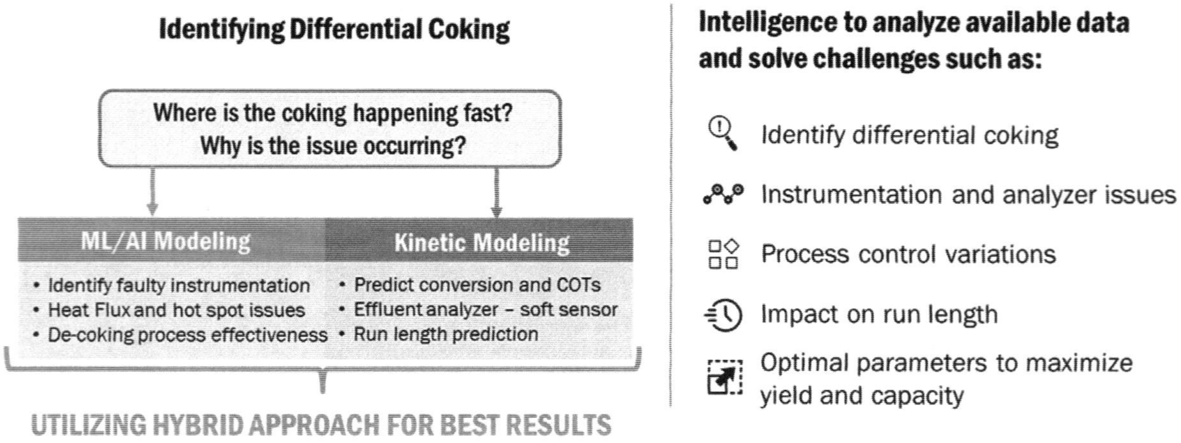

Critical adjustment variables to avoid differential coking for furnaces include firing, draft, flow, feed balancing, and dilution steam. Understanding their impact on various furnace parameters such as coil inlet/outlet pressure, tube metal temperature, and firing duty is essential for optimal operation. Integrating first principles and predictive modeling using Big Data techniques enables accurate performance prediction throughout the furnace's operational life.

The synergy between Big Data analytics and first principles modeling offers a comprehensive approach to optimize Ethylene furnace performance, leading to enhanced productivity and operational efficiency. While outcomes may vary based on feedstock availability and downstream limitations, significant production increases ranging from 5% to 20% can be achieved in ethylene units without any major CAPEX investment.

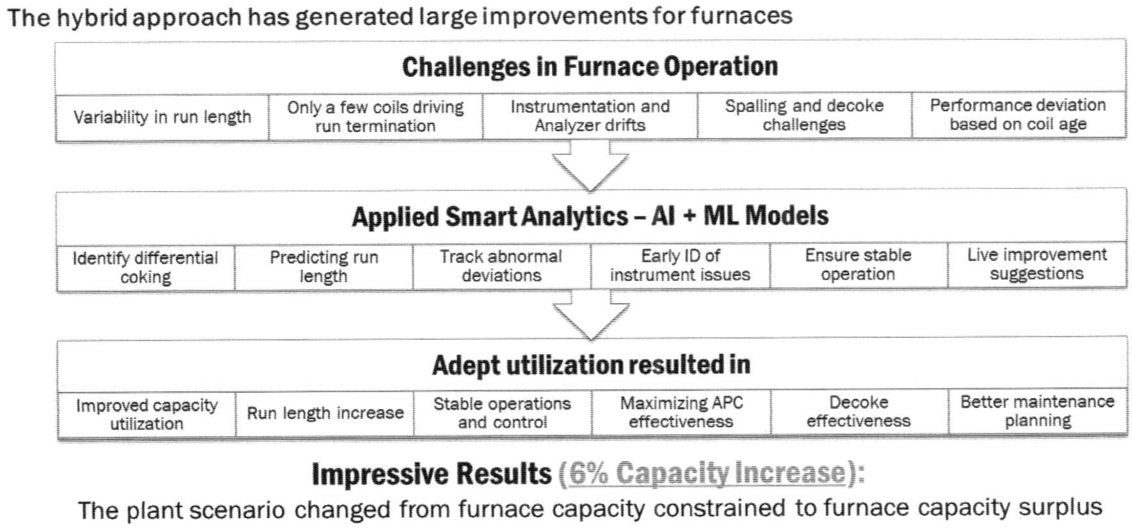

Convection Models

Convection section models are instrumental in monitoring the convection section of the ethylene process.

These models aid in improving thermal efficiency, evaluating individual bank performances, identifying fouling and flue gas channelling issues, and determining parameter settings for maximum safe production.

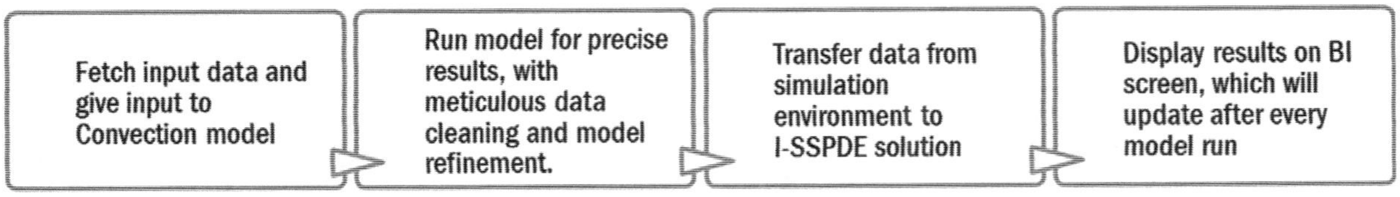

Flowchart for automating convection model runs for live insights to furnace operation.

Benefits of Convection Models:

- ***Thermal Efficiency Improvement***: The models help in optimizing the thermal efficiency of the convection section, leading to energy savings and reduced emissions.
- ***Evaluating Individual Bank Performances***: By assessing the performance of individual banks within the convection section, operational inefficiencies can be identified and rectified.
- ***Identifying Fouling and Flue Gas Channelling***: Detection of fouling and flue gas channelling issues is crucial for maintaining optimal performance and preventing operational disruptions.
- ***Determining Parameter Settings***: Setting parameters correctly ensures maximum safe production while adhering to operational and safety standards.

Convection Models play a key role in enhancing operational efficiency within the ethylene industry and through monitoring can lead to reduced energy consumption, lower emissions, improved run length, and enhanced safety measures. By leveraging these models effectively, ethylene plants can optimize their processes and achieve sustainable operational excellence.

b. Exchanger Degradation – Tracking with Soft Sensor

Exchanger is one of the key assets in any operating unit and an important aspect to ensure overall process performance improvement. Optimization of exchanger operation is critical and is feasible through application of right predictive and prescriptive analytics techniques. Exchanger performance troubleshooting can be a tedious process as multiple operating and design factors affect the actual performance. At certain times it would require smart design improvements, but at times options are available to restore the desired performance as much as possible through intelligent process adjustments. Ensuring optimal exchanger operation will require complete understanding of the design, process complexities and overall functioning of the equipment.

Soft sensor framework to model degradation in equipment, primarily focusses on the core idea to develop a multivariate model that learns the correlation among predictor variables and between predictor and dependent variable(s) (fouling indicator). The soft sensor is trained to identify a deviation in the process induced particularly by fouling, muting the effect of various kinds of other normal process variations. Further, the degree of deviation between current and clean data (without fouling) is calculated to assess the severity of fouling.

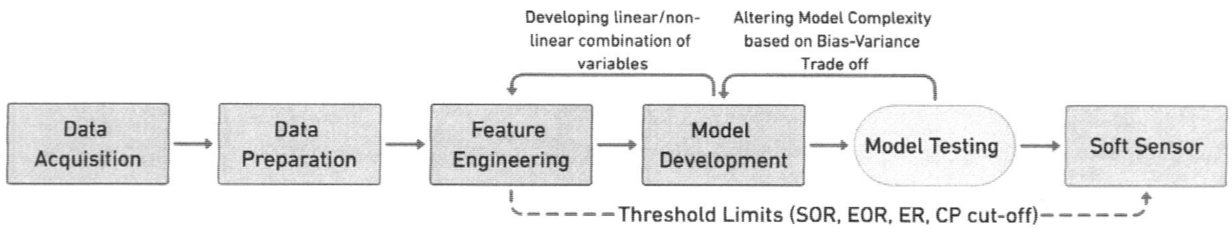

Soft sensor development methodology (general schema)

Model Development: Clean data is divided into train and test sets and multivariate regression analysis is performed on train sets to capture system characteristics. Based on model performance on test sets combined with bias-variance trade-off analysis, optimal model complexity is determined, while achieving the desired accuracy.

The complexity in the model can be introduced in the form of nonlinearity at two levels: developing non-linear combination of variables during feature engineering or building a non-linear relationship between the predictor and dependent variable(s), or a combination of both. Generally, model development iteration starts with an assumption of linearity and then based on performance, model complexity is increased.

Ethylene Plant - De-Butanizer Reboiler Example

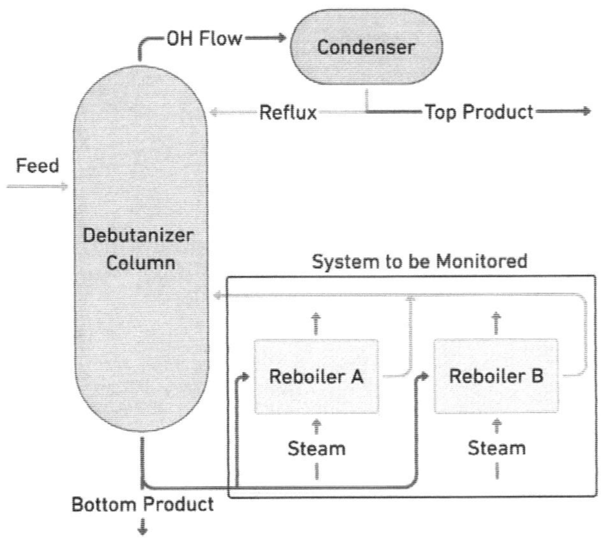

Process Variables	Description
LP Steam Flow	Flow Rate of Steam into Reboiler
LP Steam Pressure	Captures saturation level and heat transfer capacity
LP Steam Battery Limit Temperature	Alternate variable to capture steam inlet T
Column Feed Flow I	Primary Feed Stream to Debutanizer
Column Feed Flow II	Secondary Feed Stream to Debutanizer
Column Bottom Temperature	Proccess inlet T to Reboiler
LP Steam Valve Opening	Dependent Variable

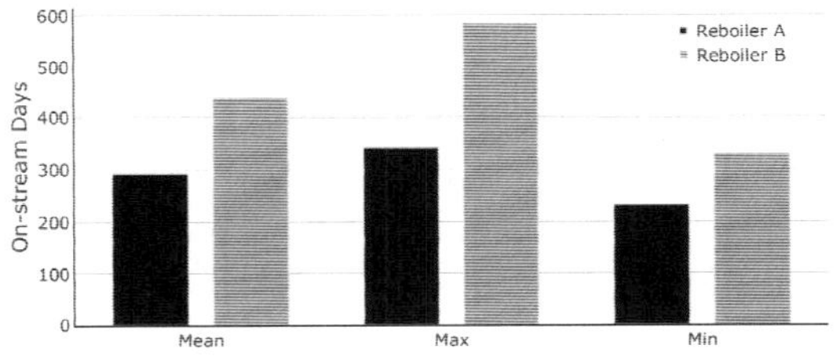

Figure: (Left) Block diagram of the process; (Top Right) List of independent and dependent variable(s) considered for soft sensor development; (Bottom Right) Run length comparison for exchanger A and B

The discussion centers on Reboiler B runs, analyzing DOF (Degree of Fouling) with Soft Sensor output. Thresholds (SOR, CP cut-off, EOR, and ER) are set based on statistical estimation and process understanding. DOF reflects equipment fouling progression qualitatively, aiding comparison within a system's historical runs. However, cross-system comparison isn't feasible.

Real-time monitoring relies on DOF, rate of DOF, and comparison plots during the Clean Phase (CP). As the system enters the Fouling Phase (FP), RUL prediction becomes crucial alongside fouling indicators. A windowed curve fitting approach is employed, selecting the best-fit curve based on mean squared error. The trend line extends to intersect EOR and ER thresholds, with a 95% confidence interval indicating prediction uncertainty. Continuous RUL prediction by the Soft Sensor aids in timely changeover planning, preventing unnecessary financial losses.

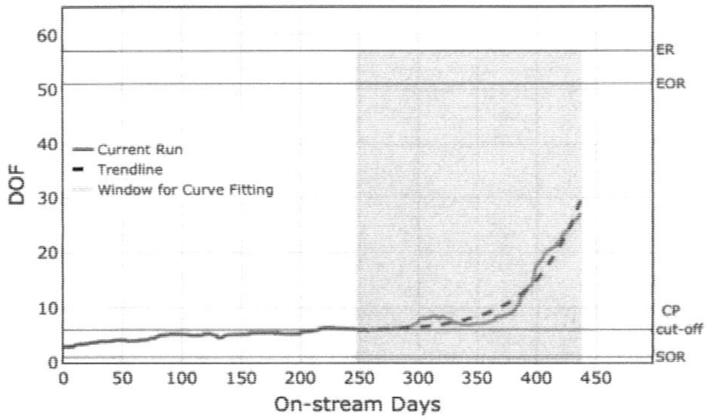

Figure: Window selection for curve fitting

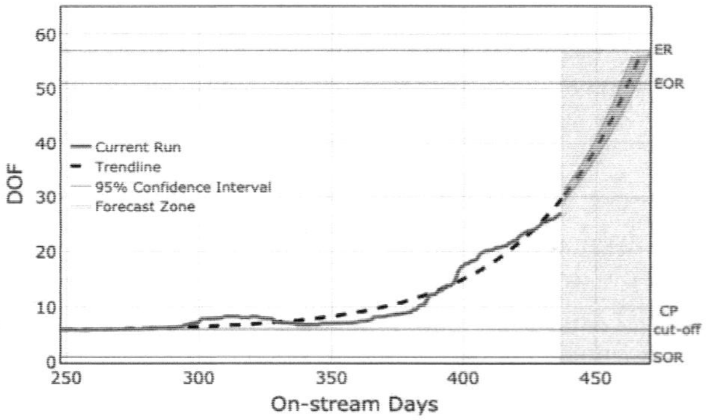

Figure: Forecasting RUL for EOR and ER conditions

c. Improving Compressor Reliability and Capacity Utilization

In the realm of ethylene operations, the health and efficiency of compressors play a pivotal role in ensuring smooth and cost-effective production processes. To address this, a comprehensive Compressor Health Monitoring System is required, focusing on the real-time assessment of compressor efficiency using advanced analytics and machine learning techniques.

The monitoring system primarily targets could be two key components: Fuel Gas Compressor (FGC) and Refrigeration Compressors. These critical assets are central to the operation of various ethylene processes, making their continuous monitoring imperative for optimal performance.

Efficiency Calculation:

The cornerstone of the monitoring system lies in the real-time calculation of compressor efficiency. By leveraging operating data such as temperatures, pressures, and feed processing parameters, efficiency is derived using a first-principle model. The model incorporates the K-value, a factor dependent on molecular weight and conversion maintained in the furnaces. This approach enables a dynamic assessment of efficiency across varied operating conditions, ensuring adaptability to changing ethylene process parameters.

Efficiency Degradation Tracking:

To pre-emptively identify potential issues, machine learning techniques are employed to track efficiency degradation patterns. Critical insights are derived, including alert generation based on consistent drops in efficiency over a predefined duration or exceeding defined thresholds. This proactive approach enables early intervention, thereby minimizing downtime and maximizing operational efficiency.

Deviation Tracker:

A deviation tracker is used to monitor critical operational parameters that could impact compressor efficiency and reliability. This includes inter-stage cooler temperature differentials, discharge temperature and pressure ratios, as well as mechanical parameters like vibrations and bearing temperatures. While mechanical impacts are not directly analyzed, early indications of potential mechanical issues are provided, allowing for timely maintenance interventions.

Overall Benefits:

The Compressor Health Monitoring System offers a multitude of benefits to ethylene operations:

- Early Issue Identification: Potential issues are identified in their nascent stages, allowing for prompt corrective actions.
- Improved Run Length: By addressing issues proactively, the system enhances the interval between shutdowns, optimizing operational uptime.
- Process Optimization: Insights derived from efficiency calculations and deviation tracking facilitate process optimization, driving overall operational efficiency.

In the relentless pursuit of operational excellence, the implementation of a Compressor Health Monitoring System stands as a testament to the convergence of cutting-edge technology and ethylene industry pragmatism. By embracing real-time analytics and machine learning, ethylene plants can safeguard their critical assets, minimize disruptions, and unlock new levels of productivity and efficiency. As we continue to refine and expand upon this monitoring framework, the horizon of possibilities for ethylene process optimization grows ever broader.

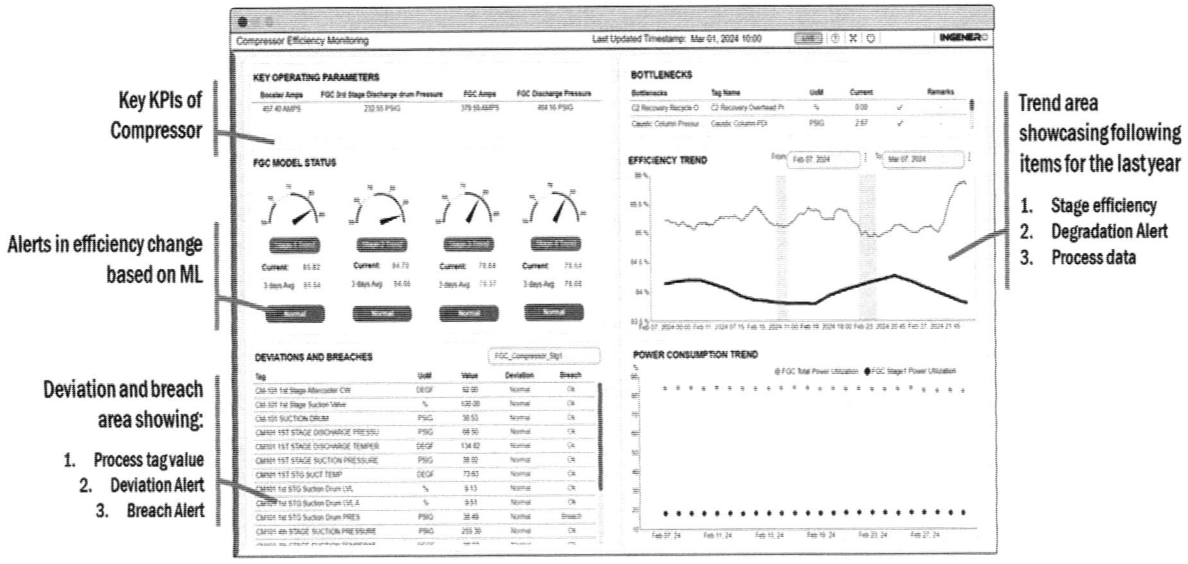

Feed Gas Compressor Performance Monitoring Display

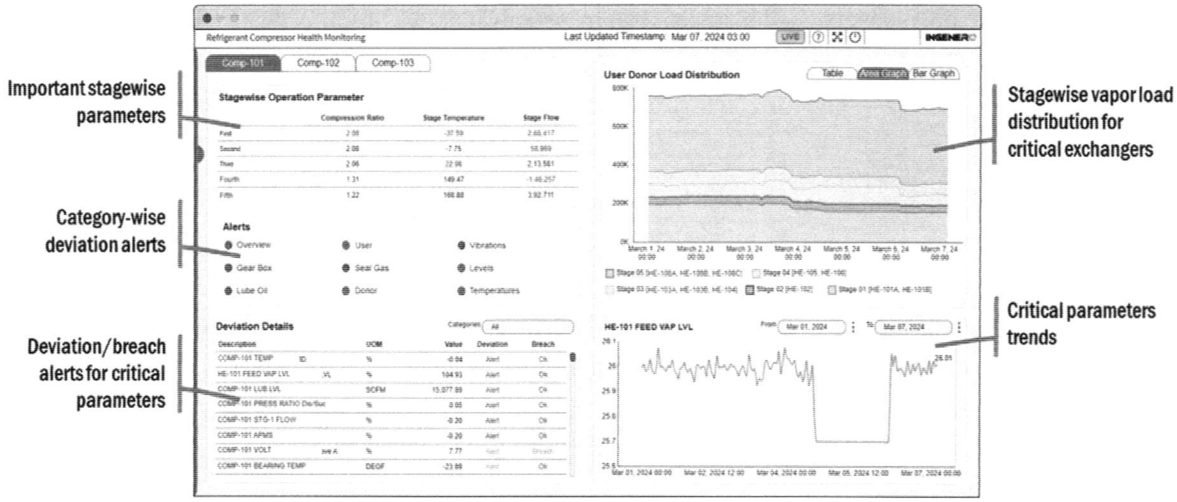

Refrigerant Compressor Performance Monitoring Display

d. Other Ethylene Applications

Showcasing the application of hybrid models in some sample areas within the ethylene industry.

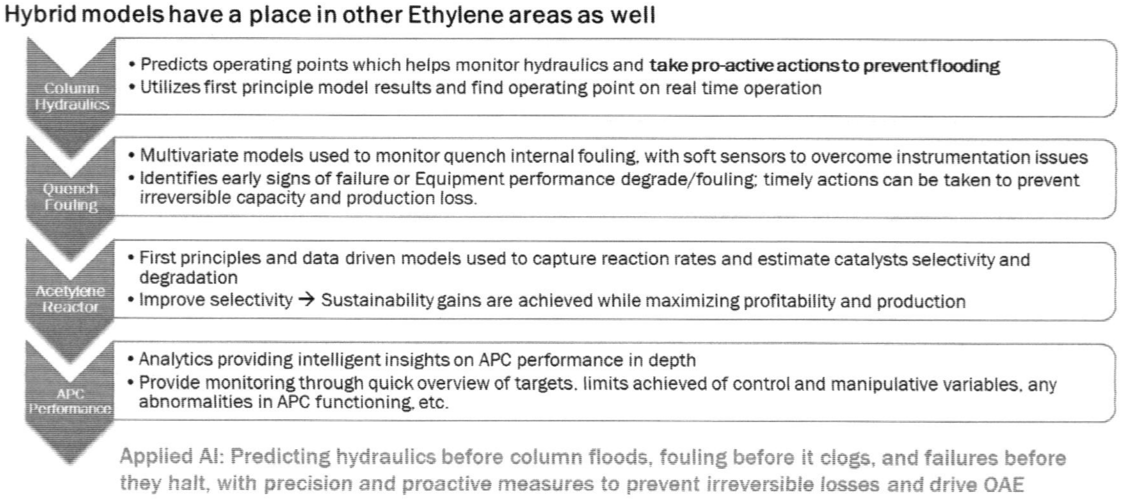

The objective is to focus on precision and proactive measures to prevent irreversible losses and enhance Overall Asset Effectiveness (OAE) through predictive modeling and anomaly detection.

5 Challenges and Considerations

Addressing the challenges and considerations in implementing a comprehensive Asset Health Monitoring System within ethylene plant operations entails navigating several key hurdles. Firstly, understanding the intricate physical processes and available data is paramount. Ethylene production involves complex interactions and variables, necessitating a deep understanding of the underlying process and the data generated by the operational processes. Secondly, integrating hybrid modeling techniques into existing plant operations poses a significant challenge. The seamless integration of advanced analytics and machine learning algorithms into traditional operational frameworks requires careful planning and execution to ensure compatibility and effectiveness. Addressing reliability issues and implementing robust performance tracking mechanisms is essential. Ethylene plants operate under demanding conditions, where equipment reliability is crucial for maintaining production efficiency. Developing strategies to track performance reliably and identify potential reliability issues proactively is critical for sustaining optimal plant performance and minimizing downtime. Successfully overcoming these challenges is instrumental in harnessing the full potential of an Asset Health Monitoring System within ethylene production facilities.

6 I-SSPDE Application

Ingenero's I-SSPDE solution, utilizing hybrid modeling as its foundation, has been successfully implemented by manufacturers worldwide, including within ethylene plants. It has played a pivotal role in empowering manufacturers globally to address challenges and elevate operational efficiency. A few illustrative examples include:

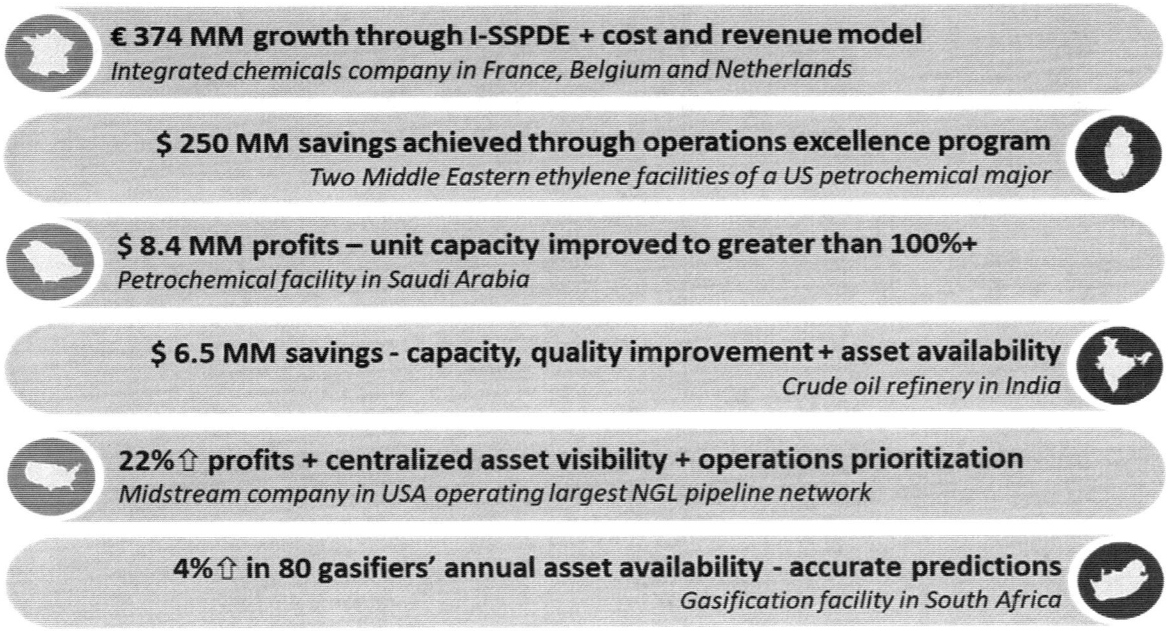

7 Key Takeaways

Equipment health and reliability plays a critical role in overall ethylene plant operation strategy and consistently maintaining high production rates. Enhanced equipment availability and reliability can be achieved through effective data analytics by proper interpretation, analysis, and use. Knowledge acquired through data can be further use for managing current problems and for determining strategy for the future. Application of data analytics through modeling techniques and systematic monitoring of equipment health is effectively utilized to identify anomalies and solutions to improve equipment reliability and performance.

The importance of Hybrid Modeling as a comprehensive approach to optimizing asset effectiveness in the ethylene processing industry cannot be overlooked. By integrating AI, ML, and first-principle models, organizations can leverage a powerful combination of modeling techniques to address complex challenges, drive improvements in asset performance, and achieve operational excellence.

The key components to this include:

- **Hybrid Modeling**: The integration of various modeling techniques, such as AI, ML, and first principle models, has proven to be extremely effective for ethylene plants. By combining these modeling approaches, raw data is transformed into actionable information and knowledge, facilitating interpretation, early issue detection, and informed decision-making.
- **Visualization and Dashboards:** Utilizing intuitive visualization tools and dashboards aids in effectively interpreting data insights. These tools enable swift action and response to operational challenges by providing a clear representation of complex data.
- **Implementation and Tracking:** Establishing robust processes for implementing recommendations derived from analytics and tracking their outcomes is crucial. This ensures that the tangible benefits of the hybrid modeling approach are realized and sustained over time, leading to improvements in the health of key equipment in ethylene units.

Results

Successful application of the above has significantly helped in improving the health of key equipment in ethylene units resulting in:

- Longer mean time between machine overhauls/ replacement → extending T/A intervals
- Improving efficiency and overall effectiveness → increasing plant production capacity depending on the severity of operating constraints

In conclusion, the transformative impact of hybrid modeling in ethylene plants, showcased how the integration of data analytics and first principles simulation can drive operational excellence, improve asset effectiveness, and optimize plant performance and reliability.

8b Improving TMT Accuracy Using IR Camera on Ethylene Cracking Furnace

Stephan Hazzam, Karsten Schulze

Tube Metal Temperatures (TMT) of radiant section coils form an important parameter in the operation of the ethylene cracker furnace as the maximum reachable value is one of the criteria to start a decoking cycle. TMT increases over the furnace run length due to a coke layer build-up. Initiating decoking too early would reduce both furnace (and thus plant) availability and coil lifetime (due to material stress during decoking). Overheating coils on the other hand would lead to material degradation and in worst-case to a direct coil rupture. Hence knowing about TMT development over time is essential to ethylene plant operations. State-of-the-art is still the manual measurement using a pyrometer and logging by some manually maintained lists. Accuracy and reproducibility of the measurement depends very much on the individual and current operating conditions during the slot of measurement.

An alternative approach is the use of thermal cameras which is presented in this paper. The technology allows the 24/7 monitoring of the TMT for the significant parts of all coils. Temperature profiles are taken every five minutes, allowing the reliable monitoring even in the light of varying process conditions. Based on the amount of data an improved prediction of the remaining furnace run time is achieved, and the generation of an expected decoking schedule is possible. In addition, different performance parameters can be retrieved, indication different heat loads on different coils or fast changing temperatures. Last but not least the captured data allow the automatic detection of hot spots and other abnormalities in the temperature distribution.

The paper explains on the implemented features and summarizes the experience. Also some guidelines are given for the implementation of this promising technology.

8c Operational Excellence in Ethylene Production Leveraging Digital Solutions

Ana Khanlari, Phd

To run an ethylene plant longer, safer, and greener digital solutions play a crucial role. In today's market conditions with low or negative margins, digital solutions are specially needed to improve overall profitability and sustainability of the operation. Digital solutions can be deployed to optimize the performance of cracking furnaces or light ends fractionation, monitor pumps or cracked gas compressors, monitor, and optimize overall plant's energy consumption, or calculate and report associated emissions at an enterprise level. In this presentation we will review several real-life examples of applying digital solutions in ethylene production and the value they have created.

SPRING24
+20TH GCPS
A Joint AIChE and CCPS Meeting

Effectively using AI/ML and APC in Ethylene

James Brigman
Ingenero Inc.
Principal and Managing Director
4615 Southwest Frwy., Suite 320
jbrigman@ingenero.com

Prepared for Presentation at
American Institute of Chemical Engineers
2024 Spring Meeting and 20th Global Congress on Process Safety
New Orleans, LA
March 24-28, 2024

Effectively using AI/ML and APC in Ethylene

James Brigman
Ingenero Inc.
4615 Southwest Frwy., Suite 320
jbrigman@ingenero.coim

Keywords: Ethylene, Digitalization, AI, ML, Digital Twin, APC

Abstract

This paper describes the use of AI/ML and APC in Ethylene production. The advantages of using APC versus AI/ML algorithms in offline and online monitoring programs is discussed. The benefits of each approach is described and the approach that Ingenero has found to be the most effective in implementation is then laid out.

1 Introduction

Simulation

Application of data analytics and simulation to address problems in petrochemical and refining operations has been an option for over 40 years. Practically all chemical engineers graduating after 1980 are exposed to first principles simulation while getting their degree. Rice University, for example, utilized Chemshare during the 1980's and then eventually switched to AspenTech's tools. Use of steady state simulation to understand the theoretical capabilities of the processing unit where they are working, the operation of distillation columns, or the sizing of heat exchangers has been a common application of simulation for engineers for the last 40 years and continues to be. Use of statistical analysis of performance using tools such as statistical process control to control manufacturing operations started around 100 years ago and was popularized by W. Edwards Deming in the 1950's and was utilized effectively through the 1980's particularly with operations still utilizing analogue controllers.

Simulation software vendors have continued to improve their products to make them easier to use with significantly better visualization and processing speed. For the last decade completed simulations of a unit or even individual equipment is referred to as a "digital twin". Having a converged digital twin of every unit in a petrochemical plant or refinery is the basic building block to understanding the performance of equipment such as reactors, furnaces, heat exchangers, distillation columns, and other separation and treatment equipment. Digital twins can be used effectively to establish theoretical capacity, optimize process design and troubleshoot issues.

However, despite the availability, the potential benefits, and the efforts of first principles software vendors to simplify the building of digital twins, effective use of the software is still questionable. Ingenero was founded on the idea that the process industry was not effectively using simulation and data analytics to its potential and that it should. Ingenero's goal was to be the best process engineering firm by applying technology to improve the petrochemical and refinery industry. From the start, it was shown that aggressively applying simulation and data analytics to petrochemical process (including Ethylene plants) to determine performance gaps and their solutions were capable of generating 20% increase in production on a regular basis. Now these initial units were probably given to Ingenero as known problem areas but the validity that utilizing digital twins and data analytics can be effectively utilized to improve operations is a given. Even today, as it was 20 years ago, most manufacturing facilities do not have a functional digital twin of their facility that is utilized. More often than not, when Ingenero goes into a facility to assist with performance improvement there is not a digital twin in place or, even after assurances that a good digital twin is in place, finding that the simulation provided is "broken" (i.e. unable to converge) and has not progressed beyond an assemblage of models around individual pieces of equipment.

AI / ML:

First principles simulation is based on iteratively solving mathematical representations of thermodynamic relationships which are typically in the form of partial differential equations. While some simpler process facilities such as for Gas Processing plants can utilize dynamic first principles models to get a real time analysis of the operation, more often than not first principles models are still too cumbersome and slow, despite the impressive gains in computing power over the last 20 years. As a result real time use of digital twins that iteratively solve partial differential equations is still often not an effective solution.

However utilizing multivariate statistical analysis and other tools to produce variable relationships that can be modeled with reduced order magnitude equations. These with the increase in processing speed and the ease of accessing years of operational data, allow data analytics methodologies including artificial intelligence and machine learning techniques to be applied. The insights from this analysis can be impressive.

Soft sensors developed on difficult to measure parameters where sensors are not available such as product composition. These calculated parameters can be utilized to provide a descriptive metric of where the plant or equipment is operating. In addition soft sensors can also be utilized to provide a prediction of where the plant or equipment will be in several hours based on the current settings. The descriptive and predictive nature of the soft sensor provides a powerful tool by providing information previously not available in real time and an understanding of the current situation and where the unit is going over the next few hours if changes are not made.

Ingenero regularly builds visual displays of the deethanizer, depropanizer, and the C2 and C3 splitters. Compositions and flooding can be modeled using so that operators can see graphically how these columns are operating.

Though large optimization is also possible in the furnaces by using AI and ML techniques. As far back as 2004, Ingenero was using Artificial Neural Net models of the furnace to optimize operations to achieve significantly improved runlengths while increasing the cracking severity. Currently using a combination of first principle modeling and the predictive fouling models, Ingenero is able to get detailed analysis typically on a coil by coil basis of the their condition and changes needed in dilution steam, fuel gas, and feed gas to achieve planned production targets while minimizing down time from decoking.

The profitability for a furnace limited plant can be absolutely impressive.

Graphic displaying $250MM over a 3 year period

However still the use of these tools remains limited despite the fanfare of digitalization and the wealth of tools developed. It is not easy to develop tools for ongoing operations. Estimates are are routinely given that digitalization efforts fail up to 80% of the time. This just shows that while it is possible, it is not easy. The quote by Oscar Wilde about experience is quite appropriate. “Experience is the hardest teacher. It gives you the test first and the lesson afterwards”.

APC:

Advanced Process Control represents a very successful application of multivariate statistical analysis that is widely used. APC utilizes a range of techniques to provide multi-variable control. However, the basis is that within a set of related process variables, relationships are established. Multivariate analysis allows the prediction of product properties in real time. This has been termed Model Predictive Control. The task of pushing performance of certain variables to defined optimal points within the defined constraints of the system can be automated. This is a closed-loop solution that takes the operator out of the task of pushing the operations.

Where there is a consistent feed composition and flow, APC can be particularly beneficial in reducing operating variability and pushing the operating unit to optimize on a particular variable such as optimizing production. Understanding where there is opportunity to utilize APC in addition to the existing DCS system based on setpoint and control algorithms can drive significant value capture.

Ingenero takes the approach with its clients to first identify where the opportunity for improvement is highest utilizing digital twins and data analytics in an offline manner. Then the degree of difficulty in terms of cost and time is considered. Getting the highest return in the shortest possible time and expense should be understood. APC is one solution that can be effectively applied in many situations to do this. In a situation where operator and engineer intuition are not part of the equation, automating the solution as part of the APC is the right move. In these situations, APC can be a relative inexpensive solution for achieving large returns.

(Two by Two Graphic showing difficulty to implement versus benefit)

The distinction that should be noted is that simulation and data analytics provide the analysis and insight while APC is a operations control methodology useful for capturing the value where appropriate. As its name implies, it is process control and not analytics to provide insight. It is a solution.

If Ingenero can find a way with the DCS system to take operational variability out, then that is approach that will be taken. However where APC offers advantages, Ingenero will help its clients with the installation and maintenance of the APC. Stabilizing operational variability provides difficult-to-measure, but meaningful, benefits such as less equipment wear and operator dependence that by themselves are difficult to track as a direct cost.

As with digital twins and machine learning applications, oversight is necessary to maintain appropriate functionality. APC, as with digitalization solutions, is reliant on soft sensors (i.e. inferential property estimation). Changes in equipment and feedstock can have significant impacts that can lead to suboptimal performance. Typical APC systems have tools to help monitor performance. Ingenero will monitor the parameters tracked within the APC monitoring tools (i.e. ranging from time the APC system is being utilized to time spent at specific set operating limits) and also continuing to compare unit performance post implementation to pre-installation and through the use of first principles simulation to thermodynamic optimal. As needed Ingenero can retune the APC parameters to achieve optimal results.

How should these tools be used effectively:

When presenting digitalization as solution to improve Ethylene operations, it is not atypical to hear either from the potential clients' APC team that "digitalization is the poor man's APC" or from the client's continuous improvement team that "digitalization is like APC on steroids". Neither answer is correct since they are distinct and while their Venn diagrams overlap they really have difference purposes.

APC is closed loop and digitalization is open loop. AI and ML applications can certainly be built to be closed loop, but as part of a digitalization solution it is open loop application of the algorithms meant to provide insight to the operations staff.

APC is great for reducing operational variability and pushing variables towards a defined optimal limit and works well within a reasonable production range. Of course it can likewise be argued that AI and ML data analytics techniques are inherently limited to the range of historical operational data. This can also be a limited range. Ingenero has found that data augmentation through the use of first principle simulation can be extremely effective even if the choice of model/algorithm applied needs to change based on specific range of application.

Ingenero has applied data analytics utilizing AI and ML in both an offline and an online manner. An offline manner is typically utilized more from the standpoint of comparing current operation to theoretical optimal and then studying the gaps. Data analytics along with simulation can be used to determine then root causes and the needs to be addressed. The online digitalization tools provide extremely rapid feedback and can compare current operation to best ever operation at the same constraints. This gives real time feedback on adjusting operating parameters. This gets a bit more comparable to purpose of APC but is providing feedback and education to the operational staff. It is using artificial intelligence to provide augmented intelligence.

In digitalization solutions for an ethylene plant AI and ML algorithms have been effectively applied for opportunity identification for profitability, reliability and sustainability. In particular

AI/ML tools implemented in Ethylene facilities as solutions that are able to address:

- Optimizing furnace runlength, yield, and coil life for increased production
- Improving the reliability and performance of key equipment including the quench tower, cracked gas compressor, and various distillation columns
- Equipment and instrumentation reliability
- Establishing awareness of current operational performance in terms of profitability and sustainability versus theoretical or best ever performance
-

Why the failures:

Digitalization and the application of AI/ML tools are in vogue and where properly applied are generating enormous value. Getting the right solutions in place is seen as necessary to compete. Yet despite this belief, successful implementation of AI/ML tools is estimated to be occurring less than 30% of the time and even then not performing up to expectations.

Based on looking at the use simulation and other statistical tools, this should not be surprising. Attempting to treat digitalization, its implementation, and its use in the same manner as simulation will result in the same limited use and applicability despite its tremendous potential. The idea that building digitalization tools is a task that can be handed off to unit engineers is the equivalent to expecting them to build, maintain, and actively use a converged digital twin of their unit. It is a rarity.

Treating AI/ML digitalization implementation and use more similarly to APC implementation can significantly increase successful application and use. Specialist are employed to handle the task separately with ongoing maintenance expected.

Recommendations:

1. Understand the urgency and value proposition:

Start with the understanding that the inability to produce to maximum design levels is an incredible loss of revenue to your firm. Production lost now is not able to be made up later. Each day of less-than-optimal production is dollars that do not go to your company. Achieving market share within the industry is a function of being able to produce the most possible at the lowest price. Lowering specific energy costs and enhancing yield is paramount and time matters. For a large petrochemical manufacturer or refiner, knocking $100MM out of production costs or dropping this amount to the bottom line is certainly possible. Ingenero in just applying a hybrid first principles and ML algorithms to deliver a digitalization solution around approximately 140 single major pieces of equipment achieved over $100MM in production gains. At a PE ratio of 20 this corresponds to an increase in worth to the shareholders of $2 Billion.

2. Have a Digital Twins on every unit:

If you do not have a digital twin, this is the first step to understanding theoretical capacity and debottlenecking your facility. Building and utilizing digital twins should be a first step on the digitalization journey. This is an extremely useful tool for pushing unit and equipment performance and troubleshooting operational problems.

However just having a digital twin is not enough. Getting value out of a digital twin requires a level of familiarity and user capability or it is just a virtual representation of the existing plant that can just as easily be understood by looking at the DCS screens. Unfortunately, if you do not have a team that has had the time to build a digital twin, you will find that you don't have a team that has the time to effectively utilize a digital twin regardless of how good the simulation is.

3. Apply the DMAIC Process:

While the need to understand the urgency and value proposition was mentioned earlier, it is the first step of the DMAIC process. Defining the value proposition is key doing this throughout an ethylene unit will generate a list of potential projects with a matching list of the value potential. While the largest value proposition will typically occur around the plant bottleneck constraints, the degree of difficulty and expenditure also needs to be considered in the approach.

However once the value proposition is understood, getting the data together to analyze is necessary. This is the "Measure" step. The importance of getting the compilation and cleaning of this data should not be underestimated. Throwing all the variables into a machine learning model does not typically improve the results but instead throws them off.

The Analysis is application of the AI and Ml algorithms and the understanding of the relationships. Having the domain expertise on the areas of application is critical to getting the right variables in your analysis. Making sure that the handles to solving the problem are understood. Applying optimization modeling to the appropriate variables may what is required.

The Improve step is something that varies by the online versus offline approach. The online approach is one of digitalization. The offline approach may utilize AI and ML but the delivery of the information is different. Credibility and belief by the operations staff is absolutely critical to getting implementation. No matter how good the solution built is; if it does not get used then it is a failure.

The Control step on APC and AI and ML is handled by monitoring and maintenance of the solutions built. Tracking the ongoing value capture and having scheduled retraining is critical for long term survival of any digitalization or data analytics solution.

4. Have a Digitalization Roadmap and make headway:

Digitalization is not easy. Knowing the value to be chased is critical for both APC and digitalization. The "build It and they will come" approach may make for cute movie but even then it depended on a fantasy coming to life. Don't have this approach with your digitalization efforts. Be realistic but aggressive and chase the value and not the blind implementation of applications or seemingly innovative tools. Configuring tools to the needs of the units is needed. The roadmap should be around the insights needed by unit type and specific value captures needed.

5. Provide Management and Personnel Transparency:

Digitalization provides the opportunity to drive accountability and transparency. This is a big part of getting use out of tools. Operations pays attention to what management pays attention to. Data analytics should provide results and action items that can be rolled up and presented to management. Transparency all the way up will drive success.

6. Don't think that you can do it with just internal resources:

The road to success is paved with the failures of previous attempts. Why do you want to be the person or internal group that must go through a hundred failures to get to the win? We live in a world where life is a bit easier than in the past. You can be a specialist and hire others to do the other tasks. You can hire someone who has done it before, understands the skill sets needed, and knows the difficulty and time required. Otherwise this a matter of seeking to be hero by believing that eventually your simulation and modeling skills will be sufficient for you to provide the solution. If the time is available this may occur assuming that you are in the position long enough and other obligations allow it. However the time required and the production lost in the interim is huge.

❑

34a Flare Operation Basics

Vance Varner

The operation of flares for ethylene production plants has changed dramatically in the past 15 years. Following the recognition that there were high emissions from many flares on a continuous basis, there has been multiple data collections, equipment testing, and new federal regulations.

This paper will cover the basics of flares – types / operation considerations / public perspective. In addition, there will be discussion on what technology direction companies are probably using for new installation. There will be discussion of what technology should / can be used for replacement of existing equipment. Finally, some potential pitfalls as the E-MACT is now in place.

Flare systems are now being, or should be, treated as another piece of process equipment that requires monitoring, maintenance, and proper operating discipline.

34b MULTI-POINT Ground FLARE Design and Operation in a Modern Ethylene Production Maximum Achievable Control Technology World

John Bellovich

New regulations imposed on Ethylene Production Facilities present new challenges in the areas of design, operation, and environmental compliance, especially with respect to Multi-Point Ground Flare (MPGF) systems. In 2020, The EPA enacted modified rules for Ethylene Production Maximum Achievable Control Technology (EMACT) to minimize pollutants emitted from these facilities as defined in the National Emissions Standard for Hazardous Air Pollutants (NESHAP). The revised requirements of the EMACT prohibit venting and flaring for all but emergency situations. Reduced emergency flaring leads to more use of what are known as "early stages" of the MPGF system. This increased use of these early stages causes more wear and tear than seen before in the same applications. Other challenges include an increased desire for facility owners to eliminate negative environmental impacts and operate the facility effectively. Efficient operation of the MPGF requires a comprehensive understanding of all potential relief cases, an intelligently designed MPGF, and continuous operator involvement. A properly designed MPGF will allow operators to continue operation through all flaring scenarios. Additionally, it is important for the MPGF system to light reliably and effectively when required. Owners may retrofit older MPGF systems to meet current requirements. Changes may be as simple as adding monitoring devices but could be as complex as adding pilots, burners, and additional stages. This article will provide a comprehensive review of the EMACT requirements and how they pertain to the operation and maintenance of a MPGF system for reduced emissions as well as how EMACT pertains to other flare systems.

34c Design and Operation of Ethylene Flares in the Post-Emact Era

Kevin Leary

With the publication of the Generic Maximum Achievable Control Technology Standards Residual Risk and Technology Review for Ethylene (EMACT) under NESHAP regulations in March 2020 the requirements for the design and operation of flares in the ethylene production industry in the US changed significantly. Compliance with the new EMACT requires operators to significantly increase the monitoring, control and reporting associated with all flares in the industry. These requirements are similar to those imposed on the Refining Industry in the Refinery Section Rule in 40CFR 63.670 which was published in December 2015. This paper will present some techniques that have been used to improve ethylene flare operation in the post-EMACT era.

The most significant change required from a flare operating standpoint is the addition of a new operating parameter called the Net Combustion Zone Heating Value or NHVcz. The NHVcz is defined as the net heating value of the flare gas just prior to combustion including the dilution effect of assist steam or assist air that is injected into the flare tip. The assisted flare must include automatic control of supplemental fuel gas to maintain the NHVcz above 270 BTU/SCF at all times.

Additionally, the EMACT required the preparation of detailed flare management plans designed to minimize flaring and to minimize any non-smokeless operation of each flare. This flare management plan established published smokeless capacities for each ethylene flare that must be maintained at all times. Any smoking event on a flare can trigger a violation of the standard requiring detailed root cause analysis and corrective actions.

Depending on the capacity, size, operation, and smokeless capacity of an ethylene flare it can be very difficult to maintain both the smokeless operation and the NHVcz within the guidelines at all times. And some operating conditions can initiate automatic supplement gas addition into the flare header that can greatly increase the utility costs associated with the flare system.

Many operators have attempted to minimize the amount of steam injected at low flaring rates in order to prevent excessive supplemental fuel usage required to maintain the minimum NHVcz required by the rule. Operating with insufficient cooling steam on a steam assisted flare tip (or cooling air on air assisted flare tip) leads to internal burning within the body of the flare tip that can cause or accelerate heat damage to the flare tip resulting in more frequent maintenance or requiring complete replacement of the flare tip.

Many design and operational changes have been developed by flare designers and operators in order to optimize the operation of ethylene flares in the post-EMACT era. This paper will present several such design and operational changes that have been used effectively to optimize ethylene flare operation. These changes / recommendations will be grouped into several different categories that have been used including:

- Major redesign and/or retrofit of the flare burners to optimize the flare burner design and improve flare operations.
- Minor retrofits to existing flare burners to improve flare operation.
- Control strategies (with or without burner retrofits) for improved flare operation.

34d Lessons Learned and Challenges Overcome with Operating Flares Under the New Emact Requirements

Eric Swisher

The U.S. Environmental Protection Agency (U.S. EPA) finalized amendments to the National Emission Standards for Hazardous Air Pollutants (NESHAP): Generic Maximum Achievable Control Technology Standards for Ethylene Production, otherwise known as "EMACT" on July 6, 2020. The EMACT revisions required affected facilities to comply with additional monitoring and compliance assurance for flares, including the implementation of flare monitoring and operational requirements of 40 CFR Part 63 Subpart CC (National Emission Standards for Hazardous Air Pollutants from Petroleum Refineries) or "Refinery MACT". Specifically, the new monitoring and compliance assurance requirements included the installation of continuous parameter monitoring system (CPMS) hardware (e.g., vent gas flow monitors, mass spectrometer, calorimeter, etc.) and the software (i.e., the means for managing the data collected by the hardware for purposes of process control and compliance demonstration) as well as additional flare operational restrictions. This paper will focus on the challenges and lessons learned for the operation of flares after July 6, 2020 in order to comply with the EMACT.

7a Decarbonizing Olefins Production

Michael Tallman

Most companies, including olefins producers, have announced ambitious goals towards carbon neutrality. Nearly all new olefins projects include requirements to reduce or eliminate carbon emissions, even though the production of olefins is a highly energy intensive process.

Accordingly, all technology licensors are working on carbon reduction technologies and strategies. In this talk various features, both commercialized and in development, will be discussed which can be applied to new crackers or as modifications to existing crackers to reduce or eliminate carbon emissions.

Additionally, carbon reduction strategies for other olefins-production technologies outside of steam cracking will also be discussed.

Finally, the potential to integrate blue / green hydrogen or blue / green ammonia facilities with olefins plants will be covered.

7b Carbon Tax Impact on Ethylene Plants

Mark Whitney

When governments start charging for every pound of carbon dioxide emitted, how will that affect the ethylene industry? That is more than idle speculation. Canada has already enacted carbon price legislation that effectively puts a tax on carbon dioxide emitted during the production of materials. The European Union is putting in a cap-and-trade system for certain products that will have the same impact as a carbon tax. Even if Washington does not add a carbon tax, every ethylene plant doing business in areas with such a tax will need to be cognizant of the implications of its carbon dioxide emissions on the final price of its products.

To put the issue in perspective, the paper will estimate the taxes using current Canadian regulations and speculate where carbon taxes worldwide are heading in the future. The impact of the carbon price on US ethylene producers will be discussed. Various methods of reducing the carbon tax will be examined using currently available technologies. Finally, a vision of the future of carbon taxes on ethylene producers will be shared.

PAPER# 7C

ADVANCES IN PURIFICATION OF REFINERY OFF-GAS STREAMS-INCLUDING A NEW LEAD-FREE ARSINE ADSORBENT

Rodrigo J. Lobo, Lead Engineer, Honeywell UOP
Stephen Caskey, Sr. R&D Manager, Honeywell UOP
Allan Hatami, Project Development Manager, Honeywell UOP
David Radtke, Principal Applied Sales Engineer, Honeywell UOP

Abstract

Refinery off-gas streams derived primarily from fluid catalytic cracking units (FCCUs) and Coker units have historically been sent to fuel gas systems, but these light hydrocarbon rich streams can be a cost advantaged feed stock for steam crackers (ethylene plants). Prior to being fed to the fractionation section of a steam cracker, removal of a variety of contaminants is required to meet ethylene plant product specifications or result in downstream deactivation of selective hydrogenation or polymerization catalysts.

Adsorbents are cost effective and efficient means of removing many of these problematic contaminants including arsine, mercury, chloride, oxygenates, sulfur compounds and more. The adsorption systems can be operated in either regenerative or non-regenerative mode. Honeywell UOP with its extensive adsorption technology expertise, field service capabilities, broad adsorbent product portfolio and commercial experience in refinery off-gas treatment, can offer solutions to these purification demands.

Honeywell UOP has developed new materials for application in refinery-off gas streams. A lead-free arsine removal adsorbent has been patented. The new arsine removal adsorbent, free of environmentally problematic lead oxide, has been field tested at a major olefins producer in Asia and is now available for use.

Background

As the chemical industry tries to maximize margins (and yield) of petrochemical products, the use of cost advantaged feeds has become common place. Feeds that in the past were considered fuel gas, are now being processed to recover valuable light components such as hydrogen, and simple olefins. In a refinery one such stream is the Refinery off gas (ROG)

Refinery off-gas typically originates from high temperature cracking processes where hydrocarbons are converted —in the absence of significant hydrogenation (such as FCC, delayed Coker). This results in a stream containing a high concentration of light olefins as well as hydrogen. Table 1 shows typical major components in an ROG stream.

Table 1: typical concentration in of refinery off gas

Component	Concentration (mol%)
Ethylene	34
Hydrogen	30
Ethane	20
Methane	8

Differences in processing schemes and the variety of feedstocks that generate the ROG lead to a wide range of contaminants (and contaminant concentrations), these contaminants can have an outsized impact on downstream processing. Table 2 shows some of the contaminants in ROG and their impact.

Table 2: typical contaminants found in refinery off-gas, and cracked gas.

Contaminants	Concentration range	Impact
Water	100s-1000s ppm	Freezing in refrigeration systems, detrimental to catalysts and adsorbents.
Acid gases (H_2S, CO_2)		Catalyst poison, corrosion
Organic sulfur (mercaptans, sulfides, disulfides, COS, CS_2)	1-100's of ppm	Product specifications, corrosion, or catalyst poisons
Mercury	100's of ppb	Can structurally compromise aluminum heat exchangers
Oxygenates		Coke formation in adsorbents and catalysts, corrosion, water formation in reactors, deactivation of catalysts
Arsine, phosphine and stibene	100's of ppb	Poison to polymerization catalysts at 0.1 ppm levels can reduce the productivity by 70%

The large variation in contaminant profiles requires that the treatment solution be tailored to each plant, and at times to each feed.

Although less contaminated, cracked gas (from steam crackers) shares similar bulk characteristics (such as high olefin, and hydrogen concentration) and similar contaminant challenges, hence, many of the solutions used for ROG, can be also used for Cracked gas.

Adsorbents for contaminant removal

The use of adsorbents is well suited to the variability in ROG and cracked gas streams, since they can be customized for the unique needs of a given process stream, and in some cases with marginal impact on the overall cost of the process.

The two major classes of adsorption process are regenerative, and non-regenerative. Non-regenerative processes frequently use materials that chemically react with the contaminant, removing it from the process stream (e.g., H_2S with a copper adsorbent producing CuS) or converting it to a more acceptable species, such as CO to CO_2. Once the adsorbent's capacity is reached, it needs to be replaced with a fresh batch of adsorbent.

To avoid performing frequent, and potentially costly, adsorbent reloads while the plant is in operation, non-regenerative adsorbent needs to have enough capacity to last multiple months or even years. Therefore, they are typically used in applications where the contaminant is at low levels (<5 ppm) (e.g. Chloride removal).

When the contaminant concentration is too high, use of non-regenerative adsorbents may become impractical (due to high reload frequency, or very large beds), a regenerative adsorbent process is required (e.g. dehydration of water saturated streams). In these processes, once the adsorbent is saturated with the contaminants, it is taken off-line and placed under new conditions that provide a driving force to return the adsorbent to its initial state. The regeneration can be effected by changes in temperature, pressure and/or by using auxiliary chemicals. Regeneration of the adsorbent introduces additional equipment to the process (compressors, heaters, etc.) making it more capital and utility intensive, compared with the non-regenerative processes. Furthermore, in most cases as the adsorbent is regenerated, the contaminant will be rejected (at much higher concentration) and may need to be further treated (e.g., rejected H2S stream can be treated by caustic washing).

Unique challenges of Refinery off-gas treatment

The high olefin and hydrogen content, along with the presence of highly unstable molecules (such as Acetylene), enables ROG streams to undergo a variety of reactions, that, if left unchecked, can be detrimental to the effectiveness of the adsorbent, economics, and safety of the process. The primary examples of concern are:

$$Polymerization/coke\ formation:\ nC_xH_{2x} \xrightarrow[acid\ catalyzed]{} C_{nx}H_{2nx}$$

$$Hydrogenation:\ C_xH_{2x} + H_2 \xrightarrow[Metal\ catalyzed]{} C_xH_{2x+2}$$

$$Metal\ reduction:\ MO_x + xH_2 \longrightarrow M + xH_2O$$

Unwanted polymerization is a concern in any olefinic streams, and any stream containing Alkynes, (and even some oxygenates), since it could result in the deposition of coke over the adsorbent. This deposition can lead to diminished performance, product quality problems (e.g., color in polymers), and deterioration of the mechanical integrity of the adsorbent, which can result in an increased pressure drop. In extreme cases, the coke deposition can be severe enough to lead to dangerous exotherms or even cause mechanical failure of vessels (Figure 1).

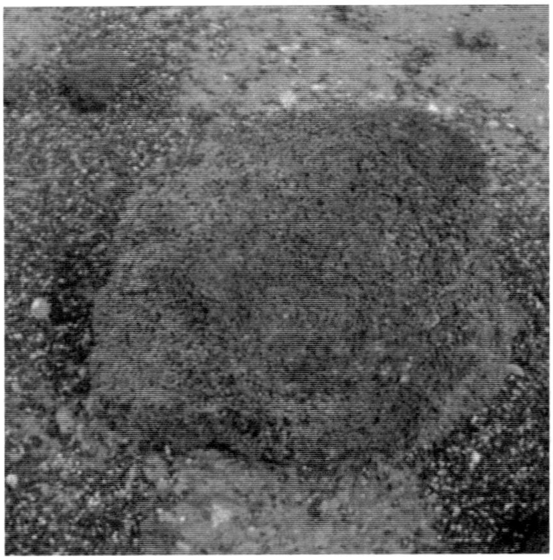

Figure 1: agglomerate of adsorbent found after vessel failure due to runaway uncontrolled exotherm.

These reactions are typically catalyzed by the presence of acid sites (even weak ones) on the surface of the adsorbent; hence if an adsorbent is to be used for ROG treatment careful control over the adsorbent design and manufacturing is necessary.

Due to its exothermic nature, Olefin hydrogenation and metal reduction can lead to a thermal runaway if left unchecked. The thermal runaway can in turn lead to feed polymerization. For example, if hydrogenation is allowed to fully occur with the composition shown in table 1 could produce an exotherm in excess of 500 °C.

Additionally, from a production cost perspective hydrogenation is undesired, as it converts valuable feedstock components into byproducts.

Transition metal oxides are ubiquitous in the contaminant adsorption field (e.g., CuO for the removal of H_2S), thanks to their ability to form resilient metal compounds (such as CuS or Cu_3As). However, in the presence of high concentration of hydrogen these metals can reduce to the metal and become active for hydrogenation, while releasing a large amount heat. This behavior limits the options to only the more oxophilic metals.

In the sections below we describe two applications in which the unique features of ROG have become critical, and strongly influence the adsorbent design.

Treatment of acid gas and other polar contaminants

Two common types of contaminants in ROG streams are Acid gases (H_2S, CO_2, and COS) and polar molecules (including oxygenates and organic sulfur species). Removal of these contaminants would traditionally involve the use of two adsorbents in a compound bed. The first, a zeolite type adsorbent; and the second a promoted alumina adsorbent. With this approach one type of contaminant (polar molecule or acid gas) is effectively removed in each section.

Despite having high selectivity for polar contaminants, the zeolite type adsorbent also interacts with olefins, producing large exotherms when a regenerated bed is first exposed to olefins. A behavior which requires special procedures to pre-load the regenerated beds, prior to being put in operation. This results in longer regeneration times and additional costs. Thus, an alternative approach is desired.

Hybrid adsorbents mitigate the temperature rise due to the heat of adsorption by the addition of another material, such as an amorphous adsorbent, in the formed particle, which acts as a heat sink to absorb some of the heat released, and significantly reduce the exotherm in the bed. A lower exotherm enables elimination of a pre-loading step, thereby resulting in safer operation.

Despite the high affinity of these materials for polar compounds, some hybrid adsorbents can be designed to have extremely low reactivity towards olefins as is the case for UOPs AZ adsorbents. This polymerization reactivity can be readily measured using an olefin reactivity test (Figure 2).

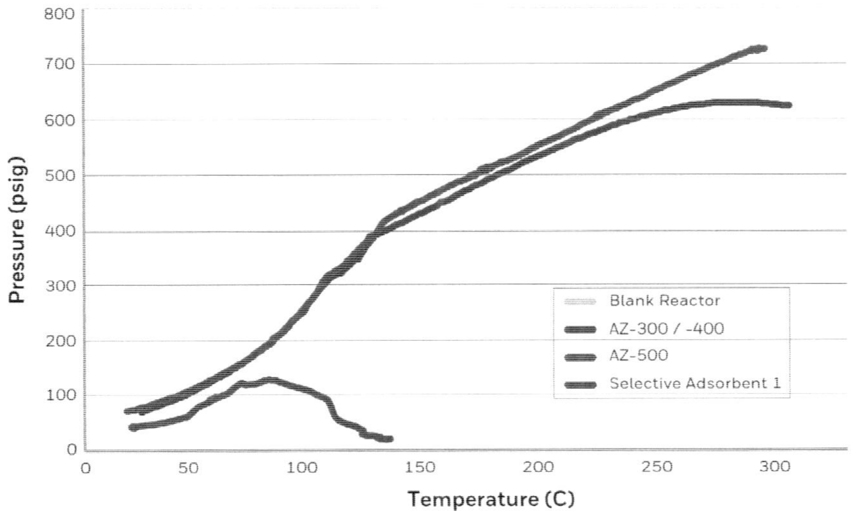

Figure 2: Results from batch isobutylene reactivity tests of hybrid adsorbent

Figure 2 shows a reactivity test in which the adsorbent and isobutylene are loaded into a closed vessel. The pressure is monitored as the temperature increases. As long as no polymerization occurs, the pressure will continue to rise with temperature. When the pressure begins to decline, it indicates that the isobutylene is being consumed due to polymerization. The higher the polymerization temperature the less reactive a material is, and therefore, less prone to coke formation. Thus, the higher polymerization temperature of the AZ-products shows greater resistance to coking than other commercial materials.

In addition to laboratory tests, Hybrid adsorbents with low reactivity, such as the AZ- family) have been tested to show lower coking tendency (Figure 3).

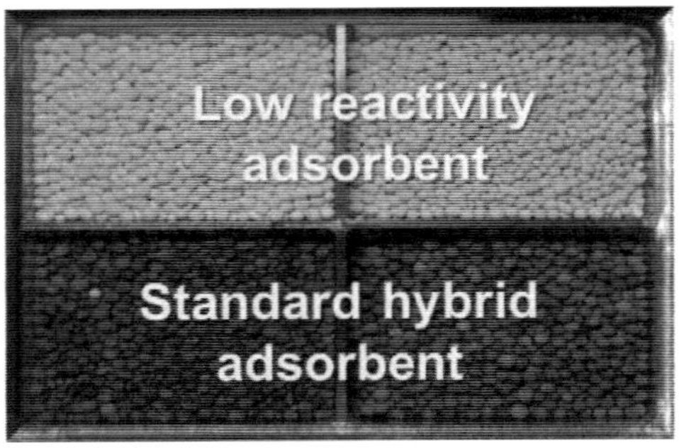

Figure 3: adsorbents after 24h of high temperature exposure to propylene. (top) Low reactivity adsorbent (used for ROG, AZ-300), (bottom), standard hybrid adsorbent.

UOPs AZ- family of low reactivity adsorbents are now installed in more than 190 plants worldwide (including in the facilities of major olefin monomer producers and polymer plants).

Developing the next generation hybrid adsorbents

Despite hybrid adsorbents having dual function, some ROG streams contain a higher amount of acid gases than the first generation of hybrid adsorbents were developed for, leading to shorter cycle times or use of compound beds with promoted alumina, resulting in shorter cycles and consequently shorter adsorbent life. Hence, the next generation of hybrid adsorbents (such as AZ-500) is intended to tackle the higher acid gas content. Figure 5 shows a comparison between two adsorbents, and the impact of using the higher acid gas capacity adsorbent operated under high acid gas removal conditions.

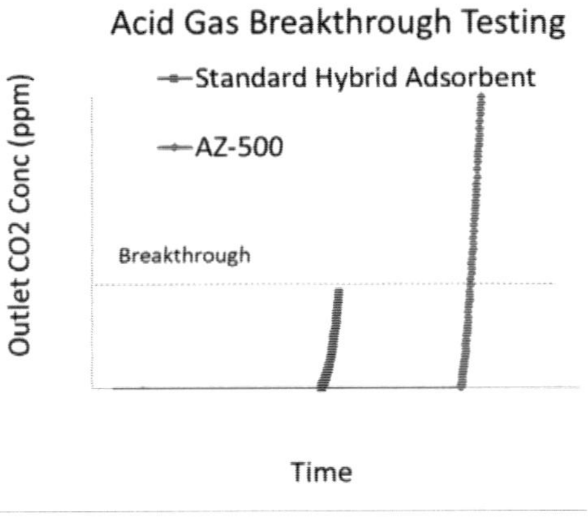

Figure 5: shows CO2 breakthrough curve of the AZ-500 and a standard hybrid adsorbent+ modified alumina adsorbent.

As shown in Figure 6 the less frequent regeneration directly leads to reduced adsorbent aging, and thus longer adsorbent lifetime.

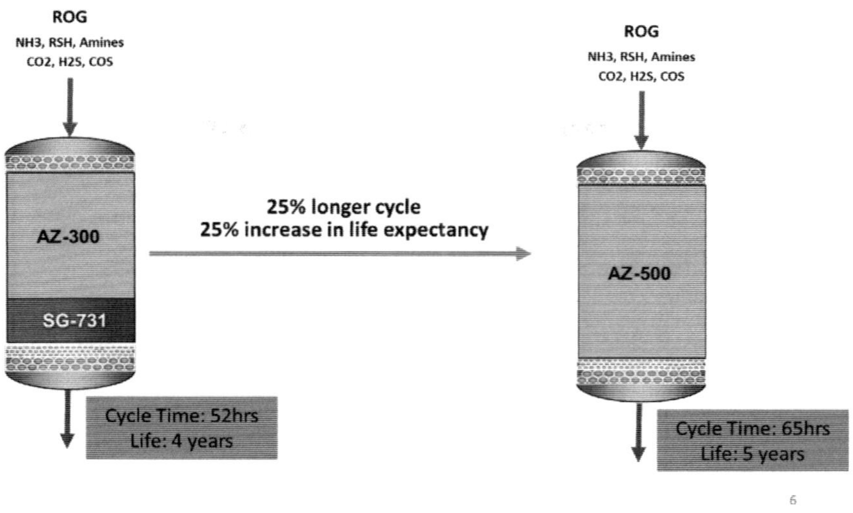

Figure 6: impact of switching to an adsorbent with higher Acid gas capacity for a ROG contaminant removal.

Arsine removal in highly reactive environments

Arsine is a highly volatile metal hydrate that is often present in ROG and some cracked gas streams due to the presence of arsenic compounds in the feedstocks processed. This chemical constitutes a very powerful poison to many of the catalysts used in the production of polyolefins[1]. The effect is so strong that even 500 ppb of arsine can have a measurable impact on the performance of polymerization catalysts. For many ROG streams it is imperative to remove AsH_3.

In streams that do not have significant levels of hydrogen, the primary choice for Arsine removal are copper oxide-based adsorbents, this removal is effected by reaction 1.

$3\,CuO + 2\,AsH_3 \rightarrow Cu_3As + As + 3\,H_2O$...............(1)

However, in the presence of hydrogen, copper oxide undergoes reduction to cause a significant exotherm. If enough hydrogen is present (>1%) the exotherm becomes a safety hazard. Furthermore, as the Cu turns into copper metal, it becomes less effective for the removal of Arsine (as seen in reaction 2)

$3\,Cu_2O + 2\,AsH_3 \rightarrow 2\,Cu_3As + 3\,H_2O$..............(2)

For decades, the solution to the reducibility problem of CuO-based adsorbents, has been to use PbO based adsorbents. PbO is well known to have no significant reduction to the metal, has a negligible olefin hydrogenation activity, and can undergo Arsine (and phosphine) removal reactions, under conditions where copper containing adsorbents are unsuitable (see reaction 3)

$3\,PbO + 2\,AsH_3 \rightarrow Pb_3As_2 + 3\,H_2O$............(3)

Although PbO-based adsorbents are highly effective, they have major drawbacks related to the Pb toxicity, and thus handling and disposal. Both loading and unloading the adsorbent require special procedures, which can be costly and time consuming to execute. Further, depending on the locality, disposal of Arsenic/Lead containing adsorbent can impose significant added costs and liability for the plant owner. Finally, some countries have moved towards restricting the import of Lead containing products altogether (including PbO-based adsorbent). For example, India now classifies products containing lead as *restricted* to import into the country according to the ITC (HS). Which imposes significant import barriers and costs. [2]

To address the handling and regulatory issues, adsorbent providers have developed materials for this service that avoid the use of Pb, with limited success. Recently, Honeywell/UOP developed a new patented Pb-free adsorbent based on manganese (GB-230), which overcomes the capacity and reactivity challenges of these applications.

Reaction 4 indicates the reaction that effects Arsine removal in a MnO-based adsorbent system, which is analogous to that Cu-based adsorbents. However, in contrast to copper, manganese is an early transition metal and more oxophilic, i.e., more resistant to reduction. Further, by virtue of the reaction stoichiometry, the molar amount of Arsine that can be adsorbed is greater, enabling greater capacity.

$$MnO_x + AsH_3 \rightarrow MnAs + H_2O \text{..........(4)}$$

Figure 7 shows the relative adsorption performance of ROG and cracked gas compatible adsorbents, Mn- and Pb-based. The results show that the manganese-based adsorbent not only has higher overall capacity (equilibrium capacity), but it has a significantly longer time before breakthrough (i.e., breakthrough capacity), which results in smaller bed requirements, or lower reload frequency.

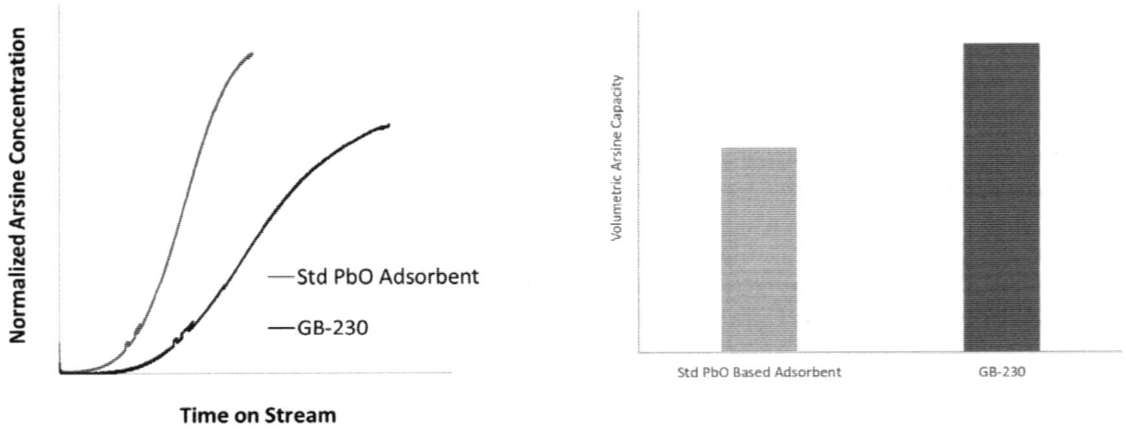

Figure 7: (left) breakthrough capacity testing results for Arsine uptake with Mn-based (GB-230) and Pb-based adsorbents; (right) relative volumetric capacity for Mn-based (GB-230) and Pb-based adsorbents.

This higher capacity (with lower toxicity) can result in smaller beds, or longer turnaround in existing units, and lower handling and disposal costs.

Reactivity and long-term stability

Over the years, multiple attempts have been made to implement Mn-based adsorbents for this application. However, the resulting materials have suffered poor long-term performance, suspected to be, in part, due to slow reduction of Mn and increase in the adsorbent's reactivity. To address the issue Honeywell UOPs GB-230 utilizes patented additives, which prevents reduction and becoming reactive towards the process stream.

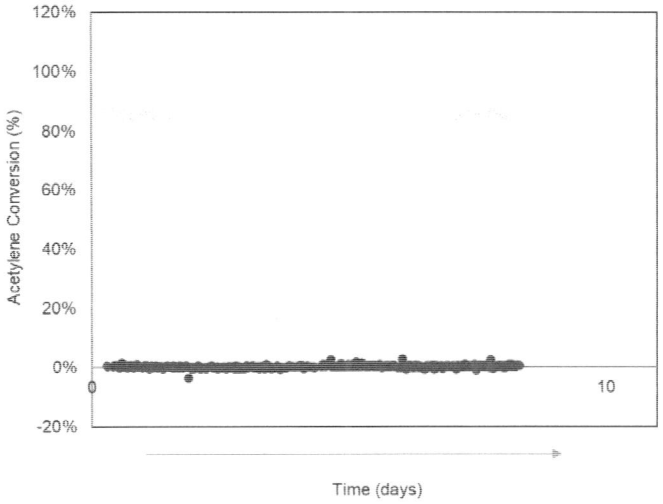

Figure 8: Acetylene conversion in the presence of H_2 and GB-230

Laboratory scale testing (Figure 8) using a flow of acetylene shows that the Mn based adsorbent has little or no activity towards hydrogenating alkynes and a low tendency to form coke.

However, because of the complexity of ROG streams, bench scale tests, although useful, are not fully representative of the conditions observed on the commercial scale. Thus, to demonstrate good performance, Mn-based (GB-230) and standard PbO adsorbent were tested in parallel cartridges in a slip stream of a commercial ROG in the plant of a major ethylene producer. The test was conducted over a period of ~5 months, to provide a realistic picture of the performance in the field.

Characterization of the exposed adsorbent showed that the Mn-adsorbent accumulated ~25% less carbon than the PbO-based adsorbent, showing that the former is significantly less reactive.

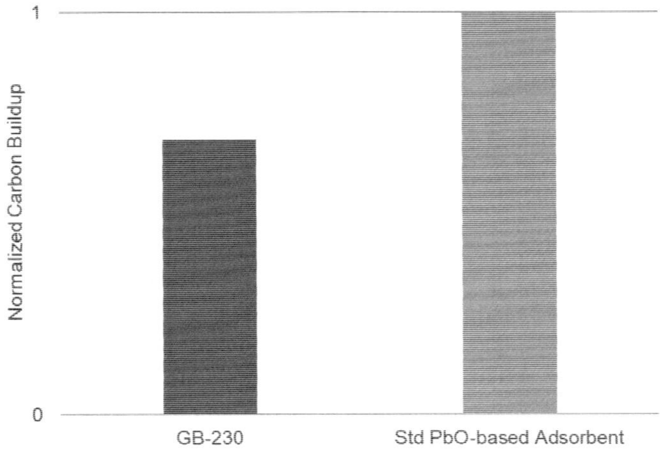

Figure 9: Relative coke amount on adsorbent after 5 months of exposure to ROG in an operating unit Mn-based (GB-230) and Pb-based adsorbents

Based on these results, one of the major ethylene producers has installed this adsorbent and the treater is currently in service for several months.

H_2S and COS capacity of the Lead-free arsine adsorbent

Both Pb and Cu, can also undergo sulfidation reactions, and can remove H_2S and COS, which are often found in ROG-like streams. This is an important feature of an arsine adsorbent, without which additional adsorbents might be required for the product to meet polymer grade specifications. Fortunately, Mn can undergo similar reactions as Pb and Cu, hence GB-230 also exhibits good capacity for these problematic contaminants, and can even be higher than some PbO adsorbents (Figure 10)

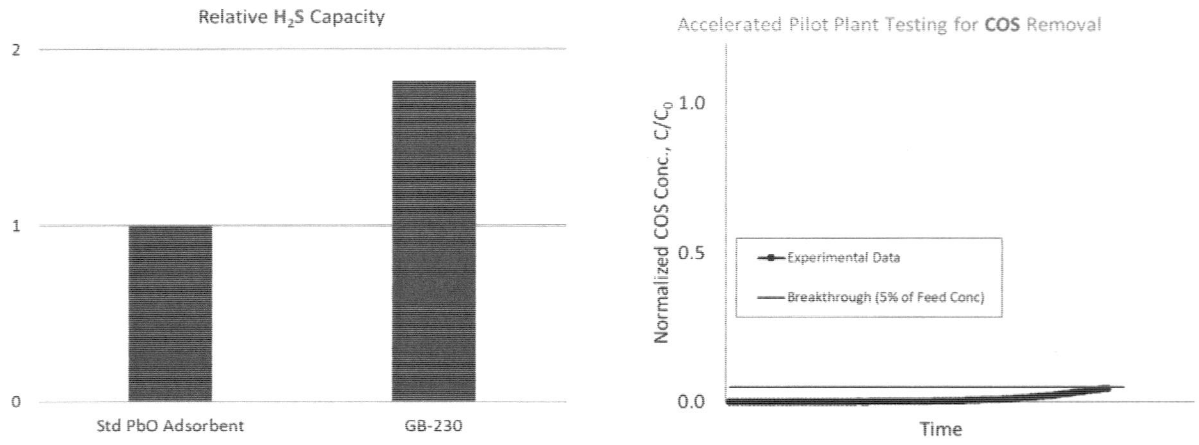

Figure 10: (left) Relative capacity for H_2S of Mn-based (GB-230) and Pb-based adsorbents; (right) COS breakthrough curve, in laboratory COS adsorption test.

Conclusions

Removal of contaminants from refinery off gas is particularly challenging due to the reactive nature of the stream, which can make adsorbents used in other hydrocarbon streams unsuitable. To achieve the cleanup, adsorbents must have negligible activity for three reactions, olefin polymerization, hydrogenation, and metal oxide reduction. A new hybrid adsorbent, AZ-500, has a higher capacity for polar compounds in addition to a significantly higher capacity for acid gases was introduced, which enables a significant reduction in the cost of olefin contaminant removal.

For Arsine and phosphine removal, a new Manganese based adsorbent that avoids the PbO material handling, and regulatory concerns is now available. This adsorbent has improved capacity for Arsine and phosphine, over PbO- based adsorbents, in addition to high H_2S and COS capacity.

References

[1]Debras, et al., US 4,861,939

[2] https://www.dgft.gov.in/CP/?opt=itchs-import-export

Manufacture of Sustainable Ethylene from Captured CO_2

James Middleton

One of the principal challenges facing the ethylene industry is improving sustainability, in particular reducing CO_2 emissions, working towards net zero. The capture of CO_2 from the flue gas of ethylene furnaces is feasible using existing carbon capture technologies, but not all plants have an outlet for captured CO_2, and tax credits for CO_2 sequestration may not make projects economically viable.

Conversion of CO_2 and hydrogen to a sustainable product provides a higher value product than captured CO_2 and is not dependent on access to an outlet for captured CO_2. LanzaTech and Technip Energies are combining their LanzaFlex™ syngas to ethanol and Hummingbird® ethanol to ethylene technologies for the manufacture of sustainable ethylene.

This presentation describes the process, a typical mass balance and utility requirements, the CO_2 footprint, and the commercial readiness of the technology.

The product from the process is polymer grade ethylene, which can be supplied to a polyethylene plant or the purification section of an existing ethylene plant.

AICHE PAPER NUMBER:

TC2C™
AN INNOVATIVE PATH FOR MAXIMUM CONVERSION OF CRUDE TO CHEMICALS

Peter Kuzma Jr.
Senior Technology Director for Ethylene
Lummus Technology

Prepared for Presentation at the 2024 Spring National Meeting
New Orleans, LA, March 24-28, 2024

TC2C™
A LOW EMISSIONS AND LOW CAPEX ROUTE FROM CRUDE TO CHEMICALS

Peter Kuzma Jr.
Senior Technology Director for Ethylene
Lummus Technology

Abstract:

With the projected decline in transportation fuels production crude suppliers will be looking for alternative outlets for their products. Simultaneously chemicals producers will be looking for new feeds to satisfy the growing demand for petrochemicals. Crude to chemicals routes promise to address both problems.

Conventional crude to chemicals refineries are capital intensive, with high utility consumptions and the resulting high carbon dioxide emissions. TC2C™ is an award winning innovative process that combines process intensification, tailored catalysts and chemistry, and advanced separation devices to reduce both CAPEX and energy consumption. Decreased energy consumption will reduce emissions directly, but the process can be enhanced by sustainable solutions such as zero carbon fuels or electrification to have net zero scope one emissions.

Introduction

Press releases for Crude to Chemicals abound, whether announcing the development of new processes, the award of new projects, or the publication of new reports on the subject. The name Crude to Chemicals encapsulates the ongoing shift in refinery-petrochemical integration, from refineries producing almost entirely transportation fuels from crude oil to refineries producing almost entirely petrochemicals feedstocks from crude oil as shown in Figure 1. The Thermal Crude to Chemicals (TC2C™) technology jointly developed by Lummus Technology, Saudi Aramco Technologies Company, and Chevron Lummus Global represents the next step on this pathway, with deep integration of the refinery and petrochemical units to the point that the boundary between the complexes disappears.

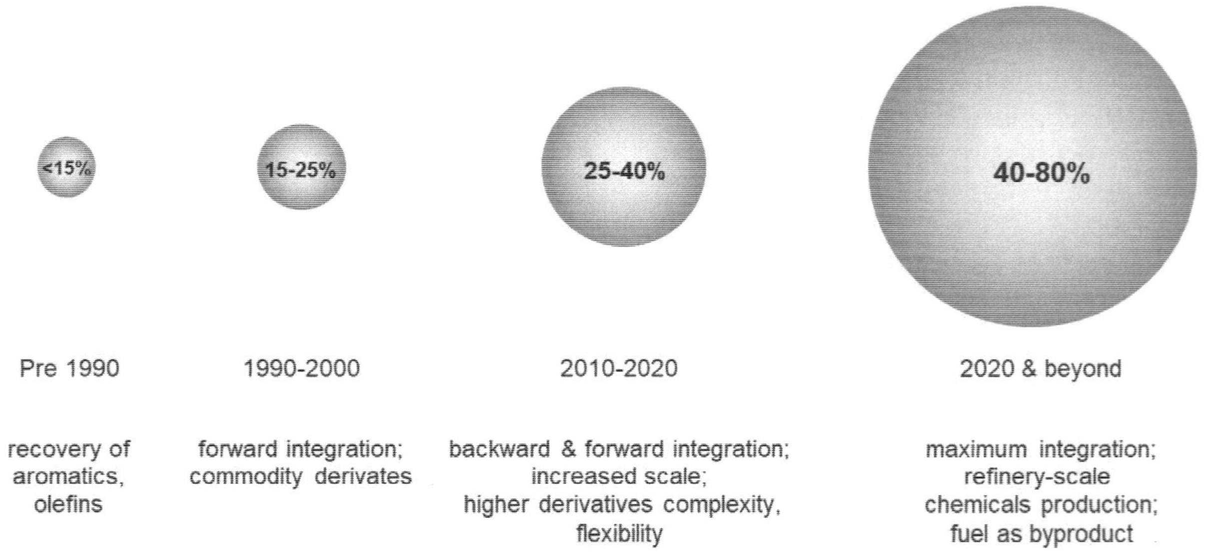

Figure 1: Historic Trend in Chemicals Yield from Crude

With the projected decrease in demand for gasoline, diesel, and other fuels larger and larger portions of crude must be converted to olefin and aromatic chemical products instead of transportation fuels. Achieving the highest yields of chemicals requires an understanding of process routes to ethylene, propylene, butadiene, aromatics, and other chemicals compounds. Integrating these process schemes is vital to ensure that the economics of crude to chemicals units remain attractive.

Achieving this deep integration between units required a thorough understanding of refinery processes, petrochemical processes, and crude chemistry and properties. The development of the TC2C technology leveraged the strengths of all three partners to rapidly develop a robust process that:

- Maximizes chemicals yields
- Reduces capital cost
- Reduces energy consumption
- Reduces CO_2 emissions

Drivers for Crude Oil to Chemicals Technologies

Two trends are combining to increase interest in maximizing conversion of crude oil to chemicals. Demand side growth in chemicals is combining with the predicted decrease in demand for transportation fuels from crude oil. Since many petrochemicals feed streams are byproducts of crude oil production the decrease in production of transportation fuels requires a shift in the output of refineries from transportation fuels to petrochemicals feedstocks.

Chemicals Demand Growth

Estimates for growth in petrochemicals vary, but the consensus is that petrochemicals demand growth will exceed GDP growth for the next twenty years or more. Figure 2 shows the IMF predictions on GDP growth overlaid with Chemicals Market Analytics estimates for petrochemicals demand growth.

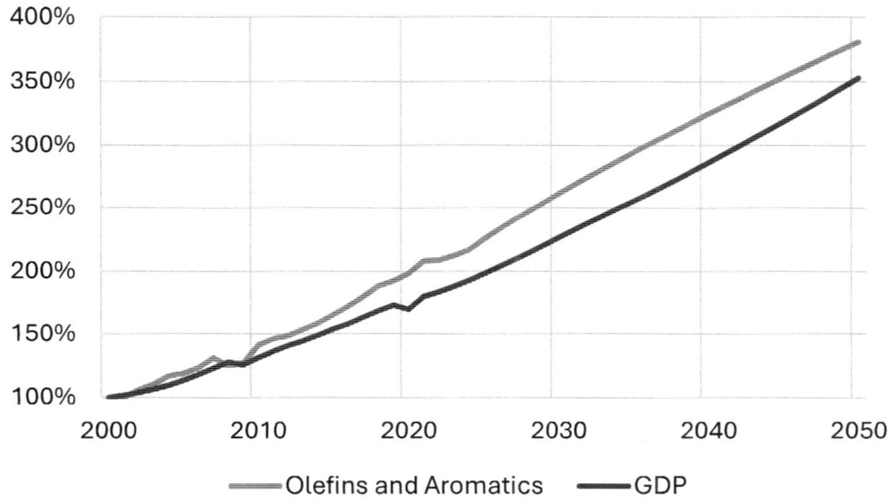

Figure 2: GDP Growth and Petrochemicals Demand Growth

The growth in petrochemicals demand naturally leads to a need for new petrochemicals production facilities. Focusing on ethylene in particular, significant growth in the 2010 decade came from a wave of new ethane cracking plants in the US, capitalizing on the availability of low cost feedstock. Projects in Europe, China, India, and other locations have also been developed based upon this cost advantaged feedstock. Moving forward, crude oil may become the next cost advantaged feedstock fueling a wave of new development.

Decline in Transportation Fuels Demand

While the exact timing varies depending upon the organization making the predictions, there is broad consensus that gasoline demand will start to decrease in the near future, with other liquid fuels demand following the same trajectory on a slightly longer timeline. Figure 3 shows projections from OPEC's 2023 World Oil Outlook, with decline in gasoline demand starting to occur between 2025 and 2028 while diesel demand beginning to decrease sooner.

Figure 4 shows this trend in terms of overall global energy supply. The contribution of different energy sources to global energy demand is shown. By 2040 the energy demand from crude oil is predicted to be only 90% of current demand, while by 2060 the demand for energy from crude oil will be only 60% of current demand.

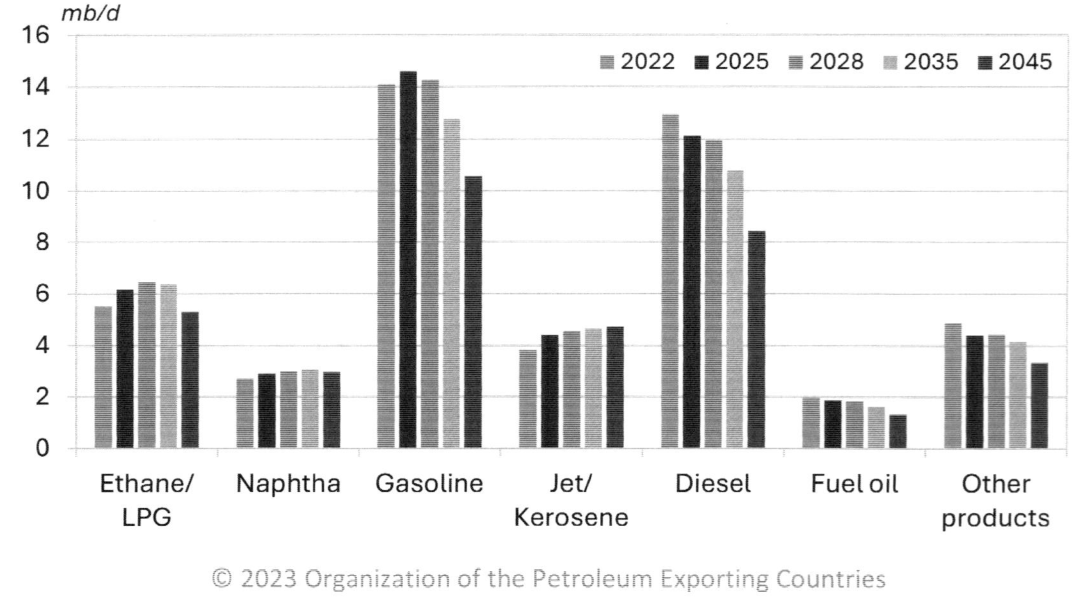

Figure 3: Demand Projections for Selected Oil Products

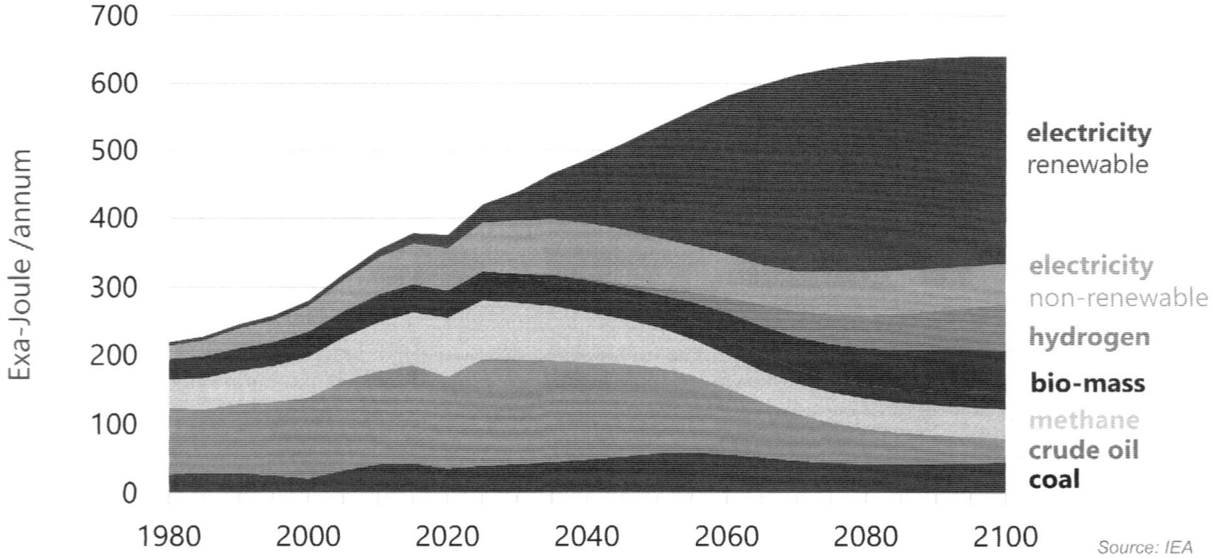

Figure 4: Global Energy Sources

The combination of demand growth in petrochemicals and demand decrease for transportation fuels in particular and energy from crude oil in general will leave crude suppliers seeking alternative outlets for their product while petrochemicals will find a decreasing supply of refinery byproducts to feed their demand growth. Direct conversion of crude oil to chemicals is a way to address the needs of both parties.

Development of TC2C™ Technology

Direct cracking of crude has been studied since the 1960s, but has not been widely pursued due to poor economics. Figure 5 shows the boiling point curves for several typical crudes and condensates. The legend shows crude name, typical API gravity, and typical microcarbon residue.

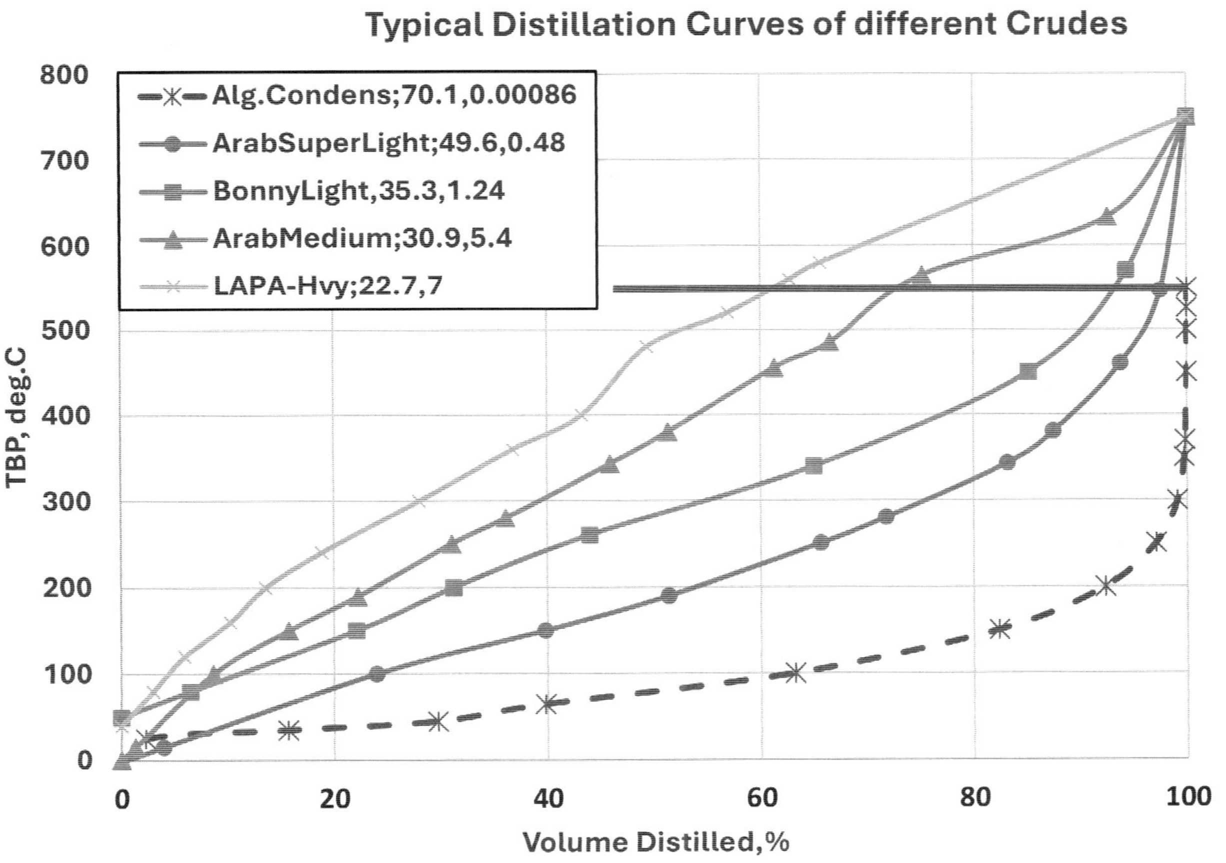

Figure 5: Boiling Point Curves for Selected Crudes and Condensates

To prevent fouling in the cracking heater convection section or rapid coking in the radiant coil the heavy boiling fraction of the crude, roughly the components boiling above 540°C, must be removed before cracking. For light crudes, like Algerian Condensate and Permian crude (not shown), The residue represents a small fraction of the feed, less than 5 wt%. As crudes get heavier the residue increases to as high as 40 wt% for the crudes shown above.

Several methods to remove the heavy tail are available, from a simple flash drum, to conventional distillation, to Lummus' heavy oil processing system (HOPS). Whatever method is used, as figure 5 shows for many crudes a significant portion of the feed must be discarded. This negates the cost advantage of crude oil for most crudes. Direct cracking is limited to a small subset of light crudes or condensates with low impurity levels.

Conventional crude to chemicals projects have been developed to address the limitations of direct crude cracking. These projects use conventional refinery operations, but shift a portion of the refinery products from transportation fuels to petrochemicals feedstocks, with chemical yields approaching fifty weight percent of crude. This allows a wider selection of crudes, but has a high capital and energy cost since full refinery units are utilized and transportation fuel specifications must still be met. Increasing chemicals yield while reducing CAPEX and OPEX required the talents and expertise of three major players in the petrochemical industry.

Development Partners and Process

In January 2018 a Joint Development Agreement was signed between Saudi Aramco Technologies Company, Lummus Technology, and Chevron Lummus Global. Each partner brought their depth of background knowledge to an accelerated development program. The goal of the program was to develop a process solution with chemicals yields greater than seventy weight percent of crude with reduced capital and energy requirements compared to conventional crude to chemicals technologies, without compromising reliability and on-stream factor.

TC2C builds on fundamental research in the areas of process technology, process chemistry, and catalyst technology. It is anchored on commercially proven technologies such as Saudi Aramco's vapor liquid separation device, Lummus Technology's steam cracking technology (particularly of streams with a higher boiling point than naphtha), and CLG's integrated hydroprocessing reactor platforms for crude conditioning and heavy oil upgrading. This extensive commercial experience enables TC2C to capitalize on flexible, commercially proven technology components.

An intensive multi-year research program ensued, building upon ongoing work in molecular speciation and process analytics to characterize crude oil and the various fractions contained in crude. New and existing catalysts and reactor platforms were tested with different crude oils. The system was tested with and without the recycle of low value streams from the steam cracker to increase overall chemicals yield. Steam cracker yields were modeled based upon the properties of the feed conditioning section products and total chemicals yield was calculated assuming recovery of olefins, butadiene, and aromatics from the steam cracker product streams.

Advanced separation devices were tested at bench scale and cold flow units were tested to evaluate the separation efficiency of multiple configurations. Laser droplet counters were used to determine the quantity and particle size distribution for each configuration tested. Testing confirmed that the devices can achieve the required separation at low capital and energy cost.

In a remarkably short period of time the partners brought the technology to market and were rewarded with the first license for the new technology less than two years after starting the development process.

Applicability of TC2C™ Technology

The TC2C technology can be designed for a wide range of crudes, from light to heavy, by varying the relative sizes of the individual processing sections. In addition to crude oils the system can process a variety of streams from refineries and steam crackers. These include low value refinery streams such as SDA pitch, high sulfur fuel oil, FCC slurry oil, LCO, and others. Recycle streams from the steam cracker include pyrolysis fuel oil, pyrolysis gas oil, and heavy gasoline product. Excess hydrogen from the steam cracker is recycled to the feed conditioning section to reduce the requirement for imported hydrogen.

As a result of the feed flexibility the TC2C technology is well suited to either grassroots or brownfield applications. In a grassroots configuration the plant receives crude oil feedstock and exports chemicals products and a small quantity of very low sulfur fuel oil. In a brownfield configuration the feedslate to the plant is crude oil plus a variety of orphaned refinery streams that the technology will upgrade to chemicals products and very low sulfur fuel oil. There have been two licenses awarded for TC2C technology, the first in a brownfield application and the second in a grassroots configuration.

TC2C™ Process

TC2C is a modern intensification of traditional processes in that it takes the key technical features of commercially proven technologies and fuses them together to meet processing objectives distinct from traditional refinery and petrochemical complexes. The processing objectives are:

- Maximizing the yield of chemicals from crude. (>70 wt% chemicals on crude accomplished).
- Minimizing the yield of transportation fuel
- Optimizing the operating costs, the global utility consumption and the carbon footprint through scalable process intensification and integration.

Key features that were needed to meet these objectives include:

- novel proprietary fractionation technologies which eliminate the need for costly conventional crude atmospheric and vacuum distillation units.
- a breakthrough hydrocracking catalyst with a tailored mesopore structure
- smartly configured fixed bed and residue hydrocracking reactors that selectively condition crude for mixed feed steam cracking
- a mixed feed steam cracker that converts a wide boiling range feed into petrochemical building blocks such as ethylene and propylene along with other chemicals

The core philosophy during the development of the TC2C process was to optimize the hydroprocessing and separation performed in the feed conditioning section to meet the operating objectives. Figure 6 shows the main units in the TC2C complex.

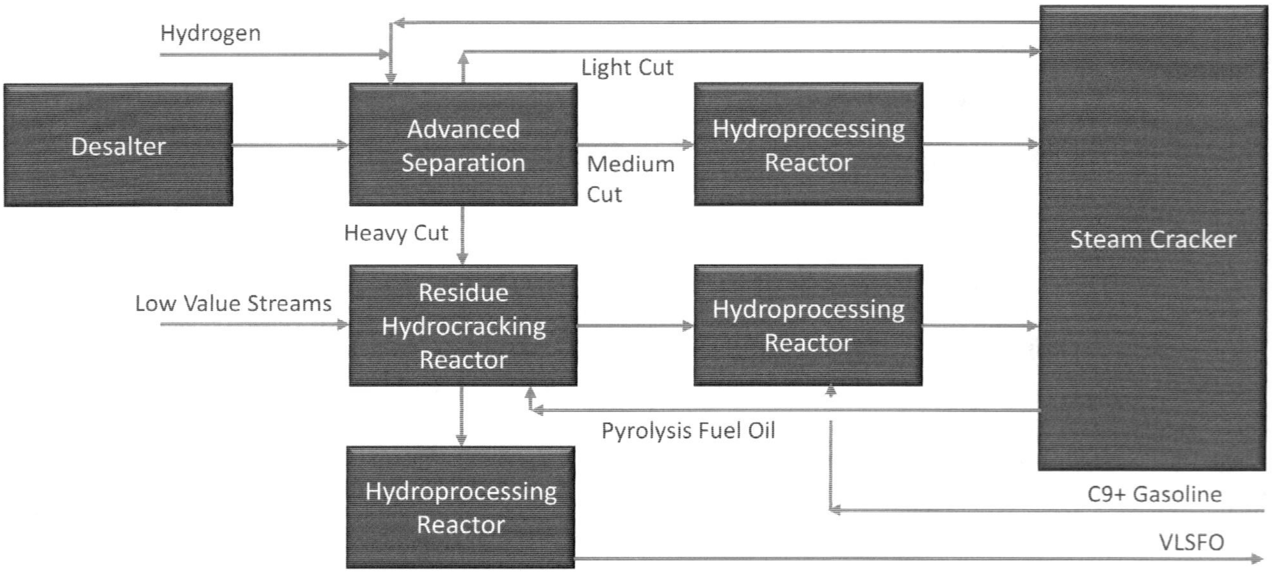

Figure 6: Block Flow Diagram for TC2C

Crude oil is first desalted in a two stage desalter. This typical refinery process is required to ensure that the feed to the system has minimal quantities of sodium present. Two stages of desalting and the quality of the separation in the advanced separation section ensure removal of sodium from the direct cracker feed since ppb levels of sodium in cracker feedstocks can damage the radiant coils and impact run length and yield. Desalting the crude ensures that contaminant load to the hydroprocessing units is minimized and avoids costly metallurgical upgrades.

The desalted crude then flows to the advanced separation section. This section of the process contains two separation devices, an integrated separation device and a hot hydrogen stripper, to produce three crude cuts, light, medium, and heavy. The patented Integrated Separation Device (ISD) includes a proprietary cyclonic inlet design to reduce entrainment of heavies into the overhead stream. The hot hydrogen stripper uses hydrogen already being fed to the complex to help separate the medium cut from the heavy cut without requiring vacuum, excessive temperatures, or large consumption of steam that must then be condensed and recovered from the product.

Cutpoints and operating temperatures have been selected to ensure long cycle times of the downstream sections while producing feed cuts suitable for the steam cracker. To minimize equipment and energy consumption the products of the advanced separation unit only meet the specifications required for the operating objectives. This reduces energy consumption and piece count.

The light cut, similar to a straight run naphtha product, is sent directly from the integrated separation device to the cracking heaters in the steam cracker. This stream typically has low levels of sulfur and other impurities and is rich in normal paraffins, making it an excellent feed to the cracker. Additional processing would offer only minor gains in hydrogen content of the stream. If the iso to normal paraffin ratio of the feed increases due to hydrotreating then the net chemicals yield from the feed can actually decrease due to increased methane yield from isoparaffins.

The medium cut from the advanced separation section, representing the heart cut from the crude, is sent to a hydroprocessing section of the TC2C unit. In most configurations of TC2C this will be mild hydroprocessing to remove impurities and slightly increase the hydrogen content of the stream before feeding the steam cracker. Mild hydroprocessing minimizes the isomerization of normal paraffins to isoparaffins, maintaining a high yield from the feed. This stream is then fed to the steam cracker.

The heavy cut from the advanced separation section is fed to the residue hydrocracking reactor section. Pyrolysis fuel oil from the cracker and any other low value refinery streams are also fed to the residue hydrocracking reactors to be upgraded to cracker feedstock. The residue hydrocracking reactor can be an ebullated bed reactor or a slurry reactor, depending upon project objectives. The converted products from the residue hydrocracking reactors are sent to an integrated fixed bed hydroprocessing reactor for final cleanup and hydrogen addition. C_9 plus gasoline from the cracker is recycled back to the integrated hydroprocessing reactor as well. Heavy product from the residue hydrocracking reactor section is sent to a heavy oil hydrotreater to remove sulfur and produce an IMO 2020 compliant very low sulfur fuel oil (VLSFO) with less than 0.5 wt% sulfur.

The severity of the integrated hydrocracker is carefully controlled. Conversion is maintained high enough to reduce the concentration of polynuclear aromatics and other undesirable molecules in the crude to a level that results in an optimized cracker feed but at a low enough level to minimize unnecessary over-conversion to naphtha and other light components. Some naphtha and lights are formed, these are sent to the cracker as well as the gas oil from the integrated hydrocracker. Limiting conversion minimizes hydrogen consumption in the reactor and minimizes the production of isoparaffins. Table 1 shows a comparison of typical once through cracking yields for straight run and hydrotreated feeds.

Typical Cracker Yields	Straight-run Feeds			HVGO from Hydrocracking
	Full Range Naphtha	AGO	VGO	
CH_4	15.40	10.90	9.20	12.60
C_2H_4	29.00	24.50	13.50	30.00
C_3H_6	13.50	13.30	12.00	13.30
C_4's	8.70	8.85	7.85	8.60
C_5-204°C (1)	21.20	19.05	16.60	19.00
PFO	5.60	18.16	37.45	10.10

Table 1: Once Through Cracker Yields

Despite the heavier boiling range of hydrocracked HVGO, the addition of hydrogen in the hydrocracker results in a stream with cracking yields similar to straight run naphtha. Hydrocracking the feed to naphtha or beyond is not required, the isoparaffin content of the hydrocracked naphtha would reduce the ethylene yield from the hydrocracked naphtha. Even the higher production of PFO from hydrocracked gas oil is not an issue for overall product yields since the PFO is recycled to the crude conditioning section and returned to the cracker as feed.

The steam cracker receives the crude conditioning section products. Naphtha and gasoil streams are sent directly to the cracking heaters, LPG products are sent to the charge gas compressor section of the cracker and separated in the recovery section. Lummus' experience with heavy feed cracking ensures that the cracking heaters achieve high yields and long run lengths on all feedstocks. Lummus' recovery section, featuring the low pressure chilling train and multicomponent refrigeration systems, provides efficient, low capital cost, and low energy consumption separation of the cracking heater effluents into final product streams.

The integration of all these technology features results in a process that yields 75 wt% to over 80 wt% chemicals from crude. The process integration and simplification in the crude conditioning section reduces piece count by 40% compared to conventional refinery processes. Minimizing processing of the crude by optimizing hydrogen addition, integrating the high pressure circuits of the hydrocracking reactor systems, and eliminating fractionation to achieve strict product specifications results in energy savings for the overall process compared to a typical crude to chemicals refinery configuration.

Case Study: TC2C™ vs Conventional Crude to Chemicals Refinery and Naphtha Cracker

TC2C offers significant advantages compared to crude to chemicals refinery configurations. These are:

- Higher chemicals yield
- Lower capital cost
- Lower utility consumptions
- Reduced CO_2 emissions

TC2C also offers significant advantages compared to stand alone naphtha crackers. These advantages include:

- Advantaged feedstock pricing
- The ability to upgrade low value streams
- Competitive energy consumption when the energy required to produce naphtha is not considered
- Lower energy consumption when the energy required to produce naphtha from crude is considered
- Competitive CO_2 emissions when the CO_2 emitted in naphtha production is not considered

- Lower CO_2 emissions when the CO_2 emitted in naphtha production is not considered

The following case study will compare the yields and energy consumption for Arab Light crude processed in a TC2C complex, Arab Light crude processed in a crude to chemicals refinery complex, and a naphtha cracker complex with typical naphtha feed. All complexes in the case study have the same configuration for downstream units to recover chemicals from the cracker C_4 and gasoline streams.

Conventional Crude to Chemicals Refinery Flow Scheme

Figure 7 shows the configuration of the crude to chemicals refinery assumed for the case study. After the desalter the crude flows to a conventional separation section consisting of crude and vacuum distillation towers to produce light, medium, and heavy products. The separation section lights (crude tower overhead) are processed in a saturated gas plant and naphtha splitter to produce LPG and naphtha feeds for the cracker. Gas oil products from the crude and vacuum tower are sent to hydroprocessing units to increase the hydrogen content of these streams and remove impurities. The naphtha, kerosene, and diesel products from the refinery are fed to the cracker. Residue from the vacuum tower is sent to a residue hydrocracking reactor and the converted products are fed to the integrated hydroprocessing system along with the gas oils from the separation section. Heavy product from the residue hydrocracking reactor section is mixed with pyrolysis fuel oil (PFO) from the cracker and sold as LSFO, with less than 3.5 wt% sulfur but more than 0.5 wt% sulfur.

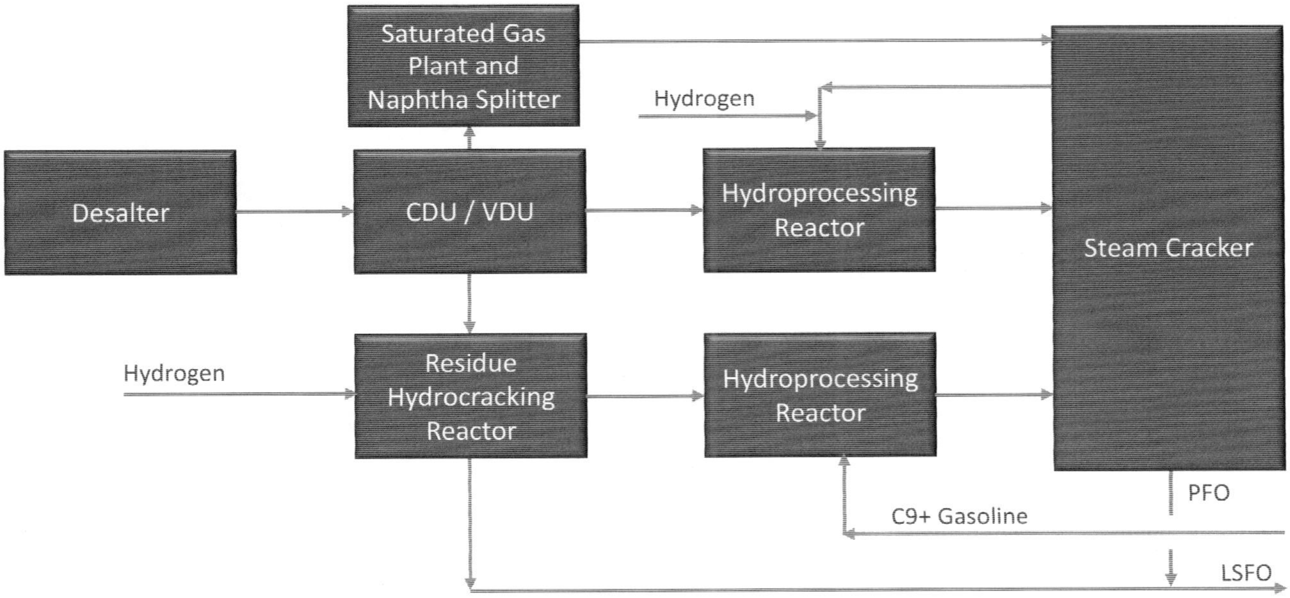

Figure 7: Block Flow Diagram for Crude to Chemicals Refinery

Cracker Byproduct Flow Scheme Used for Analysis

To accurately calculate the chemicals produced from each flow scheme the following series of units has been considered when generating overall material balances. The mixed C_4s stream from the cracker is sent to a butadiene extraction unit to recover 1,3 butadiene product. The raffinate from butadiene extraction is sent to an MTBE unit to remove isobutene from the raffinate and the MTBE is then reacted back to high purity isobutene. The raffinate from isobutene production is sent to an olefins conversion unit where normal butenes are reacted with ethylene to produce propylene. The remaining raffinate is hydrogenated and recycle cracked. The C4 product flow scheme is shown in Figure 8

The pyrolysis gasoline from the cracker is sent to a two stage pyrolysis gasoline hydrogenation unit to saturate the diolefins and olefins present in the stream. C_5s in the gasoline are recycled back to the cracker and the C_9 and heavier gasoline is sent to the hydroprocessing units in the TC2C and crude to chemicals refinery configurations, while in the naphtha cracker configuration the C_9 and heavier gasoline is a product. The C_6-C_8 cut from the gasoline hydrotreater is sent to a BTX extraction unit to recover benzene, toluene, and mixed xylenes. The non-aromatic raffinate from the extraction unit is recycled to the cracker. The pyrolysis gasoline flow scheme is shown in Figure 9.

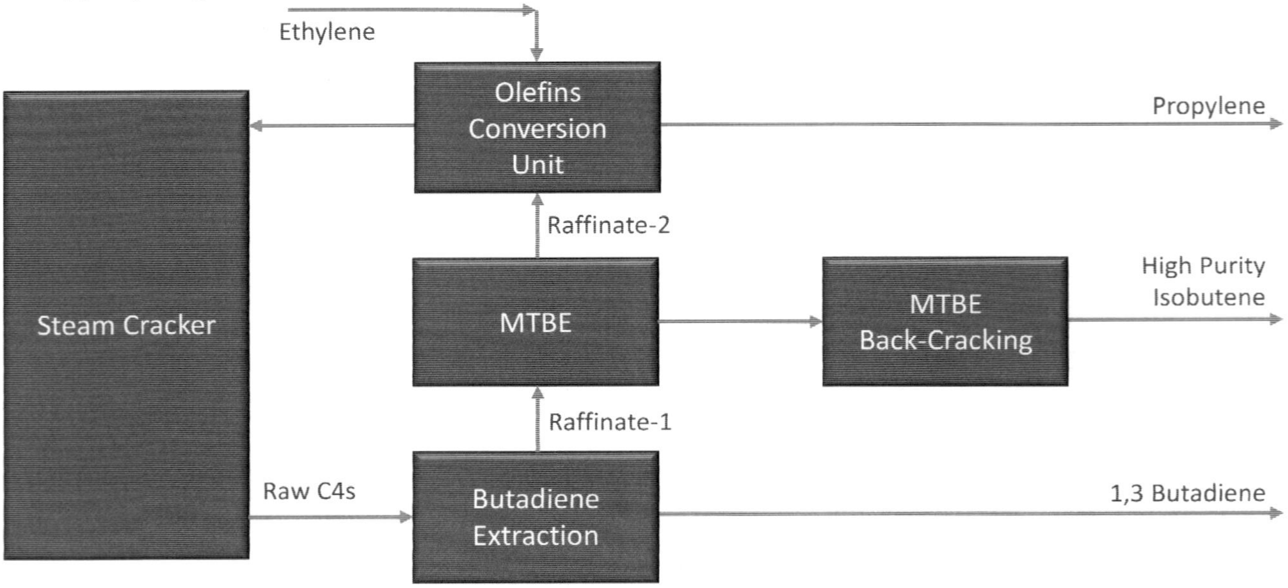

Figure 8: C4 product flowscheme

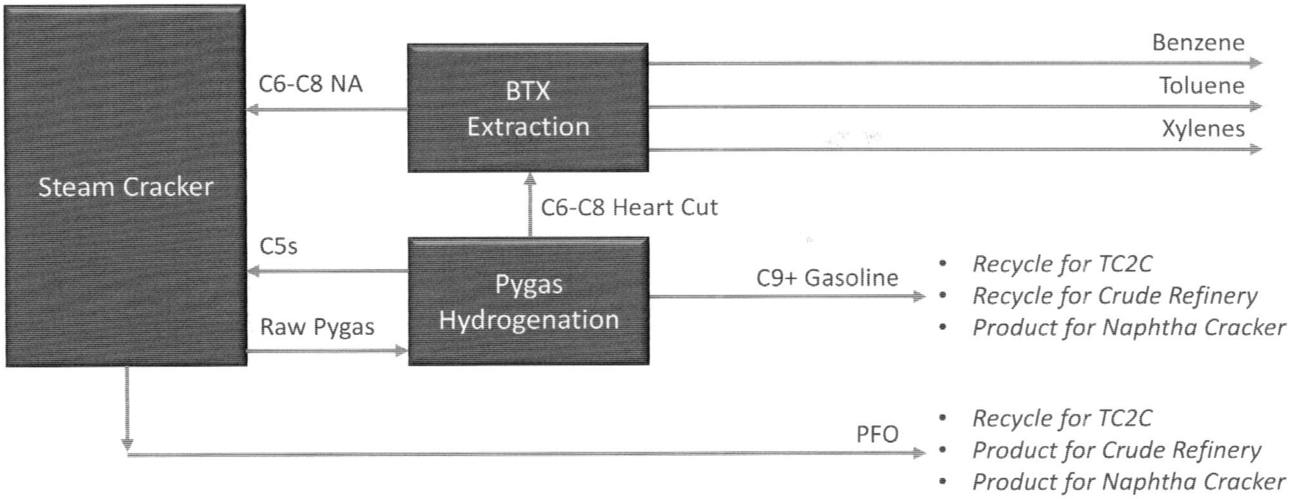

Figure 9: Pyrolysis gasoline product flow scheme

Mass Balance Comparison

Material balances for each complex are provided below. Net ethylene production for each complex is fixed at 2000 kta, feed and byproduct rates vary. Productions of ethylene and other chemicals are shown, with the yield per kg of feedstock shown in parenthesis.

As expected, the naphtha cracker has significantly lower feed rate and high ethylene yield due to the high feed quality. The chemical refinery has higher feed requirements than the TC2C configuration due to the over-processing of the feed, resulting in higher relative yields of propylene and other chemicals compared to TC2C.

The VLSFO produced from the TC2C complex has a higher value than the LSFO and PFO produced from the other complexes. Even though it does not meet the chemicals definition, the higher value of IMO 2020 compliant fuel oil means that this stream offers a significant economic benefit for TC2C compared to the other complexes.

Total chemicals production from the crude based units is higher than from the naphtha cracker, this is to be expected due to the higher ethylene yield of naphtha cracking resulting in lower feed rates. Looking at the ratios of chemicals and high value products to feed the advantage of TC2C over conventional flow schemes becomes apparent. Compared to a naphtha cracker the TC2C unit has virtually identical chemicals yield, but the VLSFO product from TC2C has significantly higher value than the PFO from the naphtha cracker. When this is considered TC2C yield of high value products is more than 8% higher than a naphtha cracker. The chemicals yield from a chemical refinery is 5% lower than the yield from TC2C as the overprocessing of the feed results in increased isoparaffin production in the cracker feeds, leading to higher methane yield. Since methane and fuel oil are not chemical products the higher flowrate of these streams represents a loss to chemicals production

The yields for the TC2C and Chemical Refinery flow schemes are based upon ebullated bed reactor in the residue hydrocracking reactor section of the complexes. If slurry technology is used in the residue hydrocracking reactors the increased conversion in the residue hydrocracking reactors will result in more than 5% higher yield of chemicals. This would result in TC2C having higher chemicals production than the naphtha cracker.

Flow in kta	TC2C Complex	Chemical Refinery	Naphtha Cracker
Crude	6262	8355	-
Naphtha	-	-	4806
Ethylene	2000 (32%)	2000 (24%)	2000 (42%)
Propylene	1396 (22%)	1664 (20%)	824 (17%)
Other Chemicals	1295 (21%)	2163 (26%)	806 (17%)
VLSFO	567 (9%)	-	-
LSFO / PFO	-	1348 (16%)	248 (5%)
Total Chemicals	4691	5827	3630
Total High Value Products	5258	5827	3630
% Chemicals / HVP	74.9 / 84.0	69.7 / 69.7	75.5 / 75.5

Table 2: Material Balance Comparison

Utilities and CO2 Emission Comparison

Due to the differences in complex capacities, comparing the utility consumption of each unit does not clearly show the differences between units. Comparison on a specific energy basis is more useful to determine the efficiency of each complex. When calculating the specific energy for the complexes three different denominators have been used, net ethylene production, total chemicals production, and total high value product flowrates. Values for the naphtha cracker have been calculated two ways, first by excluding the energy required to produce naphtha from crude oil and second by including an energy of production of 1000 kcal/kg for naphtha.

As expected, when specific energy is calculated on an ethylene only basis the higher ethylene yield from naphtha cracking results in significantly lower energy for a naphtha cracker than for the crude based complexes that must expend energy to upgrade their feed to cracker feedstock. TC2C has a significant energy savings over a crude to chemicals refinery due to the benefits already discussed. When the energy required to produce naphtha is included the specific energy of ethylene production from naphtha increases and the TC2C complex offers lower energy consumption

When propylene and other chemicals are considered the advantage naphtha crackers have decreases, but there is still an advantage to naphtha cracking. When the energy to produce naphtha is included this advantage is lost. TC2C maintains an advantage over a crude to chemicals refinery.

Considering the VLSFO from TC2C as a high value product, TC2C has an advantage against naphtha cracking even when the energy to produce naphtha is not included in the calculation. This advantage only improves once that is considered.

CO_2 emissions show a similar pattern to energy consumption, as expected. Values for naphtha cracking are shown on a cracker only basis and with the assumption of 0.3 kg CO_2 emissions per kg of naphtha. When the TC2C complex is compared to the naphtha cracker and emissions to produce naphtha are not included the naphtha cracker has an advantage when only ethylene production is considered. This advantage decreases when other chemicals are considered and the advantage shifts to TC2C when high value products are considered. If the emissions to produce naphtha are considered then TC2C offers lower emissions than naphtha cracking on every basis. TC2C has lower emissions than a crude to chemicals complex by every metric.

	TC2C Complex	Chemical Refinery	Naphtha Cracker	
			Cracker only	Including naphtha
Specific Energy				
kcal/kg ethylene	7243	10075	5156	7559
kcal/kg chemicals	3024	3458	2841	4165
kcal/kg HVP	2553	3458	2841	4165
CO_2 Emissions				
kg/kg ethylene	1.45	2.09	1.08	1.80
kg/kg chemicals	0.61	0.72	0.60	1.00
kg/kg HVP	0.51	0.72	0.60	1.00

Table 3: Energy Consumption and CO_2 Emissions

Economic Evaluation

Economics for a particular project will be dependent upon the specific economic criteria for that project, such as feed and product costs, utility costs, carbon costs or credits, discount rate, tax rate, etc.. Economic analysis was performed for each of the complexes under discussion. Typical pricing for feed, products, and energy was assumed. No cost of CO_2 emissions was included in the analysis. A discount rate of 9% was used for NPV calculations.

The naphtha cracker is significantly less expensive than the TC2C complex. This is not only due to the lack of a feed conditioning section in the naphtha cracker, but also the smaller size of the cracker due to lower throughput. The higher cost of the crude to chemicals refinery compared to the TC2C unit shows the savings from the piece count reduction in the feed conditioning section of the TC2C complex, with a 40% piece count reduction, and the smaller size of the TC2C cracker.

The feed price advantage of the crude based complexes contributes to their advantages in OPEX. The throughput of the TC2C complex is 30% higher than the naphtha cracker, but the operating margin is almost 80% higher. The high chemicals production rate and the production of high value VLSFO also improve margin for the TC2C unit. The crude to chemicals refinery margin comparison to the naphtha cracker shows increases of more than the increase in plant throughput due to the lower feedstock cost, but the improvement is not as large as for TC2C. This is due to the higher operating costs for energy and the lower value fuel oil product from the crude to chemicals refinery.

The net present value for TC2C is 80% higher than the net present value of the naphtha cracker. This is again significantly higher than just the increased throughput of the TC2C unit. This improvement in NPV represents an increase of almost 1% in the IRR for this analysis, a significant improvement in profitability. If CO_2 penalties or incentives are considered the advantage for TC2C will only increase.

Flow in kta	TC2C Complex	Chemical Refinery	Naphtha Cracker
CAPEX	100%	151%	62%
Net Margin	100%	114%	56%
Net Present Value	100%	55%	55%

Table 4: Economic Comparison

Conclusions

TC2C provides a high yield, economical, low emission route from crude to chemicals. A wide variety of crudes can be processed with yields exceeding 80 wt% chemicals from crude. This expands the potential feed slates when compared to direct crude cracking, allowing access to more cost advantaged feedstock. The hydrocracking components of the feed conditioning section allow the plant to process low value refinery streams and upgrade those to more valuable chemicals.

Process integration and reduced fractionation requirements result in capital and energy savings compared to conventional chemicals production routes. The reduced energy consumption results in reduced CO_2 emissions. The combination of low capital cost, advantaged feedstock pricing, high chemicals yields, and low energy consumption results in improved project economics compared to naphtha cracking or conventional crude to chemicals complexes.

TC2C technology draws from the strengths of the development partners to provide an efficient reliable technology for chemicals production. In less than 5 years from the start of development work the technology has had two commercial awards, one in a brownfield setting and one for a grassroots unit.

AUTHOR INDEX